Environmental Transport and Transformation of Pollutants

Environmental Transport and Transformation of Pollutants

Guest Editor

Xiaoxia Lu

MDPI

Basel • Beijing • Wuhan • Barcelona • Belgrade • Novi Sad • Cluj • Manchester

Guest Editor
Xiaoxia Lu
College of Urban and
Environmental Sciences
Peking University
Beijing
China

Editorial Office
MDPI AG
Grosspeteranlage 5
4052 Basel, Switzerland

This is a reprint of the Special Issue, published open access by the journal *Toxics* (ISSN 2305-6304), freely accessible at: www.mdpi.com/journal/toxics/special_issues/0S0PFQ6650.

For citation purposes, cite each article independently as indicated on the article page online and as indicated below:

Lastname, A.A.; Lastname, B.B. Article Title. *Journal Name* **Year**, *Volume Number*, Page Range.

ISBN 978-3-7258-6133-0 (Hbk)
ISBN 978-3-7258-6134-7 (PDF)
https://doi.org/10.3390/books978-3-7258-6134-7

Contents

About the Editor

Xiaoxia Lu

Xiaoxia Lu, Ph.D, College of Urban and Environmental Sciences, Peking University. Dr. Lu primarily engages in research on the environmental behavior and effects of pollutants. In 2012, she was selected for the New Century Excellent Talents in University Project supported by the Ministry of Education. She has led and participated in about 30 scientific research projects and has published near 100 papers. She serves as an Editorial Board Member for journals such as *Toxics, Water Environment Research, International Journal of Environmental Research and Public Health,* and *Journal of Chemical Defense Research.* She has received several awards such as the Second Prize of Natural Science Award from the Ministry of Education, China, the First Prize of Beijing Science and Technology Award, and the Teaching Excellence Award from Peking University.

toxics

MDPI

Editorial

Environmental Transport and Transformation of Pollutants

Xiaoxia Lu

Ministry of Education Laboratory for Earth Surface Processes, College of Urban and Environmental Sciences, Peking University, Beijing 100871, China; luxx@urban.pku.edu.cn

1. Introduction

In recent years, study on pollutant environmental transport and transformation has witnessed remarkable progress driven by interdisciplinary integration and technological innovation. Research focus has expanded from conventional pollutants to emerging contaminants [1]. High-resolution mass spectrometry and genomic technologies have enabled the accurate identification and quantification of trace emerging pollutants and microbes, facilitating the shift from "detection" to "mechanism exploration" [2,3]. The recognition of multi-media transport processes has deepened, with studies revealing the complex exchange of pollutants among the atmosphere, water, soil, and biota [4]. Numerical modeling techniques have advanced significantly, including the development of coupled reactive transport models for complex pollution systems and the integration of machine learning to improve prediction accuracy [5,6].

Despite these advances, critical knowledge gaps remain. For instance, the synergistic transport and transformation mechanisms of coexisting pollutants are not fully understood, particularly the coupling effects of chemical, physical, and biological processes in complex environmental matrices. The environmental behaviors of emerging pollutants are not well characterized, including their transformation pathways across different media and the ecological risks of transformation products. Existing numerical models often lack integration of multi-scale processes and real-time data assimilation, leading to uncertainties in risk assessment.

2. An Overview of Published Articles

This Special Issue comprises 12 studies (11 research articles and 1 review) covering core themes of pollutant adsorption, microbial degradation, nanoparticle fate, multi-phase partitioning, risk assessment, and toxicity. These works target and fill several key knowledge gaps identified in the field, as summarized below.

Zhang et al. (2024) (contribution 1) investigated pyrene (a PAH) and arsenite (As(III)) adsorption by micro/nano carbon black (C 94.03%, spherical, 100–200 nm) and iron oxide (hematite, irregular rod-shaped, ~1 μm long). The key findings were as follows: (1) carbon black preferentially adsorbed pyrene (24 h adsorption capacity: 283.23 μg/g; pseudo-second-order rate constant: 0.016 mg/(g·h)), while iron oxide adsorbed As(III) (24 h adsorption capacity: 3.45 mg/g; rate constant: 0.814 mg/(g·h)), with chemical reactions as the main mechanism; and (2) As(III) reduced pyrene adsorption on carbon black (effect strengthened with increasing As(III) concentration), while pyrene enhanced As(III) adsorption on iron oxide. This fills the gap in co-pollutant interaction during adsorption, guiding combined PAH-As remediation.

Liu et al. (2024) (contribution 2) explored benzene and toluene biodegradation under sulfate-reducing conditions using groundwater samples from contaminated sites. By the

https://doi.org/10.3390/toxics13121028

end of the study (day 90), about 99% benzene and 96% toluene were removed. During the study, bacterial community richness initially decreased but subsequently increased over time. Key degraders of benzene and toluene were identified as *Pseudomonas*, *Janthinobacterium*, *Novosphingobium*, *Staphylococcus*, and *Bradyrhizobium*. This fills the gap in microbial community dynamics during sulfate-driven BTEX biodegradation, providing a basis for biostimulation strategies.

Khan et al. (2024) (contribution 3) analyzed ZnO-NP aggregation in simulated water with perfluorooctanoic acid (PFOA), humic acid (HA), and electrolytes (NaCl, $CaCl_2$) over 3 weeks. ZnO-NP size increased from 162.4 nm (1 day) to >10 μm (3 weeks) and Zeta potential decreased from −47.2 mV (1 day) to −0.2 mV (3 weeks). HA and PFOA dispersed ZnO-NPs via aliphatic carbon interactions and complex structures, while electrolytes altered surface charge. This clarifies how multiple coexisting substances regulate ZnO-NP aggregation, addressing the gap in nanoparticle fate prediction.

Khan et al. (2025) (contribution 6) explored how the presence of coexisting organic pollutants (like tetrabromobisphenol-A (TBBPA)), electrolytes (NaCl and $CaCl_2$), humic acid (HA), and bovine serum albumin (BSA) in water affected the behavior of ZnO-NPs. They found that TBBPA and salts promoted aggregation via cation bridging and hydrophobic interactions, while HA/BSA enhanced dispersion by modifying zeta potential. This further expands understanding of ZnO-NP behavior in complex systems.

Yang et al. (2024) (contribution 4) studied estrogen (estrone E1, 17β-estradiol E2, estriol E3) partitioning in Taiwan's Wulo Creek (impacted by feedlot wastewater), separating samples into suspended particulate matter (SPM), colloidal, and soluble phases. The results showed that estrogens dominated in the soluble phase (85.8–87.3%), followed by colloids (12.7–14.2%); Log K_{COC} (4.72–4.77 L/kg-C) was much higher than Log K_{OC}/Log K_{POC} (2.02–3.40 L/kg-C), indicating that colloids play a critical role in estrogen transport. This addresses the gap of ignoring colloidal phases, improving ecological risk assessment accuracy.

Hou et al. (2024) (contribution 5) synthesized a covalent organic-framework-modified biochar (RH-COF) for cadmium (Cd^{2+}) adsorption. The modified material showed a 14-fold increase in Cd^{2+} capacity (from 4.20 to 58.62 mg/g) due to elevated nitrogen (from 0.96% to 5.40%) and oxygen (from 15.50% to 24.08%) content. Adsorption followed pseudo-second-order kinetics and Langmuir isotherms, with mechanisms including surface complexation, chelation, and electrostatic adsorption. This addresses the gap in low-efficiency Cd remediation by developing a high-performance adsorbent.

Kawichai et al. (2025) (contribution 7) developed a machine learning (ML) model to predict long-term $PM_{2.5}$ concentrations in upper northern Thailand (impacted by biomass burning). The best ML prediction model was selected considering root mean square error (RMSE), mean prediction error (MPE), relative prediction error (RPE), and coefficient of determination (R^2). Using 2011–2020 data (PM_{10}, CO_2, O_3, fire hotspots, air pressure, rainfall, relative humidity, temperature, wind direction, and wind speed), the random forest (RF) model outperformed others (RMSE: 6.82 μg/m^3, R^2: 0.93). This fills the gap in long-term $PM_{2.5}$ retrospective data, supporting health effect studies.

Cai et al. (2025) (contribution 9) measured 10 organophosphate esters (OPEs) in 46 homes, 12 offices, 6 dormitories, and 60 private cars in Guangzhou. Among the four microenvironments, private vehicles exhibited the highest total OPE concentrations (ΣOPEs), with an average of 264.89 ng/m^3—statistically significantly higher than the other three environments ($p < 0.05$). In homes, offices, and student dormitories, tris(2-chloroethyl) phosphate (TCEP) and tris(2-chloropropyl) phosphate (TCPP) dominated the OPE mixture, accounting for 56% and 34% of ΣOPEs, respectively. By contrast, private cars were characterized by elevated levels of TCPP (68% of ΣOPEs) and tris(1,3-dichloro-2-propyl) phosphate (TDCP, 12%). This addresses the gap in OPE data for car microenvironments.

Shi et al. (2025) (contribution 10) investigated the adsorption, diffusion, and advection–dispersion behavior of ^{99}Tc in the following three types of rocks: granite, clay rock, and mudstone shale. They found that the three types of rocks had no significant adsorption effect on ^{99}Tc. The anion exclusion during diffusion and advection–dispersion processes could block small "channels", causing some nuclide migration to lag, but accelerated the nuclide migration rate in larger "channels". In addition, parameters characterizing the size of anion exclusion in different migration behaviors, such as effective diffusion coefficient (De) and immobile liquid region porosity (θim), were fitted and obtained, guiding radioactive waste disposal.

Berrios-Rolon et al. (2025) (contribution 11) studied three low-molecular-weight polycyclic aromatic hydrocarbons (PAHs), naphthalene (NAP), phenanthrene (PHEN), and anthracene (ANT), in Puerto Rico's Cucharillas Marsh. ∑3PAH concentrations ranged from 7.4 to 2198.8 ng/L, with higher wet-season levels (mean = 745.79 ng/L) than dry-season levels (mean = 186.71 ng/L). A predominant pyrogenic origin was identified, with robust PHEN–ANT correlation (r = 0.824) confirming shared combustion-related sources. Acute ecological risk was low (HQ < 0.01), but chronic risks from PHEN/ANT were noted. This fills the gap in PAH data for tropical urban wetlands.

Berrios-Rolon et al. (2025) (contribution 8) provided a systematic review offering new perspectives on the distribution, sources, and ecotoxicological impacts of PAHs in freshwater systems. They investigated spatiotemporal variability across geographic regions, examining both anthropogenic and natural sources, as well as the mechanisms driving PAH transport and fate. Special attention was given to the ecotoxicological effects of PAHs on freshwater organisms, including bioaccumulation, endocrine disruption, and genotoxicity. They identified knowledge gaps and proposed an interdisciplinary risk assessment framework, serving as a roadmap for future freshwater PAH research.

Xu et al. (2025) (contribution 12) assessed the cytotoxic effects of three typical organic iodides (1,2-diiodoethane, 1,3-diiodopropane, and 1,4-diiodobutane) identified in shale gas extraction wastewater on human hepatocellular carcinoma (HepG2) cells. All three diiodoalkanes exhibited significant toxic effects on HepG2 cells at a concentration of 25 μM. They induced abnormal expression of genes associated with the extracellular space, extracellular matrix (ECM), and endoplasmic reticulum (ER) in HepG2 cells, and triggered excessive intracellular ROS production. This fills the gap in diiodoalkane toxic mechanisms.

3. Conclusions

This Special Issue showcases the diversity and depth of current research on pollutant environmental transport and transformation, addressing critical knowledge gaps through empirical and theoretical contributions. The studies not only advance scientific understanding but also provide practical tools for pollution remediation and risk management. As the field evolves, interdisciplinary collaboration and technology-driven innovation will remain key to tackling emerging environmental challenges and safeguarding ecological and human health.

Funding: This research received no external funding.

Data Availability Statement: Not applicable.

Conflicts of Interest: The author declares no conflicts of interest.

List of Contributions:

1. Zhang, S.; Yecrkenbieke, G.; Shi, S.; Wang, Z.; Yi, L.; Lu, X. Adsorption of Pyrene and Arsenite by Micro/Nano Carbon Black and Iron Oxide. *Toxics* **2024**, *12*, 251. https://doi.org/10.3390/toxics12040251.

2. Liu, Z.; Lin, X.; Wang, X.; Sun, M.; Ma, S.; Zhang, S. Shift in Bacterial Community Structure in the Biodegradation of Benzene and Toluene under Sulfate-Reducing Condition. *Toxics* **2024**, *12*, 423. https://doi.org/10.3390/toxics12060423.
3. Khan, A.U.H.; Liu, Y.; Naidu, R.; Fang, C.; Shon, H.K.; Zhang, H.; Dharmarajan, R. Changes in the Aggregation Behaviour of Zinc Oxide Nanoparticles Influenced by Perfluorooctanoic Acid, Salts, and Humic Acid in Simulated Waters. *Toxics* **2024**, *12*, 602. https://doi.org/10.3390/toxics12080602.
4. Yang, K.; Hung, H.; Huang, W.; Hsieh, C.; Chen, T. Multiphase Partitioning of Estrogens in a River Impacted by Feedlot Wastewater Discharge. *Toxics* **2024**, *12*, 671. https://doi.org/10.3390/toxics12090671.
5. Hou, Y.; Lin, S.; Fan, J.; Zhang, Y.; Jing, G.; Cai, C. Enhanced Adsorption of Cadmium by a Covalent Organic Framework-Modified Biochar in Aqueous Solution. *Toxics* **2024**, *12*, 717. https://doi.org/10.3390/toxics12100717.
6. Khan, A.U.H.; Liu, Y.; Naidu, R.; Fang, C.; Shon, H.K. Influence of Tetrabromobisphenol-A on the Fate and Behavior of Zinc Oxide Nanoparticles Affected by Salts, Humic Acid, and Bovine Serum Albumin in Water Systems. *Toxics* **2025**, *13*, 148. https://doi.org/10.3390/toxics13030148.
7. Kawichai, S.; Sripan, P.; Rerkasem, A.; Rerkasem, K.; Srisukkham, W. Long-Term Retrospective Predicted Concentration of PM2.5 in Upper Northern Thailand Using Machine Learning Models. *Toxics* **2025**, *13*, 170. https://doi.org/10.3390/toxics13030170.
8. Berrios-Rolón, P.J.; Cotto, M.C.; Márquez, F. Polycyclic Aromatic Hydrocarbons (PAHs) in Freshwater Systems: A Comprehensive Review of Sources, Distribution, and Ecotoxicological Impacts. *Toxics* **2025**, *13*, 321. https://doi.org/10.3390/toxics13040321.
9. Cai, Y.; Xu, M.; Ouyang, M.; Wu, Y.; Wang, R.; Zheng, K.; Ren, G. Concentrations, Compositions and Human Exposure Risks to Organophosphate Esters in Indoor Air from Various Microenvironments in Guangzhou, China. *Toxics* **2025**, *13*, 531. https://doi.org/10.3390/toxics13070531.
10. Shi, Y.; Yang, S.; Chen, W.; Zhang, A.; Li, Z.; Wang, L.; Lian, B. Migration Behavior of Technetium-99 in Granite, Clay Rock, and Shale: Insights into Anionic Exclusion Effects. *Toxics* **2025**, *13*, 760. https://doi.org/10.3390/toxics13090760.
11. Berrios-Rolón, P.J.; Márquez, F.; Cotto, M.C. Occurrence and Distribution of Three Low Molecular Weight PAHs in Caño La Malaria, Cucharillas Marsh (Cataño, Puerto Rico): Spatial and Seasonal Variability, Sources, and Ecological Risk. *Toxics* **2025**, *13*, 860. https://doi.org/10.3390/toxics13100860.
12. Xu, M.; Wu, Y.; Cai, Y.; Wang, R.; Ren, G. Cytotoxicity of Typical Diodoalkanes from Shale Gas Wastewater in HepG2 Cells. *Toxics* **2025**, *13*, 943. https://doi.org/10.3390/toxics13110943.

References

1. Wang, F.; Xiang, L.; Sze-Yin Leung, K.; Elsner, M.; Zhang, Y.; Guo, Y.; Pan, B.; Sun, H.; An, T.; Ying, G.; et al. Emerging contaminants: A One Health perspective. *Innovation* **2024**, *5*, 100612. [CrossRef]
2. Meher, A.K.; Zarouri, A. Environmental applications of mass spectrometry for emerging contaminants. *Molecules* **2025**, *30*, 364. [CrossRef] [PubMed]
3. Chettri, D.; Verma, A.K.; Chirania, M.; Verma, A.K. Metagenomic approaches in bioremediation of environmental pollutants. *Environ. Pollut.* **2024**, *363 Pt 2*, 125297. [CrossRef] [PubMed]
4. Zhang, R.; Zheng, X.; Fan, W.; Wang, X.; Zhao, T.; Zhao, X.; Peijnenburg, W.J.G.M.; Vijver, M.G.; Wang, Y. Fate models of nanoparticles in the environment: A critical review and prospects. *Environ. Sci. Nano* **2025**, *12*, 3394–3412. [CrossRef]
5. Gökçe, S.; Şengör, S.S. Reactive transport modeling of uranium in subsurface: Impact of field-scale heterogeneity and biogeochemical dynamics. *Water* **2025**, *17*, 514. [CrossRef]
6. Rad, M.; Abtahi, A.; Berndtsson, R.; McKnight, U.S.; Aminifar, A. Interpretable machine learning for predicting the fate and transport of pentachlorophenol in groundwater. *Environ. Pollut.* **2024**, *345*, 123449. [CrossRef] [PubMed]

toxics

MDPI

Article

Adsorption of Pyrene and Arsenite by Micro/Nano Carbon Black and Iron Oxide

Shuai Zhang, Gulijiazi Yeerkenbieke, Shuai Shi, Zhaoyang Wang, Lijin Yi and Xiaoxia Lu *

Ministry of Education Laboratory for Earth Surface Processes, College of Urban and Environmental Sciences, Peking University, Beijing 100871, China
* Correspondence: luxx@urban.pku.edu.cn

Abstract: Polycyclic aromatic hydrocarbons (PAHs) and arsenic (As) are common pollutants co-existing in the environment, causing potential hazards to the ecosystem and human health. How their behaviors are affected by micro/nano particles in the environment are still not very clear. Through a series of static adsorption experiments, this study investigated the adsorption of pyrene and arsenite (As (III)) using micro/nano carbon black and iron oxide under different conditions. The objectives were to determine the kinetics and isotherms of the adsorption of pyrene and As (III) using micro/nano carbon black and iron oxide and evaluate the impact of co-existing conditions on the adsorption. The microstructure of micro/nano carbon black (C 94.03%) is spherical-like, with a diameter of 100–200 nm. The micro/nano iron oxide (hematite) has irregular rod-shaped structures, mostly about 1 μm long and 100–200 nm wide. The results show that the micro/nano black carbon easily adsorbed the pyrene, with a pseudo-second-order rate constant of 0.016 mg/(g·h) and an adsorption capacity of 283.23 μg/g at 24 h. The micro/nano iron oxide easily adsorbed As (III), with a pseudo-second-order rate constant of 0.814 mg/(g·h) and an adsorption capacity of 3.45 mg/g at 24 h. The mechanisms of adsorption were mainly chemical reactions. Micro/nano carbon black hardly adsorbed As (III), but its adsorption capability for pyrene was reduced by the presence of As (III), and this effect increased with an increase in the As (III) concentration. The adsorbed pyrene on the micro/nano black carbon could hardly be desorbed. On the other hand, the micro/nano iron oxide could hardly adsorb the pyrene, but its adsorption capability for As (III) was increased by the presence of pyrene, and this effect increased with an increase in the pyrene concentration. The results of this study provide guidance for the risk management and remediation of the environment when there is combined pollution of PAHs and As.

Keywords: micro/nano carbon black; micro/nano iron oxide; pyrene; As (III); adsorption; desorption

1. Introduction

Polycyclic aromatic hydrocarbons (PAHs) and arsenic (As) are common pollutants co-existing in the environment [1,2]. PAHs are mainly produced by human activities such as the incomplete combustion of coal, petroleum, and biomass [1] or by natural activities such as volcanic eruptions and forest fires [3]. Due to their high mutagenicity and carcinogenicity, 16 PAHs have been listed as priority pollutants by the European Union (EU) and the United States Environmental Protection Agency (US EPA) [4]. As is widely present in the natural environment, mainly in the form of arsenic sulfide; however, it also has a wide range of uses in human life, e.g., in industries for alloys, pesticides, pharmaceuticals, and anti-corrosion materials [5], and it can enter environmental media via waste discharge or other approaches [6]. As generally exists in two valence states in soil and water environments: III (+3) and V (+5). As (III) is more harmful than As (V) due to its higher solubility and bioavailability [6]. The long-term intake of food or water with a high As content could cause damage to the human liver, heart, and nervous vessels [7].

Black carbon and iron oxide are common substances in soil and water environments. Black carbon describes a range of carbonaceous substances, from partly charred plant residues to highly graphitized soot, that are generated as products of incomplete combustion [8]; they have variable chemical compositions depending on their sources, sometimes being primarily elemental carbon and sometimes existing as complex mixtures of elemental carbon, organic carbon, and other non-carbon species [9]. Iron oxide is one of the most abundant minerals on Earth's surface, and it is primarily located in the shallow crust. It is an iron oxide with a chemical composition of Fe_2O_3. Micro/nano black carbon and iron oxide are likely present in soil and water environments due to various physical, chemical, and biological processes.

Adsorption is a major process that affects the migration and toxicity of pollutants in the environment. Compared to the corresponding larger particles, micro/nano particles generally have higher adsorption capabilities due to their larger specific surface areas [10,11]. Previous studies have shown that PAHs are easily adsorbed by carbonaceous substances, while As is easily adsorbed by metal oxides [12–15]. However, how the co-existence of PAHs and As influences their adsorptions is not clear yet. This knowledge gap should be given attention since PAHs and As often co-exist in the environment.

This study explored the adsorption of pyrene and arsenite (As (III)) using micro/nano carbon black and iron oxide under individual and co-existing conditions through laboratory experiments. The objectives were to determine the kinetics and isotherms of the adsorption of pyrene and As (III) using micro/nano carbon black and iron oxide and evaluate the impact of co-existing conditions on the adsorption. In this study, micro/nano carbon black was used as a surrogate for micro/nano black carbon, which is primarily an elemental carbon. Pyrene and As (III) were used as representatives for PAHs and As, respectively. This is the first time that the impact of the co-existence of PAHs and As on their adsorptions using micro/nano carbon black and iron oxide has been clarified. The results of this study provide a scientific basis for the risk management and remediation of the environment when there is combined pollution of PAHs and As.

2. Materials and Methods

2.1. Materials

Micro/nano iron oxide (Fe_2O_3, purity 99.9%) and sodium arsenite ($NaAsO_2$, purity 99%) were purchased from Innochem Science & Technology Co., Ltd. (Beijing, China). Micro/nano carbon black (industrial grade purity) was purchased from Hewns Biochemical Technology Co., Ltd. (Tianjin, China). Pyrene (95% purity) was purchased from Sun Chemical Technology Co, Ltd. (Shanghai, China). Pyrene standard solution (0.2 mg/mL in Dichloromethane) and isotopic-labeled internal standard solutions (Phe-d10 0.2 mg/mL in Dichloromethane and Pyr-d10 1000 μg/mL in Dichloromethane) were purchased from AccuStandard, Inc. (New Haven, CT, USA). Arsenious acid standard solution (18.2 μg/mL) and Arsenic acid standard solution (35.8 μg/mL) were purchased from the National Institute of Metrology (Beijing, China). n-Hexane (HPLC grade) was purchased from Thermo Fisher Scientific Inc. (Waltham, MA, USA).

2.2. Characterization of Micro/Nano Particles

Scanning electron microscopy (SEM) was employed to measure the morphology of the studied micro/nano carbon black and iron oxide particles according to the Chinese Standard JY/T 010-1996 [16]. X-ray diffraction (XRD) and Fourier transform infrared spectroscopy (FTIR) were used to study the composition and functional group of iron oxide according to the Chinese Standard SY/T 5163-2018 and GB/T 32199-2015 [17,18]. X-ray photoelectron spectroscopy (XPS) was applied to study the composition of carbon black according to the Chinese Standard JY/T 013-1996 [19].

2.3. Adsorption and Desorption Experiments

2.3.1. Pre-Experiment

Pre-experiment of adsorption was performed to determine the optimal ratio of the studied micro/nano particles on pollutants (Table S1). A series of dosages of micro/nano carbon black or iron oxide (0, 5, 10, 25, 50, 100, 500, and 1000 mg) were added to 40 mL glass centrifuge tubes, and then 40 mL of ultrapure water containing 80 μg/L pyrene or 10 mg/L As (III) was added to each tube and shaken in a temperature-controlled shaker (HZQ-2, Jinyi, Changzhou, China) at 200 r/min under 25 °C for 24 h. Thereafter, the tubes were centrifugated in a centrifuge (LD5-10, Jingli, Beijing, China) at 3000 r/min under room temperature for 15 min. The supernatant was taken to measure the concentration of pyrene or As (III). The adsorption capabilities under different dosages were calculated. It turned out that the optimal dosages for micro/nano carbon black and iron oxide were 5 mg and 10 mg, respectively. These dosages were used for further studies.

2.3.2. Adsorption Kinetics

Based on the pre-experiment, micro/nano carbon black easily adsorbed pyrene but hardly adsorb As (III), while micro/nano iron oxide easily adsorbed As (III) but hardly adsorbed pyrene. Therefore, kinetics were performed for the adsorption of pyrene using micro/nano carbon black and the adsorption of As (III) using micro/nano iron oxide. Multiple adsorption tubes were prepared by adding 5 mg micro/nano carbon black and 40 mL pyrene solution (80 μg/L) or 10 mg micro/nano iron oxide and 40 mL As (III) solution (10 mg/L). These tubes were shaken at 200 r/min under 25 °C for different times (1/12, 1/6, 1/4, 1/2, 1, 2, 4, 6, 12, and 24 h). At each time point, three tubes were taken for the analysis of pyrene or As (III) in the supernatant (after centrifuging at 1850× g for 15 min). At 24 h after adsorption, iron oxide was collected for FTIR analysis. In addition, tubes containing 40 mL pyrene solution (80 μg/L) or As (III) (10 mg/L) but no micro/nano carbon black or iron oxide were prepared as controls for losses due to volatilization or other processes.

2.3.3. Adsorption/Desorption Isotherms

For experiments of adsorption isotherms, seven combinations were set up, as shown in Table 1. The concentration of pyrene ranged from 1 to 80 μg/L (1, 2, 5, 10, 20, 40, 60, and 80 μg/L), and the concentration of As (III) ranged from 0.01 to 1 mg/L (0.01, 0.02, 0.05, 0.1, 0.2, 0.4, 0.6, 0.8, and 1 mg/L). All the adsorption tubes were shaken for 24 h (25 °C, 200 r/min), centrifuged for 15 min (1850× g), and analyzed for the target pollutant in the supernatant. The experiments were performed in triplicate.

Table 1. Setup of the experiments of adsorption isotherms.

System Label	Micro/Nano Iron Oxide (mg)	Micro/Nano Carbon Black (mg)	Pyrene [a] (μg/L)	As (III) [b] (mg/L)
IO-As	10	0	0	0.01–1
IO-As-Pyr-1	10	0	80	0.01–1
IO-As-Pyr-2	10	0	1–80	0.01–1
CB-Pyr	0	5	1–80	0
CB-Pyr-As-1	0	5	1–80	1
CB-Pyr-As-2	0	5	1–80	0.01–1
CB-IO-Pyr-As	10	5	1–80	0.01–1

[a] 1–80 refers to initial concentration of pyrene at 1, 2, 5, 10, 20, 40, 60, 80 μg/L; [b] 0.01–1 refers to initial concentration of As (III) at 0.01, 0.02, 0.05, 0.1, 0.2, 0.4, 0.6, 0.8, 1 mg/L.

Experiments of desorption isotherms were performed after the adsorption isotherms experiments were finished. Briefly, the supernatants in the tubes after centrifugation were removed, and equal volumes of ultrapure water were added into the tubes. After being shaken for 24 h (25 °C, 200 r/min), the tubes were centrifuged for 15 min (1850× g), and the supernatants were analyzed for the target pollutants.

2.4. Chemical Analysis

2.4.1. Analysis of Pyrene

For each tube, 10 mL of the collected supernatant was extracted three times with 30 mL of n-hexane (10 mL each time); the collected organic liquid was concentrated at 1 mL via rotary evaporation (N-1100, EYELA, Tokyo, Japan), and the 1 mL liquid was transferred to a 1.5 mL vial for analysis. Phe-d8 and pyr-d10 were used as recovery indicator and internal standard, respectively. A Gas Chromatography Mass Spectrometry with an electron ionization source (6890/5973N, Agilent, Poway, CA, USA) and a DB-5ms capillary column (30 m × 250 μm × 0.25 μm) (Agilent, Santa Clara, CA, USA) were used for the measurement. Helium gas with a flow rate of 2 mL/min was used as the carrier gas, and the inlet temperature was 280 °C. The column temperature program was set as follows: maintained at 50 °C for 1 min, raised the temperature to 200 °C at 15 °C/min, raised the temperature to 305 °C at 20 °C/min, and maintained at 305 °C for 3 min. The ion source temperature was 245 °C, and the fourth stage rod temperature was 150 °C. The concentration of pyrene standards ranged from 1 μg/L to 100 μg/L. The instrument detection limit of pyrene was 0.70 μg/L. A standard solution of pyrene (100 μg/L) was measured every 20 samples to control the deviation of instrument performance, and the obtained deviation was within 20%.

2.4.2. Analysis of As

For each tube, 1 mL of the collected supernatant passed through a 0.22 μm filter membrane, was diluted with ultrapure water to 10 mL, and then was analyzed using a Liquid Chromatography Atomic Fluorescence Spectrometer (LC-AFS, LC-AFS6500, Haiguang, Beijing, China) with an anion exchange column (Hamilton PRP-X100, 250 mm × 4.1 mm × 10 μm) and hollow cathode lamps. Phosphate buffer was used as the mobile phase for the liquid chromatography (LC) system, with a flow rate of 1 mL/min. A 10% HCl solution and 3.5% KBH_4 solution were respectively used as the carrier and reducing agent for the atomic fluorescence spectroscopy (AFS) system, and the main current and auxiliary current were 300 mA and 150 mA, respectively. At a flow rate of 300 mL/min and shield flow rate of 900 mL/min, and under negative high pressure of 300 V, the ionic arsenic was reduced to atomic arsenic, with different species of As being separated. The concentrations of As (III) and As (V) standards both ranged from 1 μg/L to 200 μg/L. The detection limits for As (III) and As (V) were 0.545 μg/L and 0.837 μg/L, respectively. A standard solution of As (III) and As (V) (100 μg/L for each species) was measured every 20 samples to control the deviation of instrument performance, and the obtained deviations were within 10%.

2.5. Data Analysis

The adsorption processes of pyrene and As (III) using micro/nano carbon black and iron oxide were simulated by quasi-first-order and quasi-second-order kinetic models, as shown in Equations (1) and (2).

$$\ln(q_e - q_t) = \ln q_e - k_1 t \quad (1)$$

$$\frac{t}{q_t} = \frac{1}{k_2 q_e^2} + \frac{1}{q_e} t \quad (2)$$

where t is the adsorption time (h), q and q_t are the equilibrium adsorption capacity and adsorption capacity at time t (μg/g or mg/g), and k_1 and k_2 are the adsorption rates of pseudo-first-order kinetics (h^{-1}) and pseudo-second-order kinetics ($mg(g{\cdot}h)^{-1}$ or $\mu g(g{\cdot}h)^{-1}$), respectively.

Three adsorption/desorption isotherm models were used to describe the equilibrium process of pollutants between solid and liquid phases, as shown in Equations (3)–(5).

$$q_e = K_H C_e \quad (3)$$

$$\frac{1}{q_e} = \frac{1}{q_{max}K_L}\frac{1}{Ce} + \frac{1}{q_{max}} \tag{4}$$

$$\lg q_e = \frac{1}{n}lgCe + lgK_F \tag{5}$$

where q_e and q_{max} are the equilibrium adsorption capacity and maximum adsorption capacity (μg/g or mg/g), C_e is the equilibrium concentration in the liquid (μg/L), K_H is adsorption coefficient for the Henry model (L/g), K_L is a constant for the Langmuir model (L/mg), K_F (L/mg) and n (g/L) are constants for the Freundlich model.

The model fitting and data visualization were performed using Origin2022. Data between treatments and controls were compared by t-test using SPSS 26.0, with a significance level set at 0.05. Other statistical analyses of the data were conducted using EXCEL 2016.

3. Results

3.1. Characterizations of Micro/Nano Carbon Black and Iron Oxide

The characterizations of micro/nano carbon black are shown in Figure 1. The SEM image shows that the microstructure of micro/nano carbon black is spherical-like, with irregular protrusions on the surface and relatively uniform sizes, with a diameter of 100–200 nm. The XPS spectrum shows that there are high-intensity peaks in the binding energies of the C1s and O1s orbitals on the binding energy spectrum of the sample, while there are low-intensity peaks in the binding energies of the S2p and N1s orbitals. According to the quantitative results, the proportion of each element's content (in atomic number, except for H) in the carbon black is C (94.03%), O (4.49%), N (1.11%), and S (0.37%).

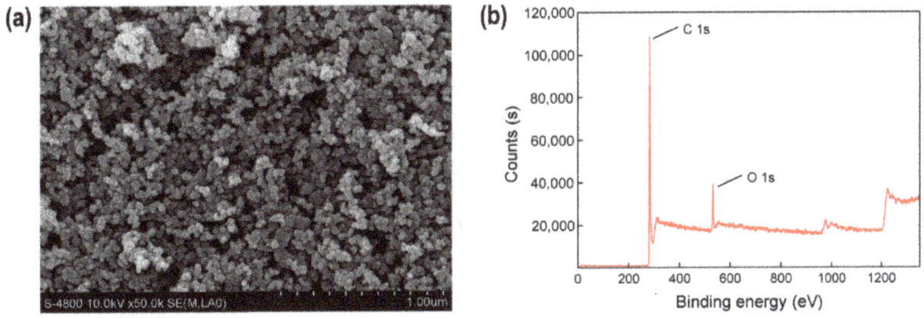

Figure 1. Characterization diagram of micro/nano carbon black. (**a**) SEM; (**b**) XPS.

The characterizations of micro/nano iron oxide are shown in Figure 2. The SEM image shows that the micro/nano iron oxide has an irregular rod-shaped structure with significant size differences. Most individual rods are about 1 μm long and 100–200 nm wide. The XRD pattern shows that there are six typical diffraction peaks that appeared in the 2θ diffraction angle of iron oxide, which are consistent with the diffraction peaks of hematite in the standard database, indicating that the iron oxide is hematite.

Figure 2. Characterization diagram of micro/nano iron oxide. (**a**) SEM; (**b**) XRD.

3.2. *Adsorption Kinetics of Pyrene and As (III) by Micro/Nano Carbon Black and Iron Oxide*

The pre-experiments showed that at a micro/nano carbon black dosage of less than 100 mg in 40 mL water containing 10 mg/L As (III), the micro/nano carbon black had little adsorption of As (III); however, when the dosage of micro/nano carbon black was larger than 100 mg, As (III) was oxidized to As (V). At a dosage of micro/nano iron oxide ranging from 0–1000 mg in 40 mL water containing 80 μg/L pyrene, the micro/nano iron oxide had little adsorption of pyrene (Figure S1).

Adsorption kinetics were studied for the adsorption of pyrene using micro/nano carbon black and the adsorption of As (III) using micro/nano iron oxide at optimal dosages of sorbents (5 mg for carbon black and 10 mg for iron oxide). The adsorption kinetics data are shown in Figure 3. Both adsorptions reached equilibrium within 24 h, and the adsorption data are fit well by both quasi-first-order and quasi-second-order kinetic models, with the latter being better (Table 2). This indicated that the mechanisms of both adsorptions were mainly chemical reactions. At equilibrium (24 h), the maximum adsorption capacity of micro/nano carbon black for pyrene was 283.23 μg/g, while the maximum adsorption capacity of micro/nano iron oxide for As was 3.45 mg/g.

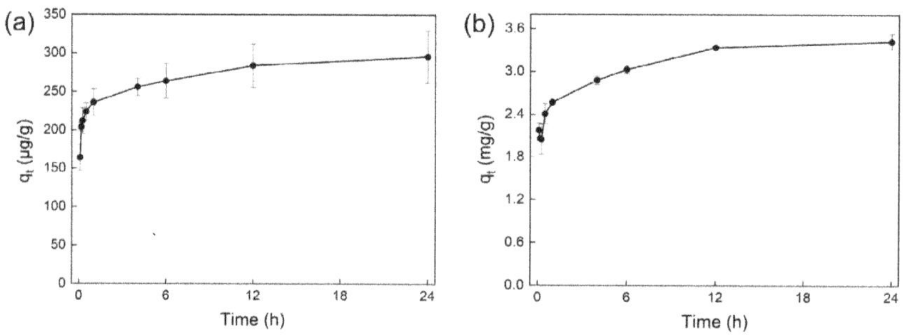

Figure 3. The adsorption kinetics of pyrene and As (III) by micro/nano carbon black and iron oxide. (**a**) Adsorption kinetics of pyrene by micro/nano carbon black. (**b**) Adsorption kinetics of As (III) by micro/nano iron oxide.

Table 2. The values of parameters in the adsorption kinetic models.

System	Pseudo-first-order kinetics			Pseudo-second-order kinetics		
	q_e (mg/g)	k_1 (/h)	R^2	q_e (mg/g)	k_2 (mg/(g·h))	R^2
IO-As	3.425	0.217	0.982	3.446	0.814	0.998

System	Pseudo-first-order kinetics			Pseudo-second-order kinetics		
	q_e (μg/g)	k_1(/h)	R^2	q_e (μg/g)	k_2 (μg/(g·h))	R^2
CB-Pyr	296.000	0.172	0.936	283.225	0.016	0.998

3.3. *Adsorption Isotherms of Pyrene and As (III) by Micro/Nano Carbon Black and Iron Oxide*

The adsorption isotherms of pyrene using micro/nano carbon black are shown in Figure 4a. Under the study conditions, the adsorption capacity of micro/nano carbon black for pyrene increased with the initial concentration of pyrene in the liquid, and the equilibrium process could be a good fit for the linear Henry model. The k-values in the different systems had the following order: CB-Pyr (28.171 L/g) > CB-IO-Pry-As (17.785 L/g) > CB-Pyr-As-2 (16.581 L/g) > CB-Pyr-As-1 (7.914 L/g), as shown in Table 3. The presence of As (III) reduced the adsorption coefficient of the micro/nano carbon black for pyrene, and this impact increased with an increased concentration of As (III). The presence of iron oxide slightly reduced the impact of As (III) due to its adsorption of As (III). Even though the iron oxide could adsorb As (III), the adsorption coefficient in the CB-IO-Pry-As system was much lower than that in the CB-Pyr system.

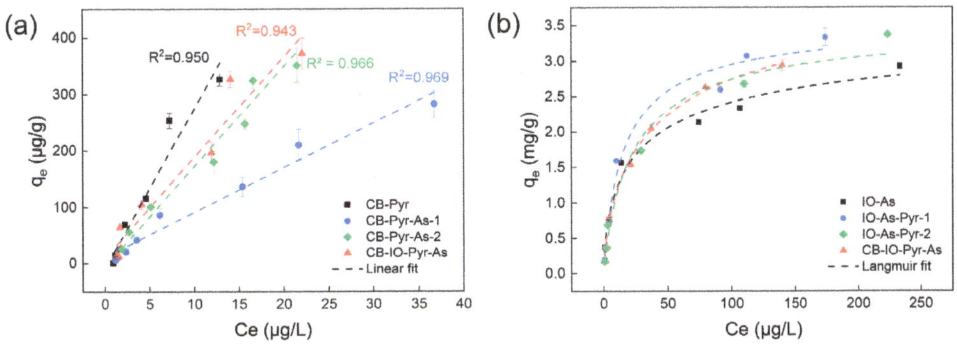

Figure 4. Adsorption isotherms and fitting model parameters of micro/nano particles for pyrene and As (III) (Details on the adsorption systems are shown in Table 1). (**a**) Adsorption isotherms of pyrene on micro/nano carbon black, The black, blue, green and red lines represent the linear fit for CB-Pyr, CB-Pyr-As-1, CB-Pyr-As-2 and CB-IO-Pyr-As, respectively; (**b**) Adsorption isotherms of As (III) by micro/nano iron oxide, The black, blue, green and red lines represent the linear fit for IO-As, IO-As-Pyr-1, IO-As-Pyr-2 and CB-IO-Pyr-As, respectively. The colors of the fit lines are corresponding to the colors of the legends.

Table 3. The values of parameters in the adsorption isotherms models.

System *	Langmuir Model			Freundlich Model		
	K_L (L/mg)	q_e (mg/g)	R^2	K_F (L/mg)	n (g/L)	R^2
IO-As	0.117	2.760	0.951	0.333	2.298	0.930
IO-As-Pyr-1	0.060	3.827	0.986	0.294	1.984	0.935
IO-As-Pyr-2	0.059	3.367	0.987	0.250	1.946	0.947
CB-IO-Pyr-As	0.079	3.366	0.954	0.294	1.967	0.951

System *	Henry Model	
	K_H (L/g)	R^2
CB-Pyr	28.171	0.950
CB-Pyr-As-1	7.914	0.969
CB-Pyr-As-2	16.581	0.966
CB-IO-Pyr-As	17.785	0.943

* In IO-As-Pyr-1, the concentrations of As (III) ranged from 0.01 to 1 mg/L, and the concentration of pyrene was kept at 80 μg/L; in IO-As-Pyr-2, the concentrations of As (III) ranged from 0.01 to 1 mg/L, and the concentration of pyrene ranged from 1 to 80 μg/L; in CB-Pyr-As-1, the concentrations of pyrene ranged from 1 to 80 μg/L, and the concentration of As (III) was kept at 1 mg/L; in CB-Pyr-As-2, the concentration of pyrene ranged from 1 to 80 μg/L, and the concentration of As (III) ranged from 0.01 to 1 mg/L.

The adsorption isotherms of As (III) using micro/nano iron oxide are shown in Figure 4b. Under the study conditions, the adsorption capacity of the micro/nano iron oxide for As (III) increased with the initial concentration of As (III) in the liquid, and the equilibrium process could be a good fit for the Langmuir model and Freundlich model, with the former being better (Table 3). The q_e values in the different systems had the following order: IO-As (2.762 mg/g) < CB-IO-Pry-As (3.366 mg/g) < IO-As-Pyr-2 (3.367 mg/g) < IO-As-Pyr-1 (3.827 mg/g), as shown in Table 3. The presence of pyrene increased the adsorption capacity of the micro/nano iron oxide for As (III), and this impact increased with an increased concentration of pyrene. The presence of carbon black barely reduced the impact of pyrene, even though the carbon black easily adsorbed the pyrene.

The adsorption capacity in the CB-IO-Pry-As system was much higher than in the IO-As system. The obtained q_e values of As (III) were comparable with those reported in the literature. However, the obtained q_e values of pyrene were much less than those reported in the literature, and the reason was that the initial concentrations in our study were much less than those in the reported studies (Table S2).

The authors understand that the adsorption capacity of a material increases with an increasing adsorbate concentration; therefore, a comparative table must be prepared with other adsorbents present in the literature for the same purpose, describing the maximum capacity obtained and the concentration used in each study.

3.4. Desorption Isotherms of Pyrene and As (III) from Micro/Nano Carbon Black and Iron Oxide

In the desorption experiment, the desorption of pyrene was not detected in any of the studied systems, but a slight desorption of As (III) was detected. Figure 5 shows the desorption isotherms of As (III) in the IO-As, CB-IO-Pry-As, and IO-As-Pyr-1 systems. The desorption capacity of As (III) increased with the initial loading of As (III) in the solid phase, and the equilibrium process could be described by both the Langmuir model and the Freundlich Model, with the former being better, as shown in Table 4. Due to residual As (III) in the liquid at the beginning of the desorption experiment, there were certain errors in the calculation of the desorption isotherm parameters. Nevertheless, it could be seen that the presence of pyrene reduced the desorption of As (III). Overall, the desorption rates of As (III) were small (less than 4%).

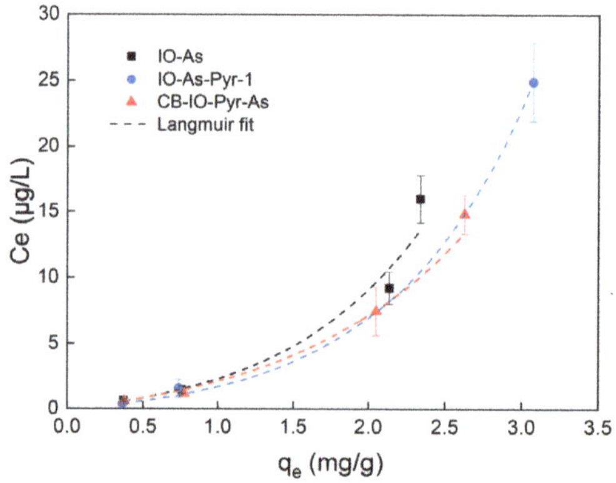

Figure 5. Desorption isotherms on micro/nano iron oxide, The black, blue and red lines represent the linear fit for IO-As, IO-As-Pyr-1 and CB-IO-Pyr-As, respectively (Details on the adsorption systems are shown in Table 1). The colors of the fit lines are corresponding to the colors of the legends.

Table 4. Langmuir and Freundlich models for desorption isotherms.

System	Langmuir Model			Freundlich Model		
	K_L (L/mg)	q_e (mg/g)	R^2	K_F (L/mg)	*n* (g/L)	R^2
IO-As	0.222	3.103	0.999	0.546	1.764	0.978
IO-As-Pyr-1	0.400	2.532	0.959	0.628	1.796	0.946
CB-IO-Pyr-As	0.181	3.946	0.990	0.592	1.709	0.972

4. Discussion

4.1. Adsorption Mechanism

This study showed that micro/nano carbon black had a strong adsorption of the pyrene but hardly adsorbed As (III), while micro/nano iron oxide had a strong adsorption of As (III) but hardly adsorbed the pyrene.

Previous studies have shown that the adsorption mechanisms of carbon materials on PAHs included hydrophobic interactions, van der Waals forces, and π bond interactions [12,20,21]. Micro/nano carbon black has a large specific surface area, and its surface hydrophobicity could create favorable conditions for the adsorption of pyrene [22]. In the production of micro/nano carbon black, high-temperature reactions of carbon-containing substances might form some aromatic functional groups, which play an important role in the adsorption of pyrene [23]. In addition to π–π interactions, N and O atoms in carbon-based materials could provide lone-pair electrons and empty orbitals, forming N–P interactions with π electrons in PAHs. At the same time, some polar functional groups on carbon adsorbents can also enhance their polarity, which enhances the adsorption of PAHs by using carbon materials in dipole interactions [13,22]. In this study, we attempted to characterize the functional groups on the surface of micro/nano carbon black using the FITR method, but due to the high blackness of the carbon material and its overall strong absorption of infrared radiation, the functional groups could not be identified. However, an XPS analysis detected certain amounts of O, N, and S, which might form polar hydrophobic functional groups. After being adsorbed by the micro/nano carbon black, the pyrene was difficult to desorb, indicating that the adsorption reaction was very strong.

Micro/nano iron oxide had strong adsorption of As (III), which was related to its large specific surface area and abundant hydroxyl functional groups on the surface. In aqueous solutions, the surfaces of metal oxides were covered by many hydroxyl groups, which originated from the metal oxides themselves or the dissociation of water molecules [14]. The surface functional groups of micro/nano iron oxide were mainly composed of bound water and –OH. After adsorbing As (III), the –OH absorption peaks at the wave numbers of 1091 cm^{-1}, 1039 cm^{-1}, and 917 cm^{-1} were significantly weakened, indicating that surface hydroxyl groups participated in the adsorption process of As (III) using micro/nano iron oxide (Figure 6). The hydroxyl groups on the surface of the iron oxide were able to form complexes with various heavy metal ions to achieve adsorption, among which arsenite was a good hydroxyl ligand [15]. The adsorption process first formed physical adsorption through hydrogen bonding between O of the arsenite and H of the hydroxyl group. The arsenite physically adsorbed on the surface of the iron oxide was further transformed into monodentate or bidentate complexes. The Gibbs activation free energy of this process was often low (less than 0), making the reaction easy to occur spontaneously [24]. Once they exchanged with the hydroxyl groups on the surface of the iron oxide through a ligand exchange to form inner spherical complexes, it was difficult to detach them from the surface of the iron oxide.

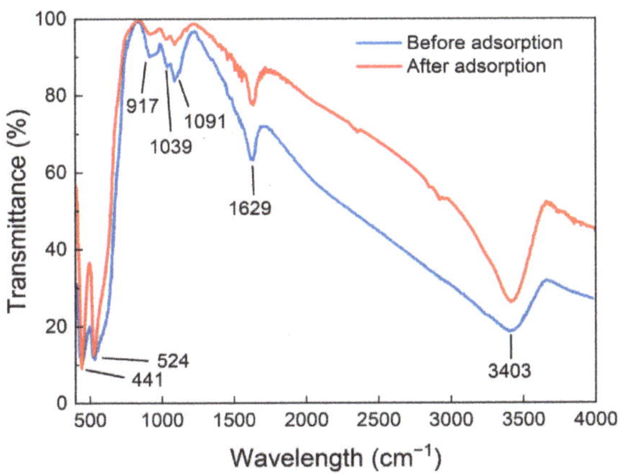

Figure 6. Changes in the FTIR spectra of micro/nano iron oxide before and after the adsorption of As (III).

4.2. The Influence of Co-Existing Pollutants on the Adsorption

Although the micro/nano carbon black did not adsorb As (III), the presence of As (III) reduced the adsorption of the pyrene when using micro/nano carbon black. The addition of micro/nano iron oxide slightly reduced the impact of As (III) on the adsorption of pyrene when using micro/nano carbon black. On the other hand, although the micro/nano iron oxide did not adsorb the pyrene, the presence of the pyrene increased the adsorption of As (III) when using the micro/nano iron oxide. The addition of the micro/nano carbon black did not reduce the impact of the pyrene on the adsorption of As (III) when using micro/nano iron oxide.

As (III) reduced the adsorption of pyrene when using micro/nano carbon black, which might be related to the pH change in the solution after the addition of arsenite. This study showed that arsenite is a weak acid that could bind with protons and hydrolyze in an aqueous solution, and it mainly exists in the forms of H_3AsO_3 and $H_2AsO_3^-$, increasing the pH of the aqueous solution [25]. A higher pH was not conducive to the adsorption of the pyrene when using micro/nano carbon black as OH- in the solution bound to the surface of the carbon black. The negative charge on the surface of the micro/nano carbon black increased the electrostatic resistance that the electron-rich pyrene needed to overcome when being adsorbed on the surface, thereby reducing the adsorption efficiency [26]. Other studies have suggested that in an aqueous solution of pyrene, which is different from a small volume of metal cations (such as Li^+), which are preferentially bound to water molecules to form stable hydration shells, larger volume metal cations (such as K^+) are more likely to preferentially bind to the pyrene through cation−π bonding, which might compete for the π bonding sites of the pyrene with micro/nano carbon black [27]. In a study by Eeshwarasinghe et al., heavy metals such as Cd, Cu, and Zn reduced the adsorption capacities of PAHs by activating carbon particles, which was related to the decrease in negative zeta potential [28].

Pyrene could increase the adsorption of As (III) when using micro/nano iron oxide, which might be due to the strong affinity of pyrene with protons dissociated from water, thereby promoting water ionization and producing more OH−. These OH− are further combined with micro/nano iron oxide to form hydroxyl radicals on its surface, providing more adsorption sites for arsenite [29,30]. Due to the large electron-rich conjugated π bond in aromatic compounds as well as the electron-deficient state of H and protons in water molecules, pyrene molecules might combine with protons to form O-H $\cdots\pi$ or H+$\cdots\pi$ forces, which are electrostatic forces similar to hydrogen bonds and promote water ionization [31]. So far, numerous theoretical calculations and experimental analyses using benzene as a model have confirmed that the molecular process of Benzenem-H_2On-H+ (m represents the number of benzene molecules) is widely present in aqueous benzene solutions [32,33]. This may provide new approaches for the remediation of environments when there is combined pollution of PAHs and As. In a study by Zhang et al., there was a synergistic effect on the adsorptions of pyrene and Cu (II) when using silica adsorbents doped with Fe (III), which might be attributed to the formation of pyrene-Cu (II) complexes [34].

We should be careful when dealing with nanomaterials to remediate the environment. Nanomaterials may cause toxicity to microbes and multi-cellular organisms [35]. In addition, the widespread application of nanomaterials in remediation has been hampered by challenges in accurately delivering them to contaminated sites due to their rapid aggregation and/or retention. An understanding of the processes that influence the environmental transportation and fate of nanomaterials is critical for optimizing environmental applications and assessing risks [36].

5. Conclusions

This study investigated the adsorption of pyrene and As (III) when using micro/nano carbon black and iron oxide under individual and co-existing conditions. The micro/nano carbon black easily adsorbed the pyrene (pseudo-second-order rate constant of 0.016 mg/(g·h) and adsorption capacity of 283.23 μg/g at 24 h), while the micro/nano iron oxide easily adsorbed As (III) (pseudo-second-order rate constant of 0.814 mg/(g·h) and adsorption capacity of 3.45 mg/g at 24 h). The mechanisms of the adsorptions were mainly chemical reactions. The adsorption isotherms of the pyrene best fit the Henry model, while the adsorption isotherms of As (III) best fit the Langmuir model. The presence of As (III) reduced the adsorption of the pyrene when using micro/nano black carbon, and this impact increased with an increase in the As (III) concentration. The presence of pyrene increased the adsorption of As (III) when using micro/nano iron oxide, and this impact increased with an increase in the pyrene concentration. The interactions between the micro/nano carbon black and micro/nano iron oxide were small. These results provide guidance for the risk management and remediation of the environment when there is a combined pollution of PAHs and As.

There were limitations in this study. The functional groups on the surface of the micro/nano carbon black could not be measured using the FITR method, and, therefore, the deeper reaction mechanism was not clarified. In addition, factors influencing the adsorptions of the pyrene and As (III) when using the micro/nano carbon black and iron oxide were not investigated. In future studies, the mechanisms and influencing factors of the impact of the co-existence of pyrene and As (III) on their adsorptions should be explored.

Supplementary Materials: The following supporting information can be downloaded at https://www.mdpi.com/article/10.3390/toxics12040251/s1, Figure S1: Adsorption of pyrene and As (III) by micro/nano carbon black and iron oxide; Table S1: Setup of Pre-Experiment; Table S2: Adsorption isotherms of As (III) and pyrene reported in the literature [37–42].

Author Contributions: Conceptualization, S.Z. and X.L.; methodology, S.Z.; validation, G.Y. and L.Y.; formal analysis, S.Z.; investigation, S.Z. and Z.W.; resources, X.L.; data curation, S.S.; writing—original draft preparation, S.Z.; writing—review and editing, X.L.; visualization, S.Z.; supervision, X.L.; project administration, X.L.; funding acquisition, X.L. All authors have read and agreed to the published version of the manuscript.

Funding: This research was funded by the Ministry of Science and Technology of the People's Republic of China (2019YFC1804200) and the National Natural Science Foundation of China (41991331, 41771528).

Institutional Review Board Statement: Not applicable.

Informed Consent Statement: Not applicable.

Data Availability Statement: The data presented in this study are available on request from the corresponding author. The data are not publicly available due to privacy.

Acknowledgments: This research was supported by High-performance Computing Platform of Peking University and Public Instrument Platform of College of Urban and Environmental Science, Peking University.

Conflicts of Interest: The authors have declared no potential conflict of interest with respect to the research, authorship, and/or publication of this article.

References

1. Zhou, Y.; Long, T.; Zhu, X.; Wang, L.; Kong, L.; Li, Y.; Shi, J. Research Progress on Distribution and Remediation Technologies for the Combined Pollution of Heavy Metals and Polycyclic Aromatic Hydrocarbons in Soil. *J. Ecol. Rural Environ.* **2019**, *35*, 964–975.
2. Xu, J.; Wu, J.; Wang, Y.; Lu, Y.; Sun, H. Effects of dissolved organic matter on the adsorption and desorption of pyrene on peat and kaolin. *Environ. Chem.* **2021**, *40*, 531–539.
3. Liu, B.; Chen, B.; Zhang, B.Y.; Jing, L.; Zhang, H.; Lee, K. Photocatalytic Degradation of Polycyclic Aromatic Hydrocarbons in Offshore Produced Water: Effects of Water Matrix. *J. Environ. Eng.* **2016**, *142*, 04016054. [CrossRef]
4. Wang, Y.; Li, F.; Liu, C.; Liang, J. Photodegradation of polycyclic aromatic hydrocarbon pyrene by iron oxide in solid phase. *J. Hazard. Mater.* **2009**, *162*, 716–723. [CrossRef] [PubMed]
5. Liao, X.-Y.; Chen, T.-B.; Xie, H.; Liu, Y.-R. Soil As contamination and its risk assessment in areas near the industrial districts of Chenzhou City, Southern China. *Environ. Int.* **2005**, *31*, 791–798. [CrossRef] [PubMed]
6. De, D.; Mandal, S.M.; Bhattacharya, J.; Ram, S.; Roy, S.K. Iron oxide nanoparticle-assisted arsenic removal from aqueous system. *J. Environ. Sci. Health Part A-Toxic/Hazard. Subst. Environ. Eng.* **2009**, *44*, 155–162. [CrossRef] [PubMed]

7. Deng, T.; Ma, P.; Li, H. As (III) Adsorption Effects of Montmorillonite, Iron Oxides and Their Complex. *J. Ecol. Rural Environ.* **2017**, *33*, 252–259.
8. Shrestha, G.; Traina, S.J.; Swanston, C.W. Black Carbon's Properties and Role in the Environment: A Comprehensive Review. *Sustainability* **2010**, *2*, 294–320. [CrossRef]
9. Long, C.M.; Nascarella, M.A.; Valberg, P.A. Carbon black vs. black carbon and other airborne materials containing elemental carbon: Physical and chemical distinctions. *Environ. Pollut.* **2013**, *181*, 271–286. [CrossRef]
10. Li, N.; Jiang, Y.; Jia, X.; Zhang, P.; Xia, T. Accumulation Characteristics and Health Risk Assessment of Cr, As and PAHs in PM_(10) and PM_(2.5) Fractions of Soils. *Ecol. Environ. Sci.* **2019**, *28*, 1700–1712.
11. Liu, D.; Xu, Q.; Zhang, L.; Wang, J.; Peng, S. Assessing the water-sediment diffusive process of PAHs in Chaohu Lake watershed: A fugacity based approach. *Acta Sci. Circumstantiae* **2018**, *38*, 930–939.
12. Sun, Y.; Yang, S.; Zhao, G.; Wang, Q.; Wang, X. Adsorption of Polycyclic Aromatic Hydrocarbons on Graphene Oxides and Reduced Graphene Oxides. *Chem.-Asian J.* **2013**, *8*, 2755–2761. [CrossRef]
13. Zhang, K.; Chen, B.; Mao, J.; Zhu, L.; Xing, B. Water clusters contributed to molecular interactions of ionizable organic pollutants with aromatized biochar via pi-PAHB: Sorption experiments and DFT calculations. *Environ. Pollut.* **2018**, *240*, 342–352. [CrossRef]
14. Wang, L.; Shi, C.; Pan, L.; Zhang, X.; Zou, J.-J. Rational design, synthesis, adsorption principles and applications of metal oxide adsorbents: A review. *Nanoscale* **2020**, *12*, 4790–4815. [CrossRef]
15. Siddiqui, S.I.; Chaudhry, S.A. Chaudhry, Iron oxide and its modified forms as an adsorbent for arsenic removal: A comprehensive recent advancement. *Process Saf. Environ. Prot.* **2017**, *111*, 592–626. [CrossRef]
16. Goss, K.U.; Schwarzenbach, R.P. Linear free energy relationships used to evaluate equilibrium partitioning of organic compounds. *Environ. Sci. Technol.* **2001**, *35*, 1–9. [CrossRef] [PubMed]
17. *JY/T 010-1996*; General Principles of Analytical Scanning Electron Microscopy. State Education Commission of the PRC: Beijing, China, 1997.
18. *SY/T 5163-2018*; Analysis Method for Clay Minerals and Ordinary Non-Clay Mineralsin Sedimentary Rocks by the X-ray Diffraction. National Energy Administration: Beijing, China, 2018.
19. *GB/T 32199-2015*; Standard Practice for General Techniques for Obtaining Infrared Spectra for Qualitative Analysis. General Administration of Quality Supervision, Inspection and Quarantine of the People's Republic of China: Beijing, China, 2015.
20. *JY/T 013-1996*; General Rules for Electron Spectroscopic Analysis. State Education Commission of the PRC: Beijing, China, 1997.
21. Paszkiewicz, M.; Sikorska, C.; Leszczyńska, D.; Stepnowski, P. Helical Multi-walled Carbon Nanotubes as an Efficient Material for the Dispersive Solid-Phase Extraction of Low and High Molecular Weight Polycyclic Aromatic Hydrocarbons from Water Samples: Theoretical Study. *Water Air Soil Pollut.* **2018**, *229*, 253. [CrossRef] [PubMed]
22. Li, F.; Chen, J.; Hu, X.; He, F.; Bean, E.; Tsang, D.C.; Ok, Y.S.; Gao, B. Applications of carbonaceous adsorbents in the remediation of polycyclic aromatic hydrocarbon-contaminated sediments: A review. *J. Clean. Prod.* **2020**, *255*, 120263. [CrossRef]
23. Rong, H.A.O.; Shaolin, P.E.N.G.; Yantun, S.O.N.G. Effects of different temperature on surface functional groups of black carbon. *Ecol. Environ. Sci.* **2010**, *19*, 528–531.
24. Farrell, J.; Chaudhary, B.K. Understanding Arsenate Reaction Kinetics with Ferric Hydroxides. *Environ. Sci. Technol.* **2013**, *47*, 8342–8347. [CrossRef]
25. Raposo, J.C.; Sanz, J.; Zuloaga, O.; Olazabal, M.A.; Madariaga, J.M. Thermodynamic model of inorganic arsenic species in aqueous solutions. Potentiometric study of the hydrolytic equilibrium of arsenious acid. *J. Solut. Chem.* **2003**, *32*, 253–264. [CrossRef]
26. Gupta, H. Removal of Phenanthrene from Water Using Activated Carbon Developed from Orange Rind. *Int. J. Sci. Res. Environ. Sci.* **2015**, *3*, 248–255. [CrossRef]
27. Pašalić, H.; Aquino, A.J.A.; Tunega, D.; Haberhauer, G.; Gerzabek, M.H.; Lischka, H. Cation-π interactions in competition with cation microhydration: A theoretical study of alkali metal cation-pyrene complexes. *J. Mol. Model.* **2017**, *23*, 131. [CrossRef] [PubMed]
28. Eeshwarasinghe, D.; Loganathan, P.; Vigneswaran, S. Simultaneous removal of polycyclic aromatic hydrocarbons and heavy metals from water using granular activated carbon. *Chemosphere* **2019**, *223*, 616–627. [CrossRef]
29. Deng, M.; Wu, X.; Zhu, A.; Zhang, Q.; Liu, Q. Well-dispersed TiO_2 nanoparticles anchored on Fe_3O_4 magnetic nanosheets for efficient arsenic removal. *J. Environ. Manag.* **2019**, *237*, 63–74. [CrossRef] [PubMed]
30. Cheng, W.; Xu, J.; Wang, Y.; Wu, F.; Xu, X.; Li, J. Dispersion-precipitation synthesis of nanosized magnetic iron oxide for efficient removal of arsenite in water. *J. Colloid Interface Sci.* **2015**, *445*, 93–101. [CrossRef]
31. Grabowski, S.J.; Sokalski, W.A.; Leszczynski, J. Is a $\pi\cdots H+\cdots\pi$ Complex Hydrogen Bonded? *J. Phys. Chem. A* **2004**, *108*, 1806–1812. [CrossRef]
32. Kryachko, E.S.; Nguyen, M.T. Low energy barrier proton transfer in protonated benzene-water complex. *J. Phys. Chem. A* **2001**, *105*, 153–155. [CrossRef]
33. Cheng, T.C.; Bandyopadhyay, B.; Mosley, J.D.; Duncan, M.A. IR Spectroscopy of Protonation in Benzene-Water Nanoclusters: Hydronium, Zundel, and Eigen at a Hydrophobic Interface. *J. Am. Chem. Soc.* **2012**, *134*, 13046–13055. [CrossRef]
34. Zhang, Z.; Hou, X.; Zhang, X.; Li, H. The synergistic adsorption of pyrene and copper onto Fe(III) functionalized mesoporous silica from aqueous solution. *Colloids Surf. A Physicochem. Eng. Asp.* **2017**, *520*, 39–45. [CrossRef]
35. Abbas, Q.; Yousaf, B.; Ullah, H.; Ali, M.U.; Ok, Y.S.; Rinklebe, J. Environmental transformation and nano-toxicity of engineered nano-particles (ENPs) in aquatic and terrestrial organisms. *Crit. Rev. Environ. Sci. Technol.* **2020**, *50*, 2523–2559. [CrossRef]

36. Bradford, S.A.; Shen, C.; Kim, H.; Letcher, R.J.; Rinklebe, J.; Ok, Y.S.; Ma, L. Environmental applications and risks of nanomaterials: An introduction to CREST publications during 2018–2021. *Crit. Rev. Environ. Sci. Technol.* **2022**, *52*, 3753–3762. [CrossRef]
37. Song, K.; Kim, W.; Suh, C.-Y.; Shin, D.; Ko, K.-S.; Ha, K. Magnetic iron oxide nanoparticles prepared by electrical wire explosion for arsenic removal. *Powder Technol.* **2013**, *246*, 572–574. [CrossRef]
38. Chowdhury, S.R.; Yanful, E.K. Arsenic removal from aqueous solutions by adsorption on magnetite nanoparticles. *Water Environ. J.* **2011**, *25*, 429–437. [CrossRef]
39. Sahu, U.K.; Sahu, M.K.; Mahapatra, S.S.; Patel, R.K. Removal of As (III) from Aqueous Solution Using Fe_3O_4 Nanoparticles: Process Modeling and Optimization Using Statistical Design. *Water Air Soil Pollut.* **2016**, *228*, 45. [CrossRef]
40. Yang, K.; Wang, X.; Zhu, L.; Xing, B. Competitive Sorption of Pyrene, Phenanthrene, and Naphthalene on Multiwalled Carbon Nanotubes. *Environ. Sci. Technol.* **2006**, *40*, 5804–5810. [CrossRef] [PubMed]
41. Yakout, S.M.; Daifullah, A.A.M.; El-Reefy, S.A. Adsorption of Naphthalene, Phenanthrene and Pyrene from Aqueous Solution Using Low-Cost Activated Carbon Derived from Agricultural Wastes. *Adsorpt. Sci. Technol.* **2013**, *31*, 293–302. [CrossRef]
42. Adeola, A.O.; Forbes, P.B.C. Optimization of the sorption of selected polycyclic aromatic hydrocarbons by regenerable graphene wool. *Water Sci. Technol.* **2019**, *80*, 1931–1943. [CrossRef]

toxics

MDPI

Article

Shift in Bacterial Community Structure in the Biodegradation of Benzene and Toluene under Sulfate-Reducing Condition

Zhengwei Liu [1,2], Xiaoyu Lin [1,2], Xinzhe Wang [1,2], Mingbo Sun [1,2], Shici Ma [1,2] and Shucai Zhang [1,2,*]

1 State Key Laboratory of Chemical Safety, Qingdao 266071, China; liuzw.qday@sinopec.com (Z.L.); linxy.qday@sinopec.com (X.L.); wangxz.qday@sinopec.com (X.W.); sunmb.qday@sinopec.com (M.S.); masc.qday@sinopec.com (S.M.)

2 SINOPEC Research Institute of Safety Engineering Co., Ltd., Qingdao 266000, China

* Correspondence: zhangsc.qday@sinopec.com

Abstract: Groundwater contaminated by benzene and toluene is a common issue, posing a threat to the ecosystems and human health. The removal of benzene and toluene under sulfate-reducing condition is well known, but how the bacterial community shifts during this process remains unclear. This study aims to evaluate the shift in bacterial community structure during the biodegradation of benzene and toluene under sulfate-reducing condition. In this study, groundwater contaminated with benzene and toluene were collected from the field and used to construct three artificial samples: Control (benzene 50 mg/L, toluene 1.24 mg/L, sulfate 470 mg/L, and $HgCl_2$ 250 mg/L), S1 (benzene 50 mg/L, toluene 1.24 mg/L, sulfate 470 mg/L), and S2 (benzene 100 mg/L, toluene 2.5 mg/L, sulfate 940 mg/L). The contaminants (benzene and toluene), geochemical parameters (sulfate, ORP, and pH), and bacterial community structure in the artificial samples were monitored over time. By the end of this study (day 90), approximately 99% of benzene and 96% of toluene could be eliminated in both S1 and S2 artificial samples, while in the Control artificial sample the contaminant levels remained unchanged due to microbial inactivation. The richness of bacterial communities initially decreased but subsequently increased over time in both S1 and S2 artificial samples. Under sulfate-reducing condition, key players in benzene and toluene degradation were identified as *Pseudomonas*, *Janthinobacterium*, *Novosphingobium*, *Staphylococcus*, and *Bradyrhizobium*. The results could provide scientific basis for remediation and risk management strategies at the benzene and toluene contaminated sites.

Keywords: benzene; toluene; sulfate reduction; biostimulation; bacterial community

1. Introduction

Groundwater contamination by petroleum hydrocarbons has emerged as a pressing environmental issue, largely stemming from oil spills and leaks from underground pipes and storage tanks during oil production, transportation, and storage at various industrial sites [1,2]. Among the suite of petroleum hydrocarbons, BTEX (benzene, toluene, ethylbenzene, and xylene) are of particular concern due to their toxicity and carcinogenicity [3,4]. Moreover, given their relatively high water solubility, BTEX compounds can migrate through natural groundwater flow systems, potentially contaminating distant drinking water sources [5].

In natural environments, microorganisms harness energy for cellular growth and maintenance by facilitating the transfer of electrons from electron donors to electron acceptors [6]. Typically, electron acceptors are elements or compounds existing in relatively oxidized states, and they mainly encompass dissolved oxygen (DO), nitrate (NO_3^-), manganese (IV) (Mn^{4+}), ferrous iron (Fe^{3+}), sulfate (SO_4^{2-}), and carbon dioxide (CO_2) [6–8]. Under most circumstances, biodegradation in aquifers primarily occurs through sulfate reduction under natural conditions. Thierrin et al. have documented BTEX biodegradation under sulfate-reducing condition resulting from a gasoline spill [9]. Thus, the potential utilization of sulfate-reducing processes for the remediation of BTEX-contaminated groundwater presents a practical and cost-effective remedial option for addressing petroleum hydrocarbon contamination [4,10–15]. Several studies have detailed the contaminant removal performance achieved through sulfate-reducing processes. For example, Chen et al. report the highest biodegradation rates under sulfate-reducing conditions compared to denitrifying, manganese-reducing, and iron-reducing conditions [4]. This superior performance is attributed to the greater abundance of SO_4^{2-}-reducing bacteria in the culture relative to other reducing bacteria. Lovley et al. demonstrate that under sulfate-reducing conditions, benzene is readily mineralized to CO_2 and water, without producing intermediate products like phenol, benzoate, or acetate [11]. Huang et al. explore the use of sulfate reduction mechanisms for simultaneous bioremediation of toluene and copper-contaminated groundwater, achieving nearly 99% removal of both contaminants over a 40-day period. The bacterial composition on day 30 included *Citrobacter*, *Klebsiella*, *Acinetobacter*, *Pseudomonas*, *Bacteroides*, *Clostridium*, *Bacteroides*, *Rhodoplanes*, *Betaproteobacteria*, *Zoogloea resiniphila*, and *Dysgonomonas* [14]. Juliana documents that combined biostimulation of iron and sulfate reduction accelerates BTEX and PAH biodegradation in diesel/biodiesel blends, maintaining low dissolved concentrations of benzene and naphthalene throughout the experiment compared to a baseline control under monitored natural attenuation. *Geobacter* spp. and *GOUTA19* spp. appear to play key roles in the anaerobic biodegradation of diesel/biodiesel blends under iron and sulfate reduction [16]. Norma compares the performance of two permeable reactive barriers with differing internal substrate configurations—one containing a sulfate solution without metals, the other with metals—for treating groundwater contaminated with acid mine drainage. Bacterial diversity was higher at the beginning and middle of the experiment in both systems [17].

However, the specific changes in the bacterial community in response to contaminant concentration and reactive phase during the biodegradation of benzene and toluene under sulfate-reducing condition remain unclear. To address this gap, actual benzene and toluene-contaminated groundwater from a petrochemical site was collected to construct artificial samples. Supplementary electron acceptor SO_4^{2-} was added to investigate the feasibility of enhanced sulfate reduction for remediating benzene and toluene-contaminated groundwater. The study aimed to evaluate the shift in the bacterial community structure during the biodegradation of benzene and toluene under sulfate-reducing condition and identify the key bacteria that are most effective in the biodegradation process of benzene and toluene under sulfate-reducing condition. Understanding the dynamics of microbial communities offers a scientific foundation for optimizing bioremediation strategies. This knowledge enables the enhancement of their activities or the targeted introduction of these microorganisms into environments where they can exert maximum efficacy, thereby improving remediation outcomes.

2. Materials and Methods

2.1. Groundwater Sampling and Analysis

Groundwater contaminated with benzene and toluene was obtained for this study. The research site is situated within a petrochemical complex in southeaster China that has been operational for more than three decades. Figure 1 illustrates the layout of the contaminated site along with the positioning of the well. The aquifer medium primarily consists of sandy clay in the examined depth range. Predominantly, the groundwater type found here is Quaternary phreatic groundwater, which extends from a depth of 3 m to 15 m beneath the surface. The natural flow direction of the groundwater flows predominantly from the northwest towards the southeast. Groundwater samples (W06) were collected using MicroPurge Low flow groundwater sampling system (SamplePRO, QED Environmental Systems Limited Inc., Hong Kong, China) after the construction of the wells.

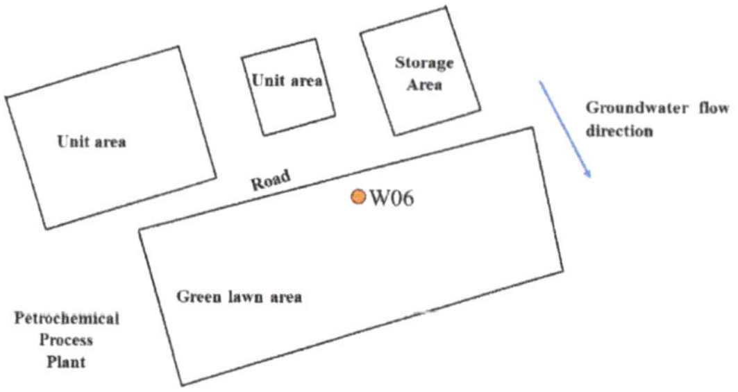

Figure 1. Locations of the groundwater sampling point.

pH, ORP, and DO were measured in situ using respective portable testers. Ion Chromatography (DIONEXTM AQUIONTM Thermo Scientific, Waltham, MA, USA) was employed to determine SO_4^{2-} and NO_3^- anions. Cations, particularly Fe^{2+} and Mn^{2+}, were quantified using a portable HACH DR3900 analyzer following the standard methods 8146 (1,10-phenanthroline photometric method) and 8034 (periodate method), respectively. The concentrations of benzene and toluene were analyzed using purge-trap and gas chromatography-mass spectrometry (GC-MS, Agilent 7890B 5977B-Atomx XYZ Analytical Instruments, Agilent, Santa Clara, CA, USA), following the procedures outlined in US EPA Method 502.2 [18].

2.2. Artificial Sample Setup

Three artificial samples, Control, S1, and S2 have been set up, and their components were shown in Table 1. In the control artificial sample, 250 mg/L $HgCl_2$ was added to inhibit microbial activity. The anaerobic artificial samples were assembled in 50 mL glass serum bottles, each filled to capacity with the contaminated groundwater (with benzene and toluene) and sulfate. The contaminant concentrations were adjusted by diluting the groundwater with deionized water to achieve the targeted concentrations. The artificial samples were prepared inside an N_2 glovebox. The artificial sample experiment was conducted over a 90-day period. On day 1, 3, 20, 40, 60, and 90, samples were collected for analysis of benzene, toluene, sulfate, ORP, and pH. On day 3, 40, and 90, samples were collected for analysis of bacterial community structure.

Table 1. Components of three artificial samples.

Artificial Sample	Components
Control	contaminated groundwater containing benzene (50 mg/L) and toluene (1.24 mg/L) + 250 mg/L $HgCl_2$+ sulfate (470 mg/L) + yeast extract (50 mg/L)
S1	contaminated groundwater containing benzene (50 mg/L) and toluene (1.24 mg/L) + sulfate (470 mg/L) + yeast extract (50 mg/L)
S2	contaminated groundwater containing benzene (100 mg/L) and toluene (2.5 mg/L) + sulfate (940 mg/L) + yeast extract (50 mg/L)

2.3. Analytical Method

At each sampling time point, three artificial sample bottles from each group were analyzed. pH and ORP were measured in situ using respective portable testers. Ion Chromatography (DIONEXTM AQUIONTM Thermo Scientific) was employed to determine SO_4^{2-} anions. The concentrations of benzene and toluene were analyzed using purge-trap and gas chromatography-mass spectrometry (GC-MS, Agilent 7890B 5977B-Atomx XYZ Analytical Instruments, Agilent, USA), following the procedures outlined in US EPA Method 502.2 [18].

2.4. Analysis of Bacterial Community Sturcture

Bacterial DNA was extracted using the MagPure Soil DNA LQ Kit (Magen Biotechnology, Guangdong, Guangzhou, China) following manufacturer's instructions to investigate the microbial diversity of the contaminated groundwater as shown in Table S1. The integrity and concentration of the extracted DNA were determined using a NanoDrop 2000 spectrophotometer (Thermo Fisher Scientific, Waltham, MA, USA) and agarose gel electrophoresis, respectively. PCR amplification of the V3-V4 hypervariable regions of the bacterial 16S rRNA gene (94 °C for 5 min, followed by 26 cycles at 94 °C for 30 s, 56 °C for 30 s, and 72 °C for 20 s and a final extension at 72 °C for 5 min) was carried out in a 25 μL reaction using universal primer pairs 343F (5′-TACGGRAGGCAGCAG-3′) and 798R (5′-AGGGTATCTAATCCT-3′) [19]. The reverse primer contained a sample barcode and both primers were connected with an Illumina sequencing adapter. The polymerase chain reaction (PCR) reaction system and reaction conditions are shown in Table S1 in the supplemental materials.

The Amplicon quality was visualized using gel electrophoresis. The PCR products were purified with Agencourt AMPure XP beads (Beckman Coulter Co., Brea, CA, USA) and quantified using Qubit dsDNA assay kit. The concentrations were then adjusted for sequencing. Sequencing was performed on an Illumina NovaSeq6000 with two paired-end read cycles of 250 bases each. (Illumina Inc., San Diego, CA, USA).

Paired-end reads were preprocessed using Trimmomatic software 0.39 to detect and cut off ambiguous bases. It also cut off low quality sequences with average quality score below 20 using sliding window trimming approach. After trimming, paired-end reads were assembled and employed the base with a higher quality score as the output using FLASH software 32.0.0.465. Parameters of assembly were: 10 bp of minimal overlapping, 200 bp of maximum overlapping and 20% of maximum mismatch rate. Sequences were performed further denoising as follows: reads with ambiguous, homologous sequences or below 200 bp were abandoned. Reads with 75% of bases above Q20 were retained using QIIME software 2023.5. Then, reads with chimera were detected and removed using VSEARCH 2.28.1. Clean reads were subjected to primer sequences removal and clustering to generate operational taxonomic units (OTUs) using Abundance-based Greedy Clustering algorithm with 97% similarity cutoff. The representative read of each OTU was selected using QIIME package. All representative reads were annotated and blasted against Silva database (Version 132) using RDP classifier (confidence threshold was 70%) [20].

Alpha diversity indices were calculated to evaluate the diversity of bacterial community structure. The Chao1 index was used to estimate community richness, while the Shannon diversity index and the Simpson diversity index were employed to assess community diversity. All the aforementioned indices were computed using Mothur 1.43.0, with the input dataset undergoing neither subsampling nor screening, given the adequate sequence depth. Beta diversity was analyzed to demonstrate the similarities or disparities among the artificial samples using Principal Components Analysis (PCA), with the application of Aitchison distances for metrics calculations.

3. Results

3.1. Physicochemical Properties and Bacterial Communitry Structure of the Collected Groundwater

The physiochemical properties of the collected groundwater used in the artificial samples are shown in Table 2. The groundwater was mildly acidic and in a reducing state, as evidenced by a pH of 5.89 and an ORP of −86 mV. The DO, NO_3^-, and SO_4^{2-} concentrations, serving as electron acceptors in microbial-mediated redox reactions, were found to be low. The primary contaminants were benzene and toluene, with benzene at 209.00 mg/L and toluene at 5.18 mg/L. Other contaminants, predominantly comprising ethylbenzene and xylene, were present at relatively low levels and were thus disregarded in the batch artificial sample study. Consequently, the limited availability of electron acceptors, the negative ORP, and the high contaminant concentrations collectively suggest that the environment is conducive to anaerobic microbial degradation of contaminants in the groundwater sample [6].

Table 2. Physiochemical properties of the collected groundwaters used in the artificial samples.

pH	ORP (mV)	DO (mg/L)	NO_3^- (mg/L)	SO_4^{2-} (mg/L)	Fe^{2+} (mg/L)	Mn^{2+} (mg/L)	Benzene (mg/L)	Toluene (mg/L)	Ethylbenzene (mg/L)	m, p-Xylene (mg/L)	o-Xylene (mg/L)
5.89	−86	0.75	0.797	2.82	8.84	2.65	209.000	5.180	0.460	0.110	0.188

The richness of bacterial communities in the collected groundwater, as measured by Chao1 values, is 1229.74. The Shannon and Simpson index values in the collected groundwater are 5.29 and 0.9391, respectively. Figure 2 shows the bacterial community structure of the groundwater at phylum, class, and genus levels.

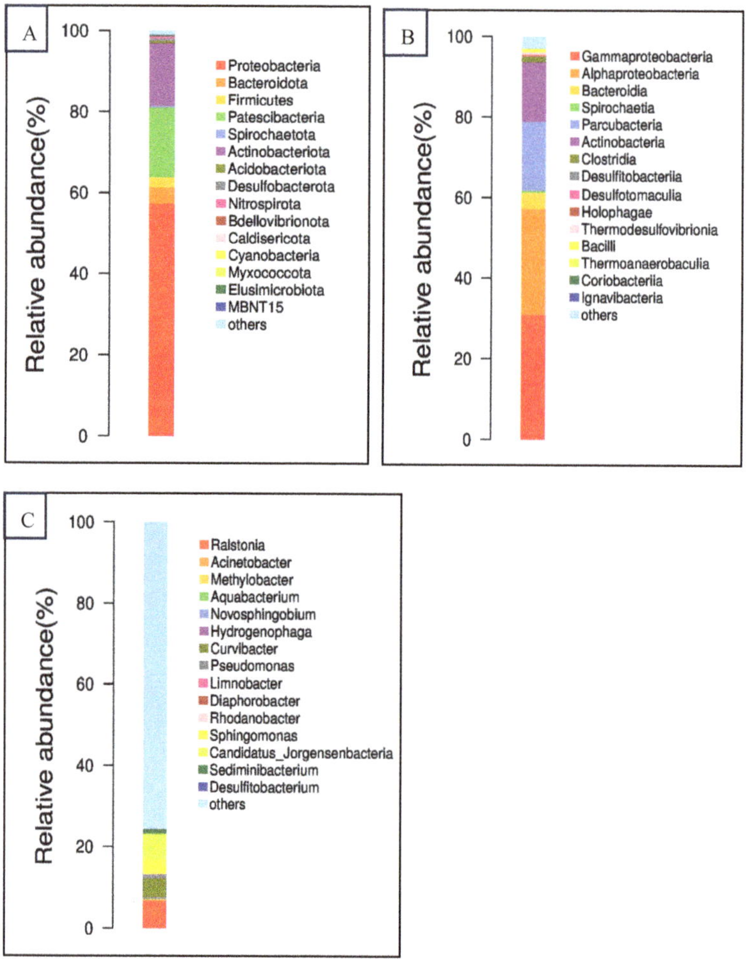

Figure 2. Bacterial communities of the groundwater sample at phylum level (**A**), class level (**B**) and genus level (**C**).

Prior to sulfate addition, the dominant phyla detected in the artificial samples were Proteobacteria, Bacteroidota, Firmicutes, Patescibacteria, and Actinobacteriota. Proteobacteria, the most abundant phylum, encompasses a diverse array of facultative or obligate anaerobic microorganisms frequently encountered in wastewater cultures and groundwater sediments [21]. These organisms have been demonstrated to possess the ability to degrade BTEX (benzene, toluene, ethylbenzene, and xylenes) compounds. Bacteroidota, typically recognized as heterotrophic aerobic and nitrifying bacteria, are capable of degrading a wide range of organic substances [22]. Firmicutes and Actinobacteria, increasingly identified in various phylogenetic lineages, are known to encompass manganese-oxidizing bacteria (MnOB) [3] and play significant roles in nitrate reduction processes. Patescibacteria, ubiquitously present in groundwater, sediments, and lakes, harbor a repertoire of organic-active enzymes that facilitate the degradation of complex organic matter into simpler molecules, as evidenced by previous research [22].

At the class level, the bacterial community in the artificial samples prior to sulfate addition was primarily composed of Gammaproteobacteria, Alphaproteobacteria, Bacteroidia, Parcubacteria, and Actinobacteria. It is noteworthy that Gammaproteobacteria and Alphaproteobacteria have been documented as the dominant classes in BTEX-contaminated groundwater ecosystems [3]. Bacteroidia, on the other hand, are believed to play a crucial role in the decomposition of complex molecules into simpler compounds, particularly in the utilization of nitrogenous substances under strictly anaerobic conditions [21].

At the genus level, *Ralstonia*, *Curvibacter*, *Sphingomonas*, and *Candidatus* genera were among the most prevalent at the genus level. Ralstonia species have been isolated from oil-contaminated soil samples [23], demonstrating their potential role in hydrocarbon degradation. *Sphingomonas*, being a strictly aerobic bacterium, is known for its capability to degrade macromolecular organic contaminants, particularly polycyclic or monocyclic aromatic compounds. It thrives in environments with contaminants, showing higher population densities when contaminants are present. *Curvibacter*, on the other hand, has been identified to degrade BTEX (Benzene, Toluene, Ethylbenzene, and Xylenes) under hypoxic conditions, which underscores its significance in anaerobic or low-oxygen environments. Each of these genera contributes to the natural attenuation of contaminants in groundwater systems, and their populations can shift in response to sulfate addition and varying contaminant concentrations, highlighting their adaptability to different environmental conditions.

3.2. Biodegradation of Benzene and Toluene in the Artificial Samples

The efficacies of concurrent benzene and toluene elimination under sulfate-reducing conditions in the artificial samples were assessed. Figure 3 depicts the changes of concentrations of benzene and toluene over the 90-daystudy. Remarkably, approximately 99% of benzene was removed in both S1 and S2 mcirocosms within 90 days. Similarly, approximately 96% of toluene was eliminated in both S1 and S2 artificial samples over the same period. In contrast, the concentrations of benzene and toluene in the Control artificial sample remained constant due to microbial inactivation. The degradation kinetics of benzene and toluene in S1 and S2 generally follow a first-order reaction pattern. In S1, the first-order degradation rate constants are as follows: 0.06744 d^{-1} for benzene and 0.08118 d^{-1} for toluene. In S2 artificial sample, the corresponding first-order degradation rate constants are 0.05252 d^{-1} for benzene and 0.03955 d^{-1} for toluene. Notably, the degradation rates of benzene and toluene in S1 and S2 artificial samples were influenced by the concentrations of these pollutants.

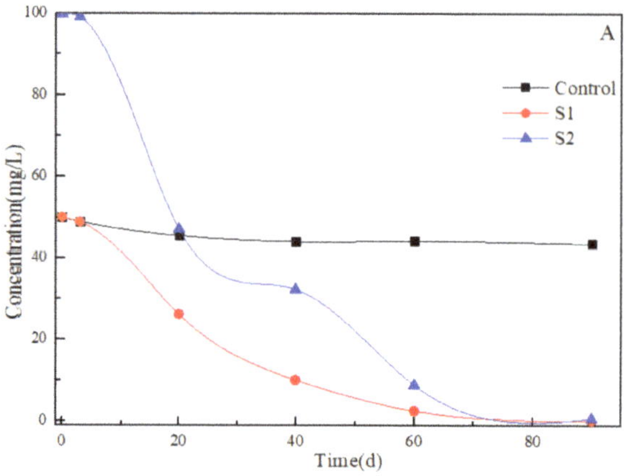

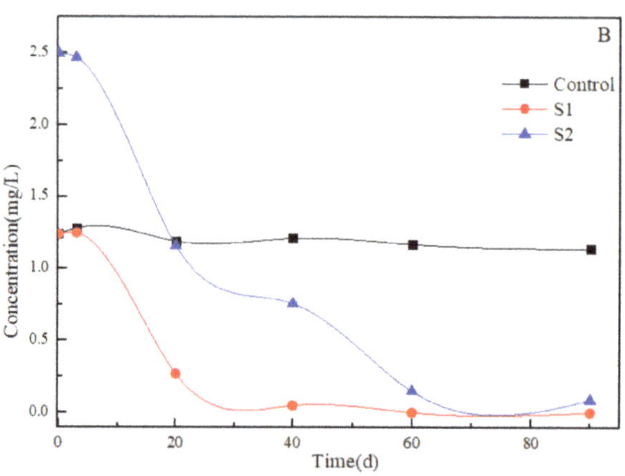

Figure 3. Changes in benzene (**A**) and toluene (**B**) concentration during the 90 days period.

3.3. Changes of Geochemcial Parameters in the Artificial Samples

Figure 4 illustrates the variations in sulfate concentration within the artificial samples. Based on the theoretical sulfate requirement for benzene biodegradation, every gram of benzene degradation necessitates 4.7 g of sulfate. In the S1 artificial sample, the sulfate concentration dwindled from an initial 470 mg/L to 352 mg/L and further to 287 mg/L after 20 and 40 days, respectively. Concurrently, in the S2 artificial sample, the sulfate concentration descended from 940 mg/L to 665 mg/L and then to 598 mg/L following 20 and 40 days, respectively. Overall, the decline in sulfate concentration approximately mirrors a 4.7-fold reduction in benzene concentration. The observed sulfate concentration alterations align with the predicted sulfate consumption. As expected, the sulfate concentration in the Control artificial sample remained unaltered throughout the experiment. It is noteworthy that reported ratios between sulfate reduction and contaminant degradation were 3.51 and 4.33 for benzene and toluene, respectively, which were strikingly close to their theoretical counterparts [24]. This congruity substantiates the significant role of sulfate reduction in mediating the biodegradation of benzene and toluene in these artificial samples.

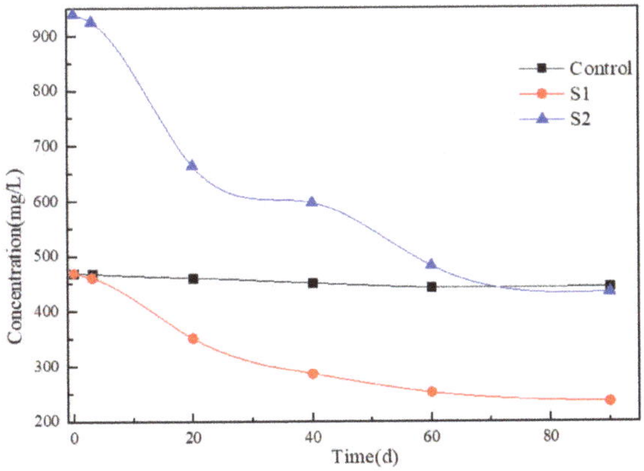

Figure 4. Changes in sulfate concentration during the 90 days period.

Thus, the observed anaerobic biodegradation of organic compounds in the artificial samples can be attributed to the sulfate-reducing process. The introduction of additional sulfate into the S1 and S2 artificial samples facilitated sulfate reduction as the predominant mechanism for benzene and toluene degradation. Essentially, microorganisms within these artificial samples utilized sulfate as the electron acceptor and benzene as the electron donor, thereby facilitating the degradation of benzene. This finding confirms the pivotal role of sulfate-reducing bacteria in mediating the anaerobic breakdown of benzene and toluene under sulfate-rich conditions.

Figure S1 presents the fluctuations in pH and ORP values throughout the artificial sample study. The pH values in S1 and S2 artificial samples were slightly higher than that of the Control group, but still remained within the neutral range, which was favorable for microbial activity. The Control artificial sample's ORP remained relatively stable, fluctuating around −50 mV. In contrast, the ORP values in S1 and S2 artificial samples experienced a rapid decrease after 3 days. Subsequently, S1 artificial sample stabilized at around −140 mV, while S2 artificial sample maintained a level of approximately −120 mV. The significant decrease in ORP values observed in S1 and S2 after 3 days indicates a shift in the redox state, likely due to increased microbial activity in the availability of electron donors and acceptors. The stabilization of ORP at lower values in these artificial samples suggests that the system has adapted to a new set of conditions, potentially indicating a more reducing environment.

3.4. Changes of Bacterial Community Structures in the Artificial Samples

As demonstrated earlier, the introduction of sulfate significantly promoted sulfate reduction as the primary pathway for benzene and toluene biodegradation in the S1 and S2 artificial samples. This observation strongly suggests the presence of microorganisms within the artificial sample environment possessing sulfate-reducing capabilities, which actively participated in the degradation of the target contaminants. These sulfate-reducing bacteria likely utilized sulfate as an electron acceptor and benzene/toluene as electron donors, driving the anaerobic degradation process observed in the study.

Table 3 displays the richness and alpha diversity estimators of the bacterial communities present in the artificial samples. In both S1 and S2 artificial samples, the richness of bacterial communities, as measured by Chao1 values, initially decreased and subsequently increased over time. The Shannon and Simpson index values, which are proxies for community diversity, demonstrated a similar trend for both artificial samples. This pattern is likely a consequence of the addition of sulfate, which altered the groundwater environment and fostered the proliferation of sulfate-reducing bacteria. As the sulfate reducers progressively became the dominant bacterial species following sulfate supplementation [4], a temporary decrease in bacterial diversity and richness was observed at day 40, however, by day 90 bacterial diversity and richness exhibited a substantial increase. Given that the contaminants have become negligible at this stage, they no longer exerted a strong selective pressure on the bacterial community, allowing for the resurgence of bacterial diversity in the later stages of the incubation under sulfate-reducing conditions.

Table 3. The richness, diversity estimators and alpha diversity estimators of bacterial community in the S1 and S2 artificial samples.

Samples	Chao1	Shannon	Simpson	Observed Species	Goods Coverage	PD-Whole Tree
S11 (3 d)	581.33	7.18	0.9674	580.50	0.9999	50.24
S12 (40 d)	133.13	2.56	0.7016	132.60	0.9999	17.44
S13 (90 d)	1429.26	8.81	0.9780	1422.90	0.9994	101.80
S21 (3 d)	513.53	6.41	0.9430	512.70	0.9999	46.19
S22 (40 d)	109.72	2.08	0.6815	108.50	0.9999	15.82
S23 (90 d)	1072.56	7.18	0.9075	1068.00	0.9996	77.54

Furthermore, throughout the entire artificial sample study, the Chao 1 value for the S1 artificial sample consistently surpassed that of the S2 artificial sample. A parallel trend was observed in the Shannon and Simpson index values. This indicated a correlation between bacterial richness and diversity and the contaminant concentration, where higher level of contaminants exerted more toxic effects on bacterial growth. This phenomenon was corroborated in the literature through comparisons of results at multiple sampling points along groundwater flow paths [25].

To illustrate the relationship between microbial structures in various artificial samples, two-dimensional PCA plots were presented in Figure 5. It is well understood that the distances between samples are crucial for analysis within a PCA plot. Typically, samples that cluster closely together indicate a higher degree of similarity in their microbial community composition, while those that are more widely spaced suggest greater disparities in community structure. Utilizing PCA with Aitchison distances, the S1 and S2 artificial samples, collected during the same reaction period at 3 d and 40 d, clustered closely together, indicating a high level of similarity in their microbial community structures. In contrast, artificial samples S13 and S23, taken after 90 days of reaction, were notably distant from each other on the PCA plot, highlighting a substantial difference in their respective microbial communities. Conversely, samples from the same group but obtained at disparate reaction stages appeared more distantly located on the PCA plot, reflecting a marked dissimilarity in their microbial compositions as time progressed. These alterations in microbial community structures correlate with fluctuations in contaminant concentration levels.

Figure 6 shows the comparisons of bacterial community structures in different samples. Following the addition of sulfate at day 3, the dominant phyla detected shifted to Proteobacteria, Bacteroidota, and Firmicutes, with a marked decrease in the relative abundance of other phyla. At day 40, the proportion of Proteobacteria reached its peak, coinciding with a substantial decline in contaminant concentrations, although still sufficient to exert a selective pressure on the microbial population. By day 90, with contaminants present at negligible levels, bacterial diversity rebounded significantly. The fluctuations in phylum-level composition parallel the trends observed in the Shannon and Simpson diversity indices, indicating that the dominance of Proteobacteria, Bacteroidota, and Firmicutes in the groundwater artificial samples during this period was accompanied by a substantial reduction in bacterial diversity.

At the class level, at day 3, following sulfate addition, Gammaproteobacteria, Alphaproteobacteria, Bacteroidia, and Bacilli continued to dominate the bacterial community at the class level, displaying a resemblance to the community structure in the absence of sulfate. At day 40, the bacterial community was predominantly comprised of Gammaproteobacteria and Alphaproteobacteria, whereas at day 90, Gammaproteobacteria, Alphaproteobacteria, and Bacteroidia regained their dominance.

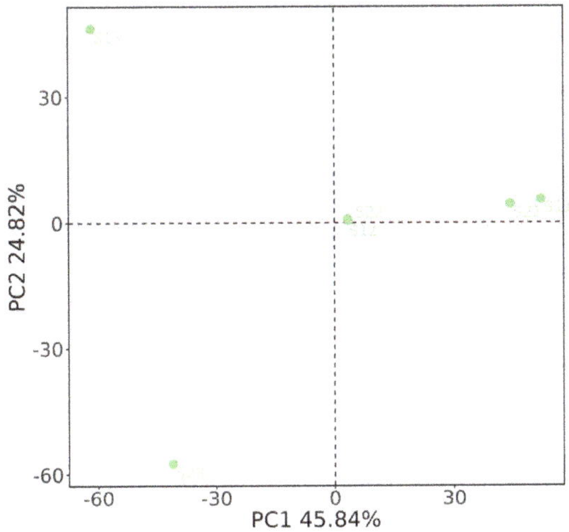

Figure 5. Two-dimensional PCA of the S1 and S2 artificial samples using the Aitchison distances.

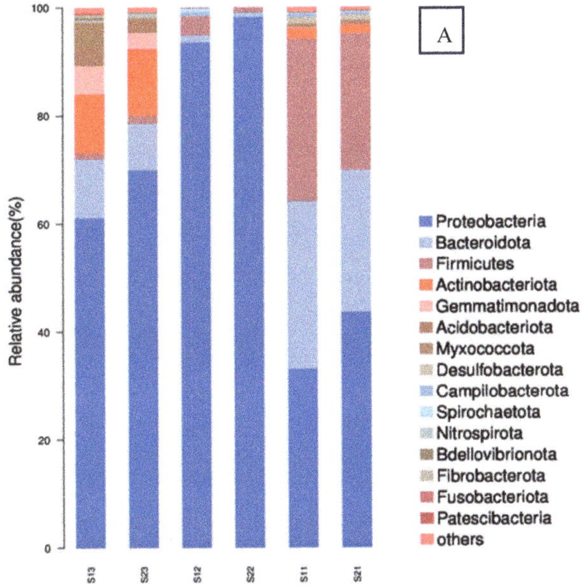

Figure 6. *Cont.*

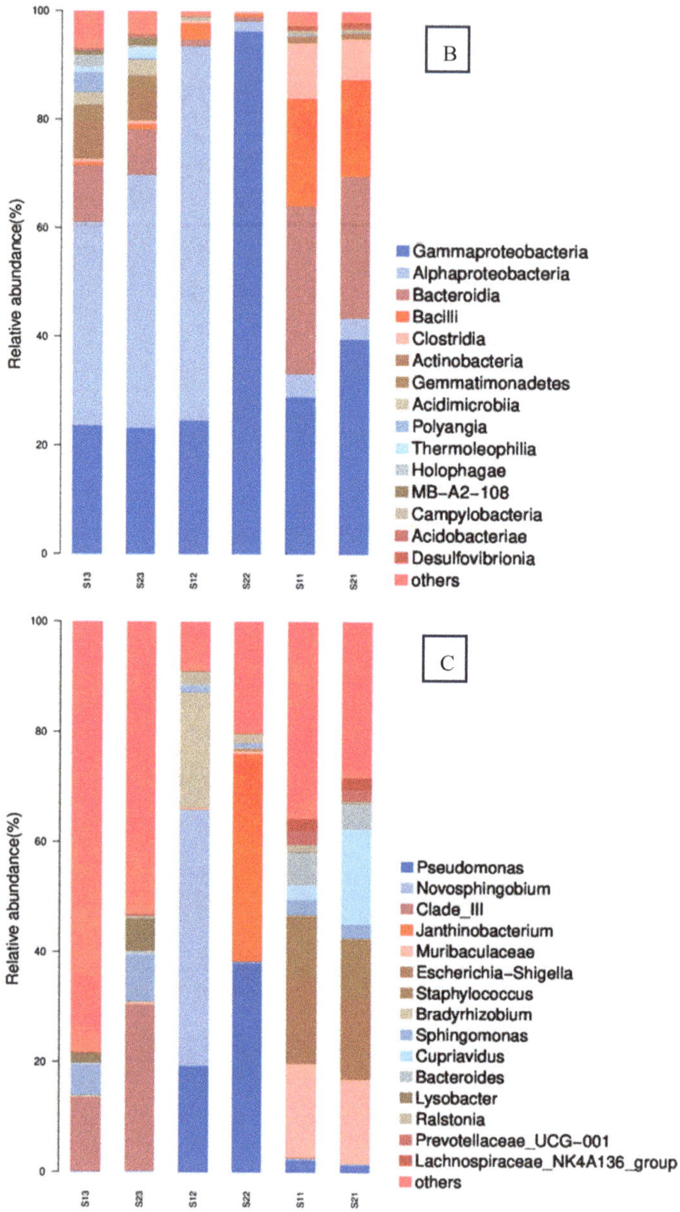

Figure 6. Bacterial communities of the treated groundwater samples at phylum (**A**), class (**B**), and genus (**C**) levels (S11 represents S1 at 3 days; S12 represents S1 at 40 days; S13 represents S1 at 90 days; S21 represents S2 at 3 days; S22 represents S2 at 40 days; S23 represents S2 at 90 days).

At the genus level, at day 3 post-sulfate addition, *Pseudomonas*, *Staphylococcus*, *Novosphingobium*, and *Bradyrhizobium* began to flourish and emerged as the dominant genera in the bacterial community. At day 40, *Pseudomonas*, *Janthinobacterium*, and *Novosphingobium* took precedence. The presence of these genera is consistent with their known involvement in aromatic hydrocarbon degradation. *Pseudomonas*, in particular, plays a pivotal role in benzene and naphthalene biodegradation under sulfate-reducing conditions. The bacteria detected in both artificial samples at day 40 are likely anaerobic, contributing to benzene and toluene degradation under anaerobic circumstances. Notably, *Pseudomonas* and *Janthinobacterium* were more dominant in the heavily contaminated S2 groundwater, while *Pseudomonas* and *Novosphingobium* were more prominent in the less contaminated S1 groundwater. By day 90, *Clade* and *Staphylococcus* topped the list of dominant genera, with a notably higher diversity compared to day 40. These findings highlight the significant shifts in the bacterial community structure over time. The bacterial community in the sulfate-treated groundwater samples differed significantly from that of the pre-sulfate addition samples. The results reveal that the bacteria present in the artificial samples at day 3, 40, and 90 did not maintain a consistent relationship, signifying that the contaminants and sulfate supplementation led to substantial changes in bacterial diversity within the artificial samples. This transformation could be due to the altered physical and chemical conditions in the artificial samples brought about by the sulfate addition and changing contaminant levels.

At the same time intervals, it's evident that the bacterial domestication, or adaptation to specific conditions in the S2 artificial sample was more pronounced compared to the S1 mcirocosm. This difference can be attributed to the higher contaminant concentration in the S2 artificial sample. Excessive contaminant levels can exert toxic effects on bacterial growth, which can lead to selection pressures that favor the survival and proliferation of certain bacterial populations better adapted to tolerate or degrade the contaminants [26]. Consequently, the more contaminated S2 artificial sample experienced a more dramatic shift in its bacterial community composition, reflecting a stronger domestication process as it adjusted to the sulfate-reducing conditions and the presence of high contaminant loads.

The original groundwater sample, which did not receive sulfate addition, exhibited a more diverse bacterial community compared to the samples at day 40 post-sulfate treatment. Conversely, the sulfate-amended samples harbored predominantly anaerobic sulfate-reducing bacteria, such as *Pseudomonas*, *Janthinobacterium*, and *Novosphingobium*, which was in line with the ample availability of sulfate as an electron acceptor in these artificial samples. Moreover, the temporal changes in bacterial community richness and diversity closely mirrored the fluctuation in contaminant concentrations. Based on the results, several bacterial genera were identified as benzene/toluene-degrading bacteria under sulfate-reducing conditions. These include *Pseudomonas*, *Janthinobacterium*, *Novosphingobium*, *Staphylococcus*, and *Bradyrhizobium*. These findings underscore the critical role played by these bacteria in the anaerobic degradation of benzene in the presence of sulfate.

Understanding the dynamics of these microbial communities provides a scientific foundation for optimizing bioremediation strategies in the following ways.

1. Enhancing activities: By adjusting environmental parameters such as pH, ORP, and sulfate availability for the growth and metabolic activity of these bacteria, it is possible to enhance their degradation capabilities.
2. Targeted introduction: Knowing these bacteria are most effective at degrading benzene can guide the targeted introduction of these species into contaminated environments. This approach can be more effective than a general inoculation with an undefined microbial mixture.
3. Monitoring and adjustment: Regular monitoring of the microbial community structure can help assess the progress of bioremediation and allow for timely adjustments to the strategy.

In conclusion, the study of these specific bacterial communities and their roles in the anaerobic degradation of benzene provides valuable information that can be used to improve the effectiveness of bioremediation strategies. This knowledge not only contributes to the remediation of contaminated sites but also enhances our understanding of the complex microbial interactions that occur in the environment.

4. Conclusions

This study investigated the shifts in the bacterial community structure during the biodegradation of benzene and toluene under sulfate-reducing conditions. In both S1 (benzene 50 mg/L, toluene 1.24 mg/L, sulfate 470 mg/L) and S2 (benzene 100 mg/L, toluene 2.5 mg/L, sulfate 940 mg/L) artificial samples, approximately 99% benzene and 96% toluene were removed from the water over a 90-day period. The contaminant concentration has a profound impact on the abundance and diversity of the bacterial communities. As the contaminant concentration decreased, bacterial abundance and diversity tended to increase, logically consistent with the notion that excessively high contaminant levels can exert toxic effects on bacterial growth. Under sulfate-reducing conditions, key players in benzene/toluene degradation were identified as *Pseudomonas, Janthinobacterium, Novosphingobium, Staphylococcus,* and *Bradyrhizobium*. Specifically, *Pseudomonas* and *Janthinobacterium* were found to dominate in the more heavily contaminated groundwater samples, while *Pseudomonas* and *Novosphingobium* were more prevalent in the less contaminated samples. The biological remediation under sulfate reduction processes provides a promising, efficient, and cost-effective strategy for remediating BTEX-contaminated groundwater. By harnessing the metabolic capabilities of sulfate-reducing bacteria and understanding their responses to varying contaminant concentrations, this approach can enhance the effectiveness of groundwater cleanup efforts.

Supplementary Materials: The following supporting information can be downloaded at: https://www.mdpi.com/article/10.3390/toxics12060423/s1, Table S1. Polymerase chain reaction (PCR) reaction system and reaction conditions. Figure S1. Changes of pH and ORP during the 90 days period.

Author Contributions: Z.L.: Conceptualization, Methodology, Validation, Formal analysis, Data curation, Writing. X.L.: Methodology, Data curation. X.W.: Formal analysis, Data curation. M.S.: Methodology, Investigation. S.M.: Formal analysis, Data curation, Writing—review & editing. S.Z.: Supervision, Project administration. All authors have read and agreed to the published version of the manuscript.

Funding: This project was funded by the Technology Development Program of SINOPEC, China (Grant No. 320131-2).

Institutional Review Board Statement: Not applicable.

Informed Consent Statement: Not applicable.

Data Availability Statement: The original data presented in the study are included in the article; further inquiries can be directed to the corresponding author.

Conflicts of Interest: All authors were employed by the company SINOPEC Research Institute of Safety Engineering Co., Ltd. The authors declare that the research was conducted in the absence of any commercial or financial relationships that could be construed as a potential conflict of interest.

References

1. Das, N.; Chandran, P. Microbial degradation of petroleum hydrocarbon contaminants: An overview. *Biotechnol. Res. Int.* **2011**, *2011*, 941810–941822. [CrossRef]
2. Varjani, S.J. Microbial degradation of petroleum hydrocarbons. *Bioresour. Technol.* **2017**, *223*, 277–286. [CrossRef]
3. Huang, H.; Jiang, Y.; Zhao, J.; Li, S.; Sarah, S.; Deng, L. BTEX biodegradation is linked to bacterial community assembly patterns. *J. Hazard. Mater.* **2021**, *419*, 5. [CrossRef]
4. Chen, X.-X.; Zhang, S.; Yi, L.-J.; Liu, Z.-W.; Ye, X.-Y.; Yu, B.; Shi, S.; Lu, X.-X. Evaluation of Biodegradation of BTEX in the Subsurface of a Petrochemical Site near the Yangtze River, China. *Int. J. Environ. Res. Public Health* **2022**, *19*, 16449. [CrossRef]
5. Christian, W.; Sabine, S.; Tillmann, L. Detection of anaerobic toluene and hydrocarbon degraders in contaminated aquifers using benzylsuccinate synthase (bssA) genes as a functional marker. *Environ. Microbiol.* **2007**, *9*, 1035–1046.
6. Todd, H.-W.; John, T.-W.; Donald, H.-K. *Technical Protocol for Implementing Intrinsic Remediation with Long-Term Monitoring for Natural Attenuation of Fuel Contamination Dissolved in Groundwater*; Air Force Center for Environmental Excellence: Brooks, TX, USA, 1999; pp. B5-4–B5-31.
7. Naraboyina, D.; Rastogi, A.-K. Remediation techniques for BTEX contamination of groundwater—A review. *Int. J. Eng. Res. Technol.* **2015**, *3*, 1–6.

8. Yu, M.; Michael, S.-W.; Shaily, M.; Nicholas, W.-J.; Kimberly, H.; Guo, S.-J.; Camilah, D.-P.; Thien, P.; Phillip, B.-G.; David, T.-A.; et al. Microbial responses to combined oxidation and catalysis treatment of 1,4-dioxane and co-contaminants in groundwater and soil. *Front. Environ. Sci. Eng.* **2018**, *12*, 2.
9. Thierrin, J.; Davis, G.-B.; Barber, C.; Patterson, B.-M.; Pribac, F.; Power, T.-R.; Lambert, M. Natural degradation rates of BTEX compounds and naphthalene in a sulfate reducing groundwater environment. *Hydrol. Sci. J.* **1993**, *38*, 309–322. [CrossRef]
10. Edwards, E.-A.; Wills, L.-E.; Reinhard, M.; Grbic-Galic, D. Anaerobic degradation of toluene and xylene by aquifer microorganisms under sulfate-reducing conditions. *Appl. Environ. Microb.* **1992**, *58*, 794–800. [CrossRef] [PubMed]
11. Lovley, D.-R.; Coates, J.-D.; Woodward, J.-C.; Phillips, E.-J. Benzene oxidation coupled to sulfate reduction. *Appl. Environ. Microb.* **1995**, *61*, 953–958. [CrossRef] [PubMed]
12. Firmino, P.-I.; Farias, R.-S.; Buarque, P.-M.-C.; Costa, M.-C.; Rodríguez, E.; Lopes, A.-C. Engineering and microbiological aspects of BTEX removal in bioreactors under sulfate-reducing conditions. *Chem. Eng. J.* **2015**, *260*, 503–512. [CrossRef]
13. Huang, W.-H.; Kao, C.-M. Bioremediation of petroleum-hydrocarbon contaminated groundwater under sulfate-reducing conditions: Effectiveness and mechanism study. *J. Environ. Eng.* **2015**, *142*, 04015089. [CrossRef]
14. Huang, W.-H.; Dong, C.-D.; Chen, C.-W.; Surampalli, R.-Y.; Kao, C.-M. Application of sulfate reduction mechanisms for the simultaneous bioremediation of toluene and copper contaminated groundwater. *Int. Biodeter. Biodegr.* **2017**, *124*, 215–222. [CrossRef]
15. Xiong, Y.; Wang, B.-Y.; Zhou, C.; Chen, H.; Chen, G.; Tang, Y.-N. Determination of growth kinetics of microorganisms linked with 1,4-dioxane degradation in a consortium based on two improved methods. *Front. Environ. Sci. Eng.* **2022**, *16*, 62. [CrossRef]
16. Müller, J.B.; Ramos, D.T.; Larose, C.; Fernandes, M.; Lazzarin, H.S.; Vogel, T.M.; Corseuil, H.X. Combined iron and sulfate reduction biostimulation as a novel approach to enhance BTEX and PAH source-zone biodegradation in biodiesel blend-contaminated groundwater. *J. Hazard. Mater.* **2017**, *326*, 229–236. [CrossRef] [PubMed]
17. Norma, P.; Alex, O.-S.; Esteban, B.; Pamela, S.; Isabel, D.; Homero, U. Performance of two differently designed permeable reactive barriers with sulfate and zinc solutions. *Sci. Total Environ.* **2018**, *642*, 894–903.
18. Volatile Organic Compounds in Water by Purge and Trap Capillary Column Gas Chromatography with Photoionization and Electrolytic Conductivity Detectors in Series. Revision 2.1. 1995. Available online: https://nepis.epa.gov/Exe/ZyNET.exe/P1018ORA.TXT?ZyActionD=ZyDocument&Client=EPA&Index=1986+Thru+1990&Docs=&Query=&Time=&EndTime=&SearchMethod=1&TocRestrict=n&Toc=&TocEntry=&QField=&QFieldYear=&QFieldMonth=&QFieldDay=&IntQFieldOp=0&ExtQFieldOp=0&XmlQuery=&File=D:%5Czyfiles%5CIndex%20Data%5C86thru90%5CTxt%5C00000039%5CP1018ORA.txt&User=ANONYMOUS&Password=anonymous&SortMethod=h%7C-&MaximumDocuments=1&FuzzyDegree=0&ImageQuality=r75g8/r75g8/x150y150g16/i425&Display=hpfr&DefSeekPage=x&SearchBack=ZyActionL&Back=ZyActionS&BackDesc=Results%20page&MaximumPages=1&ZyEntry=1&SeekPage=x&ZyPURL (accessed on 10 April 2024).
19. Nossa, C.W.; Oberdorf, W.E.; Yang, L.; Aas, J.A.; Paster, B.J.; DeSantis, T.Z.; Brodie, E.L.; Malamud, D.; Poles, M.A.; Pei, Z. Design of 16S rRNA gene primers for 454 pyrosequencing of the human foregut microbiome. *World J. Gastroenterol.* **2010**, *16*, 4135–4144. [CrossRef]
20. Kao, C.-M.; Chen, C.-S.; Tsa, F.-Y.; Yang, K.-H.; Chen, C.; Liang, S.; Yang, C.; Chen, S.-C. Application of real-time PCR, DGGE fingerprinting, and culture based method to evaluate the effectiveness of intrinsic bioremediation on the control of petroleum-hydrocarbon plume. *J. Hazard. Mater.* **2010**, *178*, 409–416. [CrossRef] [PubMed]
21. Wang, J.-L.; Zhang, Y.-L.; Ding, Y.; Song, H.-W.; Liu, T. Analysis of microbial community resistance mechanisms in groundwater contaminated with SAs and high NH4+-Fe-Mn. *Sci. Total Environ.* **2022**, *817*, 153036. [CrossRef]
22. Abbas, H.; Ahmad, A. A novel nitrile-degrading enzyme (nitrile hydratase) from *Ralstonia* sp. ZA96 isolated from oil-contaminated soils. *Biocatal. Agric. Biotechnol.* **2019**, *21*, 101285.
23. Allison, S.-D.; Lu, Y.; Claudia, W.; Goulden, M.-L.; Martiny, A.-C.; Martiny, T. Microbial abundance and composition influence litter decomposition response to environmental change. *Ecology* **2013**, *94*, 714–725. [CrossRef] [PubMed]
24. Dou, J.-F.; Liu, X.; Hu, Z.-F.; Deng, D. Anaerobic BTEX biodegradation linked to nitrate and sulfate reduction. *J. Hazard. Mater.* **2008**, *151*, 720–729. [CrossRef] [PubMed]
25. Meng, L.; Zuo, R.; Wang, J.-S.; Yang, J.; Li, Q.; Chen, M.-H. The spatial variations of correlation between microbial diversity and groundwater quality derived from a riverbank filtration site, northeast China. *Sci. Total Environ.* **2020**, *706*, 135855. [CrossRef]
26. van der Meer, J.R. Environmental pollution promotes selection of microbial degradation pathways. *Front. Ecol. Environ.* **2006**, *4*, 35–42. [CrossRef]

toxics

MDPI

Article

Changes in the Aggregation Behaviour of Zinc Oxide Nanoparticles Influenced by Perfluorooctanoic Acid, Salts, and Humic Acid in Simulated Waters

Anwar Ul Haq Khan [1,2], Yanju Liu [1,2,*], Ravi Naidu [1,2], Cheng Fang [1,2], Ho Kyong Shon [3], Huiming Zhang [4] and Rajarathnam Dharmarajan [5]

1 Global Centre for Environmental Remediation (GCER), College of Engineering Science and Environment, The University of Newcastle, Callaghan, NSW 2308, Australia; anwar.khan@uon.edu.au (A.U.H.K.); ravi.naidu@newcastle.edu.au (R.N.); cheng.fang@newcastle.edu.au (C.F.)
2 crc for Contamination Assessment and Remediation of the Environment (crcCARE), ATC Building, The University of Newcastle, Callaghan, NSW 2308, Australia
3 School of Civil and Environmental Engineering, University of Technology Sydney (UTS), City Campus, Broadway, Sydney, NSW 2007, Australia; hokyong.shon-1@uts.edu.au
4 Electron Microscope and X-ray (EMX) Unit, The University of Newcastle, Callaghan, NSW 2308, Australia; hui-ming.zhang@newcastle.edu.au
5 Australian Centre for Water and Environmental Biotechnology (ACWEB), The University of Queensland, Brisbane, QLD 4072, Australia; r.dharmarajan@uq.edu.au
* Correspondence: yanju.liu@newcastle.edu.au; Tel.: +61-(02)-4913-8738

Abstract: The increasing utilization of zinc oxide nanoparticles (ZnO-NPs) in many consumer products is of concern due to their eventual release into the natural environment and induction of potentially adverse impacts. The behaviour and environmental impacts of ZnO-NPs could be altered through their interactions with environmentally coexisting substances. This study investigated the changes in the behaviour of ZnO-NPs in the presence of coexisting organic pollutants (such as perfluorooctanoic acid [PFOA]), natural organic substances (i.e., humic acid [HA]), and electrolytes (i.e., NaCl and $CaCl_2$) in simulated waters. The size, shape, purity, crystallinity, and surface charge of the ZnO-NPs in simulated water after different interaction intervals (such as 1 day, 1 week, 2 weeks, and 3 weeks) at a controlled pH of 7 were examined using various characterization techniques. The results indicated alterations in the size (such as 162.4 nm, 1 day interaction to >10 μm, 3 weeks interaction) and zeta potential (such as −47.2 mV, 1 day interaction to −0.2 mV, 3 weeks interaction) of the ZnO-NPs alone and when PFOA, electrolytes, and HA were present in the suspension. Different influences on the size and surface charge of the nanoparticles were observed for fixed concentrations (5 mM) of the different electrolytes. The presence of HA-dispersed ZnO-NPs affected the zeta potential. Such dispersal effects were also observed in the presence of both PFOA and salts due to their large aliphatic carbon content and complex structure. Cation bridging effects, hydrophobic interactions, hydrogen bonding, electrostatic interactions, and van der Waals forces could be potential interaction forces responsible for the adsorption of PFOA. The presence of organic pollutants (PFOA) and natural organic substances (HA) can transform the surface characteristics and fate of ZnO-NPs in natural and sea waters.

Keywords: zinc oxide nanoparticles; perfluorooctanoic acid; humic acid; electrolytes; adsorption; zeta potential; aggregation

1. Introduction

Zinc oxide nanoparticles (ZnO-NPs) are among the most abundantly synthesized metal oxide-based nanoparticles, with an estimated annual global market of USD 3600 million and a global yield of 10 Mt [1]. This is due to their popular application in cosmetics, electronics, medical dressings, paints, textiles, UV filters, and other products [2–4]. Among the several causes of toxicity due to exposure to ZnO-NPs in water systems, the three major mechanisms for the toxic effects to the ecosystems are (1) photocatalytic activity and generation of reactive oxygen species, (2) release of dissolved zinc ions, and (3) attachment of the ZnO-NPs to the cell wall through electrostatic interactions, damaging the DNA structure and causing oxidative stress, etc. However, due to their small size, large surface-to-mass ratio, and strong ability to pose toxic effects (i.e., an impact on microbial and aquatic communities, such as DNA damage, oxidative stress, and soil and plant transfer) in the environment, ZnO-NPs can pose a problem for ecological receptors in water, soil, and human health [5–8]. The occurrence of ZnO-NPs in environmental samples and wastewater treatment plants (WWTPs) is well documented [2,9,10]. The presence of ZnO-NPs also hinders the degradation and removal of phosphorous and nitrogen in wastewater biofilms and activated sludge [8,11,12]. Understanding the fate and behaviour of ZnO-NPs once they are released, transferred, or interact with certain environmental factors is critical for evaluating their potential risks. Environmental factors, such as ionic strength, natural organic substances, pH, light, and polymeric substances, can significantly influence the colloidal stability and toxicity of ZnO-NPs [13,14]. For instance, the presence of hexabromocyclododecane or polybrominated diphenyl ethers as organic pollutants in water systems can alter the size and surface charge of ZnO-NPs [15,16]. Alterations in the surface potential and dispersion of humic acid-adsorbed/coated ZnO-NPs in aquatic environments have been reported [17]. Studies have also noted the unstable (aggregation) behaviour of nanoparticles (such as ZnO-NPs) in salt water (high ionic strength) due to a reduction in/compression of the thickness of the electrical double layer followed by a reduction in the energy barrier [18–20]. Normally, organic contaminants tend to interact in water media due to their hydrophobic nature, but they are more likely to interact and sorb onto the surface of ZnO-NPs (due to their high surface area). The possible interactions include van der Waals, electrostatic, hydrophobic, and π-π interactions followed by ligand exchange, hydrogen bonding, and molecular bridging effects [21–23]. These interactions can greatly alter the fate and behaviour of contaminants through aggregation, dispersion, surface charge alteration, surface coating/adsorption, and changes in crystallinity and purification.

Studies have also reported the interaction mechanisms of engineered nanoparticles with endocrine disrupting chemicals, such as polybrominated diphenyl ethers [16,22,24,25] and hexabromocyclododecane [15] as emerging environmental chemicals. Per- and polyfluoroalkyl substances (PFAS) are a class of endocrine disrupting compounds that may disrupt human thyroid hormone systems with possible negative impacts on pregnancy followed by fetal–child development [26]. The presence of such endocrine disrupting compounds could also influence the behaviour of ZnO-NPs. This study examined the changes in the behaviour of ZnO-NPs during interaction with PFOA, a representative PFAS, under various environmentally relevant conditions in water media.

PFOA is largely used in water-resistant products, such as carpets, paints and coatings, waterproof clothing, firefighting foams, nonstick cookware, and hydraulic fluids [27]. Concerns are rising regarding the release of PFOA into the environment due to its bioaccumulative and toxic nature [27–29]. PFOA has been identified in the influent and effluent of WWTPs and biosolids [29–31]. PFOA concentrations in influents of various WWTPs in Canada, Spain, China, Singapore, and North America have been reported to range from 2.2 ng/L to 6.6×10^4 ng/L, while they ranged from 1.3 ng/L to 1.6×10^5 ng/L in effluents [29]. Similarly, a range of concentrations of PFOA in the biosolids of many WWTPs in North America, Switzerland, Spain, Australia, China, Kenya, Canada, Singapore, and Finland has been documented, from 0.03 ng/g to 158 ng/g [29].

The national loads of PFOA in the effluents of 14 WWTPs in Australia were estimated to be 65 kg/year and 2 kg/year for biosolids [29]. Various drinking water sources in many countries now contain PFOA [32], including Australia (0–9.7 ng/L), Brazil (0.81–2.8 ng/L), India (<0.005–2 ng/L), China (<0.1–45.9 ng/L), Germany (<10–68 ng/L), Japan (2.3–84 ng/L), and the USA (<5–30 ng/L). Several hundred nanograms per liter of PFOA have been reported in surface waters [33]. Sediments and biota can also contain PFOA at concentrations ranging from pg/g to a few ng/g [32,34]. The maximum concentrations of PFAS, including PFOA, in marine (onshore) waters have been detected (up to 58 ng/L), while 0.11 ng/L PFOA have been found in offshore waters [32]. PFOA concentrations ranging from 1 to 13 ng/L in Palermo's (Sicily) coastline seawaters have also been detected and reported [35].

The adsorption of PFOA may lead to an accumulation of organic compounds on the surface of ZnO-NPs when these nanoparticles are released into natural waters (e.g., from sunscreen). In this study, PFOA and its interactions with ZnO-NPs were investigated, as well as the factors enabling these interactions. In particular, the physico-chemical properties, colloidal stability, particle size and surface charge alterations of the ZnO-NPs before and after interaction with PFOA under various simulated water conditions were assessed, such as in the presence of electrolytes (sodium chloride, NaCl, and calcium chloride, $CaCl_2$) and HA. The concentration of PFOA in natural waters is much lower than some of the concentrations considered in this study, which enabled characterization of the changes in ZnO-NPs. The findings of this study will be useful for assessing the influence of environmental water conditions on the exposure of ZnO-NPs and their co-contaminants.

2. Materials and Methods

2.1. Materials and Chemicals

The characteristics of the purchased ZnO-NPs, HA, and electrolytes have been reported in our previous paper [19]. The ZnO-NPs < 100 nm particle size, 544906-50G, were purchased from Sigma Aldrich Australia. The HA was purchased from Sigma Aldrich (53680-50G, humic acid technical) Australia. Most of the nanoparticles were <100 nm in size, and some were larger than 100 nm due to aggregation. PFOA was purchased from Sigma–Aldrich ($C_8HF_{15}O_2$, molecular weight: 414.07 g/mol, and purity: 96%), Australia, and subjected to characterization for the purposes of this study.

2.2. Interaction between PFOA and ZnO-NPs

The ZnO-NPs stock suspension was prepared by adding 0.1 g of nanoparticles to 1 L of Milli-Q water, followed by sonication for 10 min. Various concentrations of PFOA, from 0, 0.5, 1, 10, 50, 100, 200, and 500 μg/L to 1, 5, 10, and 50 mg/L, were prepared in Milli-Q water containing the ZnO-NPs (0.1 g/L) suspension. The higher concentrations of PFOA aimed to amplify the effects of interaction to be detectable by the Zetasizer and particle size analyzer and help to understand the mechanism of interaction. Such alteration effects are difficult to detect with PFOA at environmentally relevant concentrations due to limitations in characterization techniques. The nanoparticles were analyzed (particle size, zeta potential, dissolution, adsorption, XRD, Raman spectroscopy, FTIR, and TEM) before and after interaction with PFOA in solution at various time intervals, such as after 1 day, 1 week, 2 weeks, and 3 weeks of interaction, to assess the changes in the behaviour of the interacting nanoparticles in comparison with that of the pure nanoparticles.

2.3. Influence of Electrolytes on PFOA and ZnO-NPs' Interaction

The influence of mono- and divalent electrolytes (such as NaCl and $CaCl_2$) on the behaviour of ZnO-NPs was observed in the presence of various concentrations of PFOA (such as 10 μg/L and 500 μg/L) as an organic pollutant. The ZnO-NP stock suspension was prepared by adding 0.1 g of nanoparticles to 1 L of Milli-Q water, followed by sonication for 10 min. PFOA concentrations (10 and 500 μg/L) were also prepared in Milli-Q water containing the ZnO-NPs (0.1 g/L) suspension. Fixed concentrations (i.e., 5 mM) of monovalent and divalent salts (NaCl and $CaCl_2$) were used to investigate the effect of electrolytes on the stability of ZnO-NPs alone and in the presence of PFOA. The changes in the particle size and zeta potential of ZnO-NPs after various time intervals, i.e., after 1 day, 1 week, 2 weeks, and 3 weeks of interaction, including PFOA adsorption on the surface of ZnO-NPs in the presence of salts, are examined, which are presented and discussed in Section 3.2.

2.4. Influence of HA on PFOA and ZnO-NPs' Interaction

The effect of HA, a natural organic substance, on the size and zeta potential of the ZnO-NPs was investigated with and without 10 and 500 μg/L PFOA. The HA powder was dissolved in 0.1 M NaOH solution to prepare the stock solution. Various concentrations of HA with ZnO-NPs (0.1 g/L) suspensions in the absence and presence of PFOA were prepared. Changes in the size, shape, and charge of nanoparticles with and without PFOA and HA were investigated. The amount of PFOA adsorbed onto the surface of the ZnO-NPs was also analyzed after 1 day and 2 weeks of interaction. The pH of all the samples was maintained at 7 by using a buffer solution (thermos scientific buffer solution pH 7, USA) composed of potassium dihydrogen phosphate. The prepared suspensions were analyzed using a Malvern Panalytical Zetasizer, UK. A Zetasizer was used at room temperature (i.e., 20 °C) using a disposable folded capillary cell. The nanoparticles in the suspension were directly dropped onto TEM Cu grids to exam any morphological changes in the ZnO-NPs (such as ZnO-NPs in water suspensions and in the presence of PFOA, HA) for TEM analysis. ZnO-NPs were also obtained for FTIR, XRD, and Raman analysis by separating them from the suspension using a high speed centrifuge.

2.5. Influence of Electrolytes and HA Together on PFOA and ZnO-NPs' Interaction

The influence of HA and electrolytes together on the size and zeta potential of the ZnO-NPs was investigated with and without 10 and 500 μg/L PFOA and fixed concentration (5 mM) of various electrolytes (such as NaCl and $CaCl_2$). Various concentrations of HA with ZnO-NPs (0.1 g/L) suspensions in the absence and presence of PFOA and electrolytes were prepared. Changes in the size, shape, and charge on the surface of nanoparticles with and without PFOA, electrolytes, and HA were investigated. The amount of PFOA adsorbed onto the surface of the ZnO-NPs was also analyzed after 1 day and 2 weeks of interaction, with results discussed in Section 3.4.

2.6. Characterization Methods

FE-SEM (Zeiss Sigma VP Field Emission Scanning Electron Microscope, Germany) was performed at 15 kV after sputter coating (10 nm platinum layer), and TEM (JEM 2100 LaB6 High Resolution Transmission Electron Microscope, Japan) was performed at 200 kV to investigate the morphology. Copper grids (Lacey carbon film, 300 mesh) for TEM imaging were obtained from PST (ProSciTech), Australia. The micromeritics TriStar II (Microtrac, USA) and an XRD system (Empyrean Malvern PANalytical, Malver, UK) were used to calculate the nanoparticle surface area, pore size distribution, and phase-dimensional identification. FTIR (Agilent Technologies, Cary 600 Series) was used to determine the functional groups present before and after interactions (wavenumber range: 400–4000 cm^{-1} and number of scans: 16). Raman spectra were recorded using a WITec confocal Raman microscope (Alpha 300 RS, Germany) equipped with a 532 nm laser diode (<30 mW) under an objective lens (×100 or others) at room temperature.

A Malvern Panalytical Zetasizer was used to determine the changes in charge and size of the ZnO-NPs alone and in the presence of PFOA, HA, and different electrolytes in a pure water state (Milli-Q water). Disposable folded capillary zeta cells (DTS1070) for dynamic light scattering were obtained from Malvern Instruments Ltd. in Malvern, UK. An inductively coupled plasma/optical emission spectroscope (ICP–OES, PerkinElmer's NexION 350x) was used to determine the concentrations of dissolved zinc in the water samples before and after interactions with PFOA. An Agilent liquid chromatography–mass spectrometry (LC–MS) was employed to determine the adsorbed amount of PFOA before and after interaction.

3. Results and Discussion

3.1. Interaction between PFOA and ZnO-NPs

Particle size: Alterations in the particle size and surface charge of the ZnO-NPs were observed following interaction with various concentrations of PFOA. With the addition of various concentrations of PFOA to the ZnO-NPs (after 1 day of interaction), the particle size of the ZnO-NPs at the peak of the particle size distribution curve (PSDC) remained between 157 and 186 nm, and the particle size ranged from 106–955 nm (Figures 1a and S1a). The slight variations in the particle size at the peak of the PSDC and within the particle size ranges after 1 day of interaction may be due to less interaction time between the nanoparticles and the organic compound (PFOA). It is suggested that interactions between ZnO-NPs and PFOA as organic pollutants could be time-dependent after their coexistence was ascertained.

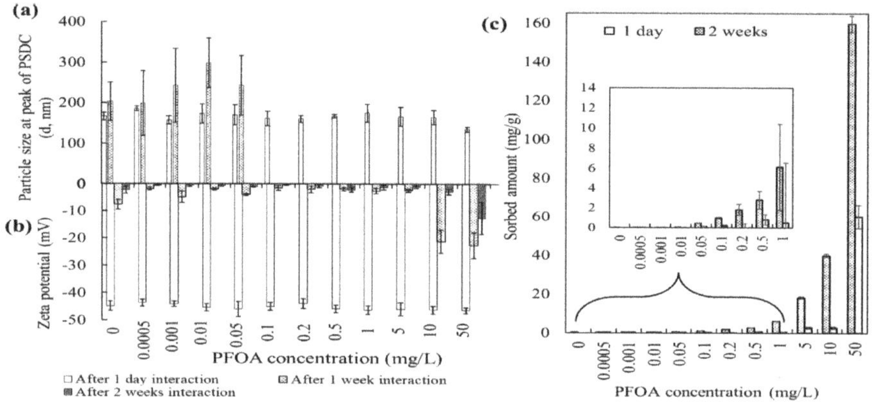

Figure 1. The analysis of particle size (**a**), zeta potential (**b**), and sorption of PFOA by ZnO-NPs (**c**), after ZnO-NPs interaction with PFOA at different times. Zeta potential values are minus.

A significant increase in the particle size of the ZnO-NPs was observed after 1 week of interaction with PFOA. The size of the ZnO-NPs at the peak of PSDC increased from 166 nm to 203 nm, and similar trends were observed in the presence of PFOA up to 50 µg/L PFOA (Figures 1a and S1b). However, the size ranges of ZnO-NPs in the presence of PFOA from 100 µg/L to 50 mg/L were not measurable via a Zetasizer, with a maximum size ranging from 0.3 nm to 10 microns. It could be assumed that the size of the ZnO-NPs after 1 week of interaction with PFOA (from 100 µg/L to 50 mg/L) is more than 10 microns. The alteration in the size distribution of the ZnO-NPs was influenced by the presence of PFOA through aggregation and the magnitude of the surface coating/adsorption of organic substances [14,22,36]. Overall, an increase in the particle size of the ZnO-NPs was observed alone and in the presence of PFOA from 1 day to 3 weeks of interaction. This increase might be due to the presence of large (agglomerated) and/or sedimenting particles resulting from particle–particle interactions, electrostatic interactions, and hydrophobic interactions. The ZnO-NPs were monodispersed (particles of uniform size) from the time of the nanoparticle suspension preparation to a few days later. However, after 1 week of interaction, nonuniform (polydisperse) behaviour of the nanoparticles was observed. Overall, particles from 1 week to 3 weeks were found in their polydispersed form, which was also reflected by their surface charge.

Zeta potential: The zeta potential (Figure 1b) of the ZnO-NPs did not significantly change after 1 day of interaction with the addition of different concentrations of PFOA, while a significant decreasing trend was observed for samples after 1 and 2 weeks of interaction. The alterations in the charge potentials suggested that PFOA coating/adsorption on the surface of the ZnO-NPs decreased in the magnitude of surface charge values. The zeta potential values also suggested that the ZnO-NPs became less stable in solution and tended to agglomerate (increasing size, Figures 1a and S1b) with increasing concentrations of PFOA. This aggregation might be attributed to the higher molecular weight and greater surface coating of PFOA on the ZnO-NPs via hydrophobic and van der Waals interactions [36–38]. The hydrophobic tail of PFOA has a decreased tendency to interact with water molecules, while the hydrophilicity and charged head of PFOA could lead to a greater tendency to attach to the ZnO-NP surface. The aggregation suggested the unstable nature of the ZnO-NPs after interactions with PFOA [15,16].

Adsorption: The adsorption analysis of PFOA in solution confirmed the interaction between ZnO-NPs and PFOA after 1 day and 2 weeks of interaction. Figure 1c shows the increased adsorption of PFOA molecules on the surface of the ZnO-NPs after 2 weeks of interaction compared to after 1 day. Various interaction mechanisms could be involved in the process of PFOA adsorbing onto the surface of ZnO-NPs, which increased with time. An increase in the particle size and an overall decrease in the net charge potential of the surface nanoparticles were observed (Figure 1a,b). Once the ZnO-NPs are in water, the formation of hydroxide layers ($Zn\,(OH)^{+}_{(aq)}$) on the surface of the nanoparticles due to hydrolysis is a common process, as water molecules can be adsorbed (both chemically and physically) onto the surface of the particles [22,36,39]. This resulted in the formation of a quantity of positive charges on the surface of the nanoparticles, attracting $PFOA^{-}$ molecules to adsorb to the surface of the ZnO-NPs by electrostatic interactions. After that, the hydrophobic tail of $PFOA^{-}$ combined on the surface of the ZnO-NPs adsorbed the free $PFOA^{-}$ molecule in the solution by hydrophobic interactions (hydrophobic PFOA molecules tend to accumulate more on the surface of the ZnO-NPs in aqueous media) [40] and further increased the adsorption amount of PFOA after 2 weeks' time interval compared to that after 1 day. This can also be corroborated by the decreased surface charge of the ZnO-NPs and the increase in particle size (Figure 1a,b).

Dissolution: The dissolution of ZnO-NPs in terms of dissolved zinc (mg/L) after 1 day and 2 weeks of interaction with PFOA was determined via ICP–OES (Figure S3). The concentration of dissolved zinc in the 0 µg/L sample (i.e., ZnO-NPs in buffered water) slightly increased after 2 weeks of interaction compared to that in the 1 day interaction. The dissolved zinc concentration increased in the presence of PFOA, which further decreased after 2 weeks of interaction compared to that for 1 day. This could be due to the aggregation/agglomeration and settlement of particles in the test tube, where coprecipitation occurred and reduced the dissolved amount of zinc in solution. The dissolution rate is often claimed to be directly proportional to the specific surface area of the material, meaning that the smaller the particles are, the faster dissolution occurs [41]. Agglomeration has been proposed as a rationale for slow dissolution [42]. Moreover, aggregation/agglomeration may have resulted in decreased dissolution [42] as the particle size increased and the settlement of particles at the bottom of the test tube was observed. The surface properties of ZnO are quite complicated due to the presence of polar and nonpolar crystallographic planes [43]. The interaction of PFOA molecules with nonpolar planes leads to agglomeration and less dissolution. A similar dissolution trend was observed in the presence of electrolytes and HA, as shown in the next sections.

TEM, XRD, FTIR, and Raman analysis: Alterations in the morphology of ZnO-NPs were observed after 1 day of interaction with PFOA using TEM (Figure 2a). The original (virgin) ZnO-NPs (Figure 2a) were in an agglomerated/spongy form with a particle size ≤ 100 nm. Both rod and spherical morphologies were observed. The lattice pattern and glittering spots/rings suggest the crystalline structure of the ZnO-NPs [15,44]. The dissolved zinc concentrations in Milli-Q water at various pH values for the same batch of ZnO-NPs were reported in our previous research [15], which indicated minimal dissolution at pH ≥ 7. The changes in crystallinity after the interaction of ZnO-NPs with water were also examined using TEM analysis. TEM showed a diffraction pattern (Figure 2a) and the crystalline/lattice arrangement of the original ZnO-NPs [15,44]. The mixture was dull and cloudy after interacting with buffer containing Milli-Q water. A more disordered/amorphous structure of the ZnO-NPs (Figure 2a) was observed after interaction with PFOA. Similarly, compared with those of the pure powder, the diffuse and cloudy bright spots (Figure 2a) represented the impure morphology of the ZnO-NPs after interaction with PFOA. The elemental composition obtained from TEM analysis is shown in Figure S2a. The structural parameters, such as lattice spacing and the crystallite size of the

ZnO-NPs after 1 day of interactions (ZnO-NPs in Milli-Q water and ZnO-NPs + 50 mg/L PFOA) were also measured using TEM analysis (Figure S2b). Lattice spacing of 0.26 nm was found in the lattice structure of ZnO-NPs. However, the crystallite size ranging from 4.29 nm to 8.39 nm were observed after 1 day of ZnO-NPs interactions with 50 mg/L PFOA (Figure S2b).

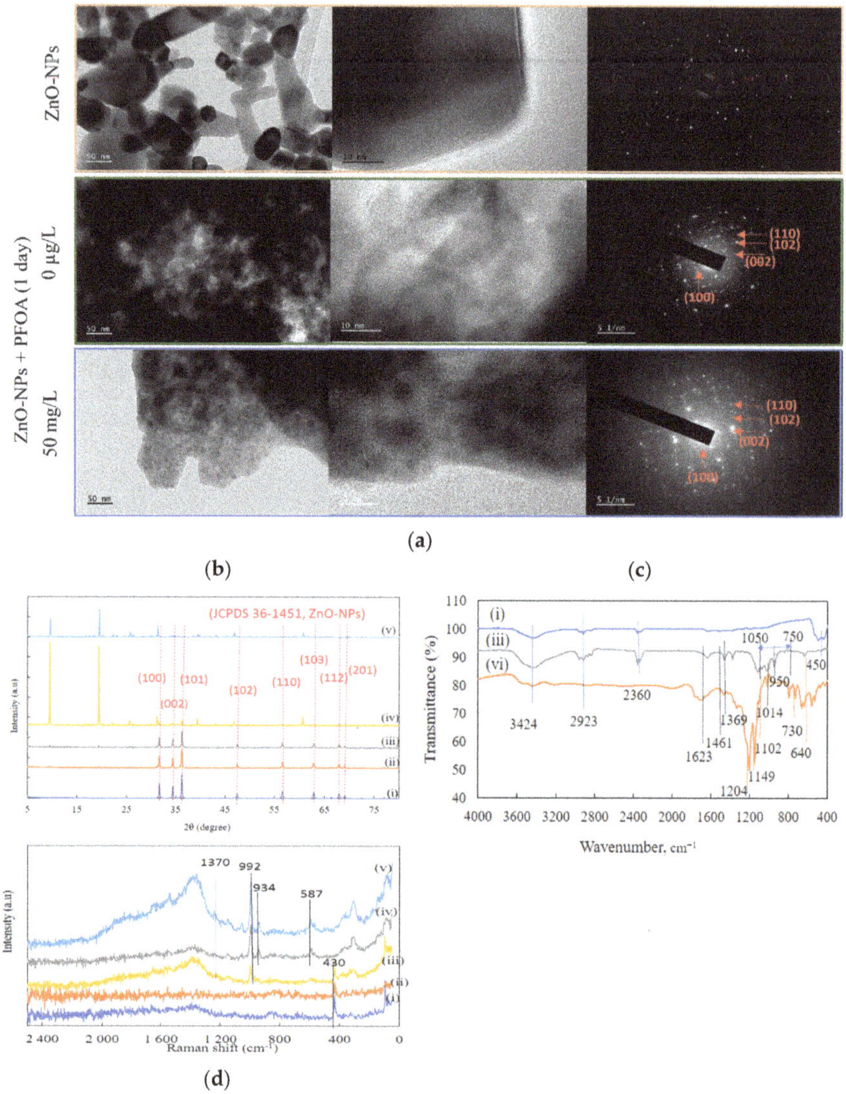

Figure 2. The characterization analysis for ZnO-NPs before and after interaction with PFOA (50 mg/L): (**a**) TEM; (**b**) XRD; (**c**) FTIR; (**d**) Raman; samples include (i) ZnO -NPs; (ii) ZnO-NPs in buffered Milli-Q water (pH 7) after 1 day; (iii) ZnO-NPs in buffered 50 mg/L PFOA (pH 7) after 1 day; (iv) ZnO-NPs in buffered Mili-Q water (pH 7) after 3 weeks; (v) ZnO-NPs in 50 mg/L PFOA (pH 7) after 3 weeks; (vi) PFOA powder.

X-ray diffraction (XRD) analysis of powder ZnO-NPs and ZnO-NPs in buffered Milli-Q water and with 50 mg/L PFOA after 1 day of interaction is shown in Figure 2b. Sharp peaks at 2θ values of 31.84°, 34.6°, and 36.5° are observed for the ZnO-NPs, which represent the (hexagonal wurtzite) crystal structure of the nanoparticles with three perfect alignments: (1 0 0), (0 0 2), and (1 0 1). These alignments match the defined standard powder diffraction (JCPDS, No. 36-1451) [22,45]. This result revealed no change in the crystal phase of the ZnO-NPs after 1 day of interaction. However, after 3 weeks of interaction (Figure 2b), the intensities of the peak alignments at (1 0 0), (0 0 2), and (1 0 1) were suppressed. Two new peaks at 2θ values of 9.62° and 19.35° were observed, which could be due to the formation of new compounds, such as zinc hydroxide dihydrate $Zn_5(OH)_{10}{\cdot}2H_2O$ [46] and zinc phosphate nanocrystalline materials [47], respectively. The intensities of the peaks at 2θ values of 9.62° and 19.35° were lower for the ZnO-NP sample with 50 mg/L PFOA than for the ZnO-NP sample in buffered water, which could be due to the coating of PFOA molecules onto the surface of the ZnO-NPs (Figure 2b).

FTIR analyses of the ZnO-NPs, PFOA, and ZnO + PFOA were also conducted (Figure 2c). The peak at 450 cm^{-1} suggested the presence of Zn–O [15,48], which is in the range of metal oxides (400–600 cm^{-1}) [15,49]. A very low-intensity peak at 450 cm^{-1} was also observed for the sample of ZnO-NPs + PFOA after interaction, indicating the presence of ZnO-NPs after interaction with PFOA molecules. The peak at 640 cm^{-1} is due to characteristic bands of organic halogen compounds (such as C–F stretching) [50], which were identified in both PFOA and ZnO-NPs + PFOA after interacting with PFOA. The vibrational peaks at 1102, 1149, and 1204 cm^{-1} may be due to C–F stretching, with the peak at 1102 cm^{-1} being identified in the spectrum of the ZnO-NPs after interaction with PFOA. C–H stretching, O–H stretching, and C=O stretching was also identified at 1369, 1461, and 1623 cm^{-1}, respectively [50,51]. These peaks were identified in both the PFOA and ZnO + PFOA samples, indicating the link between PFOA and ZnO-NPs after their interaction. The peak at 2360 cm^{-1} could be due to carbon dioxide from the atmosphere. The peak at 3424 cm^{-1} is due to stretching of the water band [15,50]. The peak detected between 750 and 1050 cm^{-1}, at 950 cm^{-1}, could be due to K-potassium and P-phosphorous (from a buffer solution used to maintain pH 7) stretching with O and C in the ZnO-NPs after interaction with PFOA [50]. The FTIR results revealed the presence of bonds in the ZnO-NPs after interaction with PFOA, suggesting that there was an association between PFOA and the ZnO-NPs. This is consistent with the elemental analysis results indicating the presence of F and C in the ZnO-NPs.

The Raman spectrum (Figure 2d) also confirmed the presence of ZnO-NPs (such as at 430 cm^{-1}) in the pure ZnO-NPs samples in buffered water and in the presence of PFOA after 1 day of interaction. After 3 weeks of interaction, new peaks at 587, 934, 992, and 1370 cm^{-1} were observed for both samples (such as ZnO-NPs in water and with PFOA) including ZnO-NPs with PFOA after 1 day of interaction (Figure 2d). Zn−O stretches fall in the region between 350 and 600 cm^{-1}. The spectral range of Zn−OH bonds (which are also called OH linkages) is between 600 and 1200 cm^{-1}. The asymmetric stretches (with a high infrared intensity) are in the range 470−550 cm^{-1}, whereas the symmetric stretches were observed at 368 and 382 cm^{-1} in the Raman spectrum. Bands below 350 cm^{-1}, as observed in the Raman spectrum, are attributed to lower-energy lattice modes [46].

3.2. Influence of Electrolytes on PFOA and ZnO-NPs' Interaction

The salinity of surface water and groundwater can vary considerably. Salinity is one of the most significant abiotic factors affecting the growth, metabolism, immunity, and survival of aquatic species in farming environments. Due to global climate change, evaporation of seawater, variations in local rainfall, and the whereabouts of ocean currents, environmental salinity in coastal areas alters frequently and violently. Under environmental stresses, physiological mechanisms are adaptively modulated to sustain body homeostasis, which can further impact the normal biological functions, comprising the immunity of the aquatic species [52]. The presence of salts could influence the interaction between ZnO-NPs and coexisting contaminants, which determines the environmental fate of ZnO-NPs. The ionic strength, pH, and other organic materials present in the solution could influence the surface charge and stabilization of ZnO-NPs. Fixed concentrations (i.e., 5 mM) of monovalent and divalent salts (NaCl and $CaCl_2$) were used to investigate the effect of electrolytes on the stability of ZnO-NPs alone and in the presence of PFOA. The changes in the particle size and zeta potential of ZnO-NPs after various time intervals, i.e., after 1 day, 1 week, 2 weeks, and 3 weeks of interaction, are presented and discussed below.

Particle size: Alterations in the size of ZnO-NPs were observed alone (i.e., ZnO-NPs in buffered Milli-Q water and PFOA) and in the presence of salts with PFOA after 1 day, 1 week, 2 weeks, and 3 weeks of interaction (Figures 3a and S4). The particle size of the ZnO-NPs in buffered water increased after 1 week of interaction compared to that after 1 day, e.g., 162.4 nm after 1 day to 206 nm after 1 week of interaction; these findings are consistent with the results shown in Figure 1a (166 nm after 1 day and 203 nm after 1 week of interaction) and Figure S1a,b. Similarly, an increase in the size of the ZnO-NPs was observed in the presence of 10 and 500 μg/L PFOA. A similar effect on the correlation between particle size and the surface charge of ZnO-NPs was observed for some samples, such as ZnO-NPs with 10 μg/L PFOA and ZnO-NPs with 500 μg/L PFOA (Figure 3a,b).

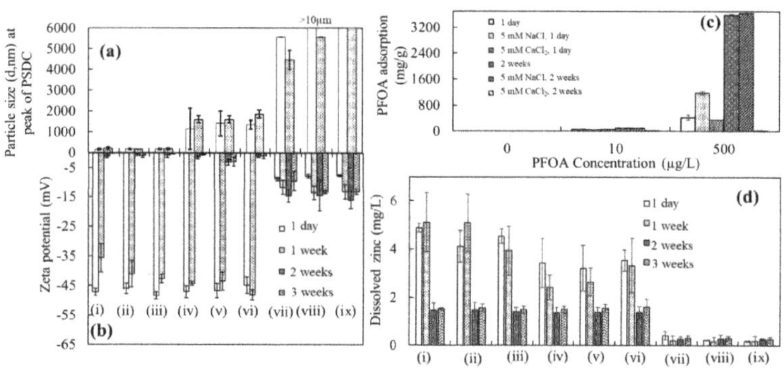

Figure 3. Analysis of particle size (**a**) and zeta potential (**b**) for ZnO-NPs after interaction with PFOA and electrolytes; analysis of PFOA sorption (**c**) and dissolved Zn (**d**) in the solution after interaction; samples include: (i) ZnO-NPs; (ii) ZnO-NPs in 10 μg/L PFOA; (iii) ZnO-NPs in 500 μg/L PFOA; (iv) ZnO-NPs in 5 mM NaCl; (v) ZnO-NPs in 10 μg/L PFOA and 5 mM NaCl; (vi) ZnO-NPs in 500 μg/L PFOA and 5 mM NaCl; (vii) ZnO-NPs in 5 mM $CaCl_2$; (viii) ZnO-NPs in 10 μg/L PFOA and 5 mM $CaCl_2$; (ix) ZnO-NPs in 500 μg/L PFOA and 5 mM $CaCl_2$. Zeta potential values are minus.

For instance, in Figure 1a, the size at the peak of the PSDC of ZnO-NPs with 10 μg/L PFOA is 173 nm after 1 day and 298.5 nm after 1 week, with a surface charge of −45.4 mV after 1 day and −1.9 mV after 1 week. The size of ZnO-NPs was also increased affecting the surface charge of ZnO-NPs at the same concentration (Figure 3a,b); the larger the particle size, the greater the decrease in the magnitude of the surface charge values and vice versa. A similar observation (such as a drop in magnitude of surface change with the increase in the particle size) with surface charge was observed for different batches of samples of the same concentration (such as ZnO + 10 and 500 μg/L PFOA, Figures 1a,b and 3a,b). After 2 and 3 weeks, the instability trend was similar, with the particle size being outside the machine range. This result showed that, initially, the stability of ZnO-NPs could vary based on Brownian motion (the random movement of particles due to bombardment by the solvent molecules that surround them). Normally, dynamic light scattering involves the measurement of particles suspended within a liquid and their shape. If the shape of a particle changes in a way that affects the diffusion speed, then the hydrodynamic size and surface charge may also change. This trend remained consistent with these concentrations (such as ZnO + 10 and 500 μg/L PFOA) in the next sections.

An increase in the particle size with a decreased surface charge was also observed in the presence of 5 mM NaCl with ZnO-NPs and with ZnO-NPs and PFOA after various time intervals (Figure 3a,b). The ions in the medium and the total ionic concentration may affect the particle diffusion speed by altering the thickness of the electric double layer (the Debye length, K^{-1}). The resulting extended double layer of ions around the particles due to electrostatic interactions leads to a reduction in the diffusion speed and results in a larger, apparent hydrodynamic diameter. However, the presence of hydrophobic PFOA molecules, which can accumulate more easily around the surface of nanoparticles due to their hydrophobic nature than when suspended in water, may also alter the size and charge of the nanoparticles.

Generally, the stability of nanoparticles in the presence of electrolytes strongly depends on the capping agents used for stabilization [53,54]. Variations in the size of the ZnO-NPs with the co-occurrence of PFOA and electrolytes could be linked to adsorption, electrostatic interactions, hydrogen bonding, van der Waals effects, and cation bridging (Schematic 7). For instance, in the case of 5 mM NaCl, few alterations in the particle size were observed at the peak of the PSDC (Figure 3a). However, the presence of 5 mM monovalent salts (NaCl) balanced the net electrostatic interactions between highly electronegative fluorine atoms and salt due to charge screening/shielding effects and London interactions [54,55].

The influence of divalent cations on the size of ZnO-NPs was different in the presence of PFOA. Alterations in the size and surface charge of the ZnO-NPs were observed by interacting the ZnO-NPs with 5 mM $CaCl_2$ alone or in the presence of PFOA after 1 day, 1 week, 2 weeks, and 3 weeks of interaction (Figures 3 and S4). This increase in the size of the ZnO-NPs could be due to the accumulation of divalent cations, which results in a decrease in the diffusion speed and generation of larger particles compared to those of NaCl. It could also be expected that in the presence of 5 mM $CaCl_2$, the electrostatic repulsion between the negatively charged ZnO-NPs surface and PFOA molecules (which have a negative charge due to the anionic nature and high electronegativity) was reduced because of bridging interactions between the negatively charged surfaces of ZnO-NPs, divalent cations, and PFOA molecules. Consequently, the gathering of positively charged divalent cations enhanced the nanoparticle size due to bridging effects and made the nanoparticles more unstable, leading to agglomeration.

Cations, such as Ca^{2+}, could be the cause of the bridging phenomenon between carboxyl groups. Consequently, the adsorption of PFOA may be hindered in certain aqueous environments enriched with the aforementioned cations due to the decrease in electrostatic interactions between PFOA and the protonated surface [56–58]. The same behaviour was observed for the ZnO-NPs in the following experiments.

Zeta potential: The zeta potentials of the ZnO-NPs (0.1 g/L nanoparticle dispersion, experimental batch one) were −47.2 and −35.7. −1.1, and −0.2 mV after 1 day, 1 week, 2 weeks, and 3 weeks of interaction, respectively (Figure 3b). The decrease in the surface charge of the ZnO-NPs could be caused by the aggregation of nanoparticles resulting from van der Waals forces, hydrogen bonding, and hydrophobic interactions based on the aging factor. A similar decreasing trend in the magnitude of the surface charge was observed in the presence of PFOA (Figure 3b). The presence of mono- and divalent salts also altered the surface charge of the ZnO-NPs after various durations of interaction with PFOA.

The presence of 5 mM NaCl did not significantly change the zeta potential of the nanoparticles after 1 day of interaction alone or in the presence of PFOA (Figure 3b). It could be argued that the large extent of PFOA adsorption on the surface of nanoparticles balanced the overall electrostatic interaction forces between PFOA molecules and the monovalent salts based on shielding effects and London interactions [54,55]. However, the surface charge of the ZnO-NPs (in the presence of 5 mM NaCl) decreased to −0.5 mV after 3 weeks of interaction. This may indicate that aging affects the surface charge of nanoparticles, allowing more attachment of monovalent cations to the negatively charged surface of ZnO-NPs via electrostatic forces of attraction and van der Waals interactions. However, the same decreasing trend after various numbers of interactions was also observed in the presence of PFOA molecules (Figure 3b).

Similarly, a decrease in the magnitude of the zeta potential of the ZnO-NPs was observed in the presence of divalent cations (5 mM $CaCl_2$) alone and in the presence of 5 mM $CaCl_2$ at various concentrations (such as 10 and 500 μg/L) of PFOA (Figure 3b). The electrostatic repulsion between the ZnO-NP (negatively charged) surface and PFOA molecules was reduced in the presence of 5 mM $CaCl_2$. The gathering of positively charged ions on the surface was responsible for the decrease in zeta potential (less negative zeta potential) (Figure 3b), which increased the particle size due to bridging effects and increased instability, leading to agglomeration (Figures 3a and S4). The aggregation behaviour of ZnO-NPs in the presence of salts suggested that aging during the interaction of ZnO-NPs with salts affects the particle diffusion speed by changing the thickness of the Debye length due to the gathering of ions, resulting in agglomeration alone and in the presence of organic pollutants.

Adsorption: PFOA was analyzed to investigate the effects of salts (such as 5 mM NaCl and 5 mM $CaCl_2$) on the adsorption (interaction) of PFOA (10 and 500 μg/L) on ZnO-NPs after 1 day and 2 weeks of interaction (Figure 3c). Increased adsorption (interaction) of PFOA molecules was identified with increasing concentration and interaction time (more sorption after 2 weeks than after 1 day). A similar trend with increased adsorption of PFOA was observed in the presence of 5 mM NaCl. A possible explanation for this result could be that the increase in the ionic strength of monovalent ions (Na^+) due to NaCl caused an increase in electrostatic attraction between the negatively charged ZnO-NP surface and the negatively charged PFOA molecules due to the presence of monovalent ions (Na^+) in between serving as a bridging carrier to support bridging interactions. However, less adsorption of PFOA on ZnO-NPs was observed with a 5 mM $CaCl_2$ concentration in the solution for both time intervals. Both $CaCl_2$ and NaCl affected the adsorption of PFOA on ZnO-NPs, potentially due to the electrostatic force of attraction. However, in the case of $CaCl_2$, the bridging effect of divalent (Ca^{2+}) cations between ZnO-NPs and PFOA may further lead to a reduction in PFOA adsorption on the ZnO-NPs [57,59]. This finding is consistent with one study showing that the adsorption of PFOA decreases with increasing ionic strength [57].

Dissolution: The dissolved zinc (mg/L) in the ZnO-NPs in buffered water and after the interaction of PFOA with the ZnO-NPs in the presence of salts (such as 5 mM NaCl and 5 mM $CaCl_2$) were measured using ICP–OES (Figure 3d). The particle size increased due to agglomeration and sedimentation after several weeks of interaction, decreasing the specific surface area and resulting in restrained dissolution. However, in the case of $CaCl_2$, less dissolution was measured than in all the other samples, which may be related to more agglomeration due to the bridging effect of Ca^{2+}. The smaller the size of the ZnO-NPs, the more easily dissolution occurred compared to the dissolution of larger particles [18]. The attachment and penetration of nanoparticles inside the pores of low-density polyethylene tubes cannot be ignored.

3.3. Influence of HA on PFOA and ZnO-NPs' Interaction

Particle size: The sizes of the ZnO-NPs in buffered water, treated with various concentrations of PFOA (such as 10 or 500 μg/L), treated with various concentrations of HA (1, 5, or 10 mg/L), and mixed with each of the other substances, were analyzed after 1 day, 1 week, 2 weeks, and 3 weeks of interaction at pH 7 (Figures 4a and S5). Overall, an increase in the particle size of the ZnO-NPs was observed alone and in the presence of PFOA from 1 day to 3 weeks of interaction. This increase might be due to the presence of large (agglomerated) and/or sedimenting particles resulting from particle–particle interactions, electrostatic interactions, and hydrophobic interactions. The ZnO-NPs were monodispersed (particles of uniform size) from the time of nanoparticle suspension preparation to a few days. However, after 1 week of interaction, nonuniform (polydisperse) behaviour of the nanoparticles was observed. Overall, particles with sizes ranging from 1 week to 3 weeks were obtained in their polydispersed form, which was also reflected by their surface charge (Figure 4b).

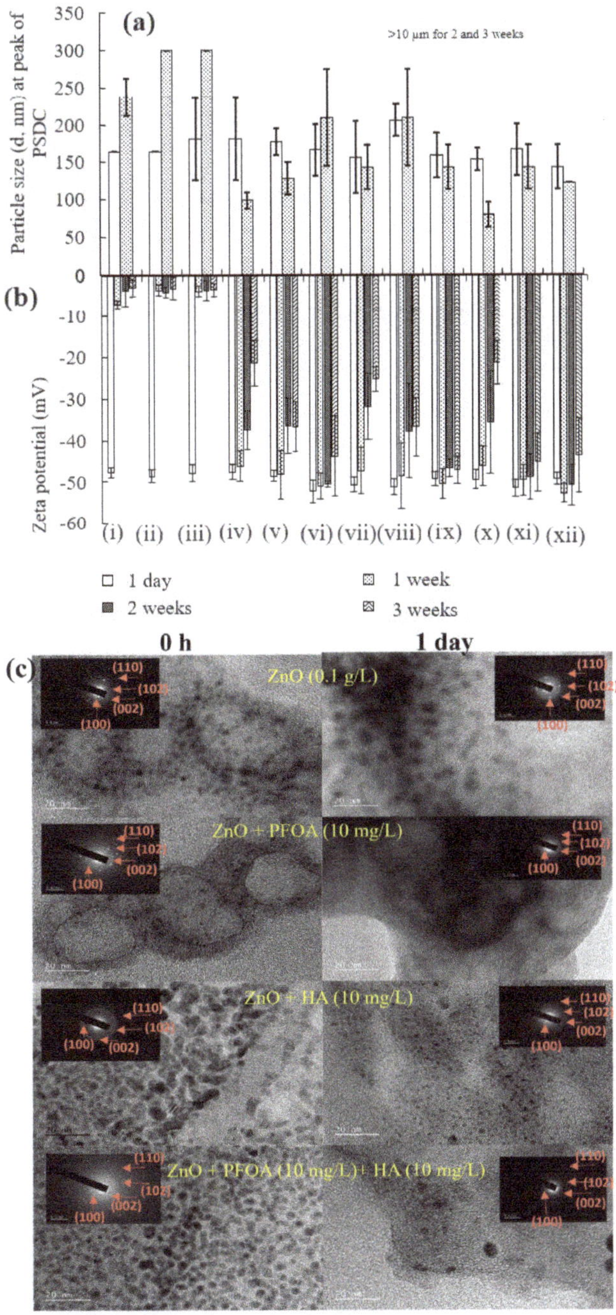

Figure 4. The analysis for ZnO-NPs after interaction with PFOA and HA; particle size (**a**), zeta potential (**b**), and TEM (**c**); samples include: ZnO-NPs (i), ZnO-NPs in 10 µg/L PFOA (ii) and 500 µg/L PFOA (iii); ZnO-NPs in 1 (iv), 5 (v) and 10 (vi) mg/L HA; ZnO-NPs in 10 µg/L PFOA and 1 (vii), 5 (viii) and 10 (ix) mg/L HA; ZnO-NPs in 500 µg/L PFOA and 1 (x), 5 (xi), and 10 (xii) mg/L HA. Zeta potential values are minus. The scale bar of all SAED images is 5 1/nm.

The size and distribution range of the ZnO-NPs in the presence of 1, 5 and 10 mg/L HA decreased after 1 week of interaction compared to after 1 day (Figures 4a and S5). However, after 2 and 3 weeks of interaction, the samples might be very polydispersed, and the particle size was not suitable for measurement by a dynamic light scattering analyzer (a scattered fraction of the samples was observed). Similar behaviour of the ZnO-NPs was observed in the presence of various concentrations of PFOA (i.e., 10 and 500 μg/L) with HA (Figures 4a and S5). It is quite possible that HA (a large aliphatic network of carbon molecules) capped the effective edges of the nanoparticles, which ultimately caused their dispersion.

The presence of both PFOA and HA altered the size of the ZnO-NPs differently than the presence of individual PFOA or HA. Figures 4a and S5 illustrate the size of the ZnO-NPs at the peak of the particle size distribution curve (PSDC (d, nm)) in the presence of various concentrations of HA and PFOA after various durations of interaction. The particle size and range of the ZnO-NPs increased (agglomerated particles) alone and in the presence of PFOA and decreased (polydispersed) at various concentrations (1, 5, and 10 mg/L) of HA. This dispersion behaviour of the ZnO-NPs may be associated with the presence of organic acids (i.e., HA), which may impact engineered ZnO-NPs by reducing their aggregation behaviour [19,60].

The number of specific affinity sites and the affinity coefficient of specific sites for organic pollutants are deemed to be the main influential parameters on the adsorbent capacity to deal with pollutants. Enhanced nanoparticle (ZnO) stability in suspension media by adsorbed dissolved organic matter can increase the total number of specific affinity sites, which supports the adsorption of organic pollutants (PFOA) on the surface of dissolved organic materials rather than on the nanoparticle surface. Simultaneously, the adsorbed organic matter may also produce new affinity sites and/or block the nanoparticle affinity sites to alter their capacity to adsorb pollutants. Dissolved organic matter, which is not adsorbed by nanoparticles, may also first adsorb pollutants and, second, curtail further adsorption of pollutants on the nanoparticle surface [15,38].

Zeta potential: The zeta potentials of the ZnO-NPs in buffered water, with PFOA, and with 1, 5, and 10 mg/L HA were measured after 1 day, 1, 2, and 3 weeks of interaction, respectively (Figure 4b). The magnitude of the surface charge of the ZnO-NPs alone and in the presence of PFOA decreased (from 1 day to 3 weeks) from −47.7 to −3.0 mV for the ZnO-NPs, from −48.6 to −3.1 mV for the ZnO-NPs with 10 μg/L PFOA, and from −47.9 to −3.5 mV for the ZnO-NPs with 500 μg/L PFOA. This behaviour confirmed the increase in the size of the nanoparticles due to agglomeration/sedimentation, which resulted in a reduced net charge (less negative) on the surface of the nanoparticles. However, the presence of HA decreased the magnitude of the change in the surface charge of the ZnO-NPs compared to that of the non-HA-containing samples. However, a high concentration of HA remained dominant in restraining the decrease in the magnitude of the zeta potential of the samples compared to that of lower HA concentrations (such as 1 mg/L HA) (Figure 4b) [18].

The aforementioned electrical potential data revealed that the aggregation behaviour of the pure ZnO-NPs in aqueous systems could be due to electrostatic interactions, van der Waals forces, and hydrophobic interactions. The environmental aging of nanoparticles alone and in the presence of organic pollutants, such as PFOA, could decrease the surface charge of the nanoparticles, increasing their sedimentation in environmental waters by decreasing their stability in aqueous systems. The presence of HA altered the surface charge in the reverse pattern compared to that of pure ZnO-NPs with and without the presence of PFOA. The HA substances covered the surface/effective sites of the nanoparticles because of their high aliphatic carbon content, which resulted in the least possibility of PFOA adsorbing on the nanoparticle surfaces. This also leads to the dispersion of the nanoparticles.

Adsorption: The PFOA in solution was measured to examine the sorption of PFOA with the ZnO-NPs alone and in the presence of various concentrations of HA after two weeks of interaction (Figure S6). An increase in the amount of adsorbed PFOA (10 and 500 μg/L) was calculated for ZnO NPs without HA after 2 weeks of interaction. However, the adsorption of PFOA decreased with the addition of high HA concentrations (such as from 1 to 5 and 10 mg/L HA). Dissolved humic acids can foul the adsorption of organic chemicals to microporous activated carbon through direct competition for adsorption sites and pore blockage [61].

Perfluoroalkyl acids, such as PFOA, contain a negatively charged hydrophilic head group and a hydrophobic–oleophobic perfluoroalkyl chain. Accordingly, a variety of mechanisms might be involved in the adsorption of PFOA in response to the surface properties (such as the charge and hydrophobicity) of adsorbents. The surface of HA is dominated by graphitic carbons, which are highly hydrophobic and have large electronic polarizability. Adsorption of PFOA molecules to HA is expected to be driven mainly by hydrophobic effects, which are combinations of entropic gradients and van der Waals (mainly dispersion) interactions between the adsorbate and adsorbent, whereas electrostatic forces play only a minimal role here. The low adsorption affinity of high concentrations of HA for PFOA is likely due to the low electronic polarizability of these molecules, thus decreasing potential van der Waals interactions despite the large electronic polarizability of graphitic carbons [62].

It could also be assumed that the highly aliphatic structure of HA dispersed the ZnO NPs, providing fewer active sites for the attachment of PFOA molecules. This also supported the results obtained (such as decreased zeta potential values for ZnO NPs alone and in the presence of 10 and 500 μg/L PFOA compared to samples with HA) in Figures 4a,b and S6.

Dissolution: The dissolution of ZnO-NPs alone or in the presence of various concentrations of HA was observed in Milli-Q water at pH 7 controlled by using buffer solution (Figure S7). The presence of zinc in its dissolved or ionic form is potentially toxic to microorganisms, such as microflora [63,64]. The dissolution of ZnO-NPs can be influenced by the presence of other compounds in water [65], such as HA. The dissolved zinc concentration (mg/L) from ZnO-NPs alone or from ZnO-NPs combined with PFOA was calculated with the addition of various concentrations of HA in this study (Figure S7). After 1 day of interaction, the dissolved zinc concentration was greater in the presence of various concentrations of HA than in the absence of HA, and this trend was observed even after 1 week of interaction. It could be assumed that HA dispersed the nanoparticles after a long interaction time (such as 2–3 weeks), decreasing this dispersion effect by dominating the electrostatic forces, van der Waals forces, and hydrophobic interactions. This could also be caused by complexation (for zinc ions) with anionic HA followed by its large complex structure. Therefore, our findings are the same as those hypothesized by [63], i.e., that HA binds zinc ions.

TEM and XRD analysis: The samples from ZnO-NPs alone or in the presence of PFOA or HA were analyzed using TEM morphology and elemental mapping after immediate preparation (such as after 0 h of interaction) and after 1 day of interaction (Figure 4c). ZnO-NPs were more aggregated after 1 day of interaction than after 0 h, which is consistent with previous findings (Figure 1a). TEM revealed an increase in the size of the ZnO-NPs in the presence of 10 mg/L PFOA (only this concentration was selected for TEM analysis to confirm the presence PFOA, such as fluorine in mapping) after 1 day of interaction compared to 0 h. Elemental mapping further confirmed the presence of Zn, O, F, P, and K (Figure S8). The dispersion patterns of ZnO-NPs, alone and in the presence of PFOA, due to the presence of highly aliphatic and complex structures of HA, can be observed (Figure 4c) when comparing images from 0 h and 1 day. The SAED images did indicate less crystallinity on the nanoparticles after 1 day of interaction, which matches the results obtained from the zeta analysis. Elemental mapping further confirmed the presence of the expected elements. Figure S8 shows the elemental composition comparisons of the ZnO-NPs with contaminants at different intervals.

Post-photocatalysis characterization, namely XRD (Figure S8b), of ZnO-NPs, HA, and ZnO-NPs with and without the presence of 10 mg/L PFOA and 10 mg/L HA was performed to further examine the impact of PFOA and HA on the crystallinity and purity of ZnO-NPs. It was observed that there was no change in the crystal phase of the ZnO-NPs after 1 day of interaction. However, after 1 week of interaction (Figure S8b), the intensities of the peak alignments at (1 0 0), (0 0 2), and (1 0 1) were suppressed. Four new peaks at 2θ values of 9.68° and 19.36°, 22.58°, and 25.6° were observed, indicating signs of alterations into the crystallinity and purity of the ZnO-NPs influence by the adsorption of co-contaminants and aging factors.

3.4. Influence of Electrolytes and HA Together on PFOA and ZnO-NPs' Interaction

A mixture of salts and dissolved organic matter could influence the particle size and surface charge differently, which was investigated and explained in this section (Figure 5).

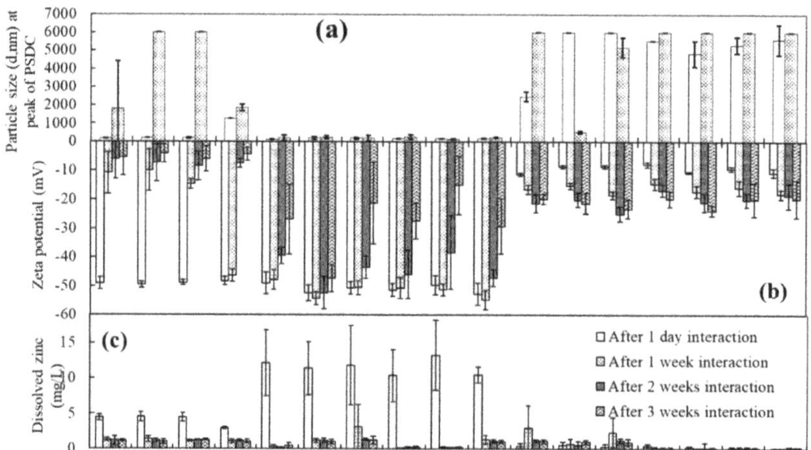

PFOA	0	10	500	0	0	0	10	10	500	500	0	0	0	10	10	500	500
HA	0	0	0	0	1	10	1	10	1	10	0	1	10	1	10	1	10
NaCl	0	5	5	5	5	5	5	5	5	5	0	0	0	0	0	0	0
$CaCl_2$	0	0	0	0	0	0	0	0	0	0	5	5	5	5	5	5	5

Figure 5. The particle size (**a**), zeta potential (**b**), and dissolved Zn (**c**) analysis for thr interaction of ZnO-NPs with PFOA under various conditions; concentrations for PFOA, HA, NaCl, and $CaCl_2$ are μg/L, mg/L, mM, and mM respectively. Zeta potential values are minus".

Particle size: The sizes of the ZnO-NPs at the peak of PSDC (d, nm) without and with the presence of PFOA and with 5 mM concentrations of monovalent and divalent electrolytes and various concentrations of HA after 1 day, 1 week, 2 weeks, and 3 weeks of interaction are illustrated (Figures 5 and S9). With the increasing size of ZnO-NPs alone or with PFOA and the addition of 5 mM concentrations of monovalent salt (NaCl) and divalent $CaCl_2$, the size of the ZnO-NPs increased as the interaction time increased, such as from 1 day to 1 week. The size was out of the machine range (range: 0.3–1.0 × 10^4 nm) after 1 week due to the aggregation of nanoparticles, and only a few fragments were measured after 1 week of interaction. It is assumed that the particle size at the peak of the PSDC would be larger than 10,000 nm. In the presence of HA, initial dispersion was observed with increasing HA concentration in the presence of salts and PFOA; however, after a few weeks, the same behaviour of aggregation/agglomeration/sedimentation was observed, with the particle size being nonuniform and not measurable, while there was a decreasing zeta potential.

These findings revealed that the influence of humic substances on the interactions between PFOA and ZnO-NPs is somewhat complicated, especially when electrolytes are present. Counteractions for PFOA between HA and ZnO-NPs could reduce the effective interactions of ZnO-NPs by decreasing the amount of PFOA available for sorption. Moreover, natural organic matter can cover the surface of nanoparticles and thereby reduce their affinity for organic pollutants [38,57,66].

Zeta potential: A decrease in the magnitude of the surface charge of ZnO NPs alone or in the presence of PFOA or salt was observed after 1 day to 3 weeks of interaction, as shown in Figure 5b. This indicated that the ZnO-NPs exhibited a similar agglomeration (size increase) behaviour (Figure 5) with a diminished surface charge, as described in the aforementioned sections. The addition of various concentrations of HA in the presence of PFOA did not significantly change the zeta potential compared to that of samples with HA. This confirms that interactions between HA molecules (the dispersion of HA-coated ZnO-NPs) are more dominant at high concentrations (such as 10 mg/L HA) than the electrostatic interactions between negatively charged nanoparticles surrounded by monovalent cations (Na^+). Conversely, divalent cations interacted more strongly with negative surface charges, and the overall zeta potential decreased in magnitude in the presence of both divalent cations (Ca^{2+}) (Figure 5b). However, the dispersion effect of HA on ZnO-NPs was also observed when PFOA and divalent electrolyte ($CaCl_2$) salts were present (Figure 5b), as noted in the findings. It could be inferred that salinity and natural organic substances play significant roles in the transport of ZnO-NPs and their associated organic pollutants (PFOA) from fresh water to the ocean, especially in estuary regions.

Adsorption: Adsorption of PFOA was examined for ZnO-NPs alone or in the presence of various concentrations of HA in the presence of 5 mM NaCl and $CaCl_2$ after two weeks of interaction (Table S1). An increase in the amount of adsorbed PFOA (10 and 500 μg/L) was calculated for ZnO NPs without HA after 2 weeks of interaction. However, the adsorption of PFOA decreased with the addition of high HA concentrations (such as 1 and 10 mg/L HA) (Table S1). It could be assumed that the highly aliphatic structure of HA dispersed the ZnO NPs, providing fewer active sites for the attachment of PFOA molecules. This result also supported the results obtained, i.e., decreased zeta potential values for ZnO NPs alone and in the presence of 10 and 500 μg/L PFOA compared to samples with HA in Figure 5b.

Dissolution: The dissolved zinc concentration (mg/L) from ZnO-NPs alone, with PFOA and salts, and with the addition of various concentrations of HA, was measured in this study (Figure 5c). After 1 day of interaction, the dissolved zinc concentration was greater in the presence of various concentrations of HA than in the absence of HA. It could be assumed that HA dispersed the nanoparticles; however, after a long interaction time (such as 1, 2, or 3 weeks), the dispersion effect decreased due to the dominant electrostatic forces, van der Waals forces, and hydrophobic interactions.

TEM analysis: The morphological behaviour of ZnO-NPs with PFOA, HA, and $CaCl_2$ after 0 h and 1 day intervals was examined via TEM (Figure S10). The particles were aggregated after 0 h in the presence of 5 mM $CaCl_2$. However, after 1 day of interaction, shaded (due to $CaCl_2$) and dispersed (due to HA) patterns of the nanoparticles can be observed. The bright diffraction signals are due to the crystalline ZnO-NPs, including the presence of $CaCl_2$ crystals. To examine the morphological changes associated with high concentrations of $CaCl_2$, 10 mM $CaCl_2$ (after 0 h of interaction) was added, which generated clusters/agglomerates. However, after 1 day of interaction, the nanoparticles were dispersed by coating them with large HA molecules (Figure S10d).

FTIR: The interactions of ZnO-NPs with PFOA in the presence of HA and electrolytes were investigated via FTIR analysis (Figure 6). The metal oxide (ZnO) absorbance ranged from 600 to 400 cm^{-1} [15,49,67], which indicated the presence of interacting ZnO-NPs. This difference was detected in all the samples, while the intensity of the absorbance peak depth and location varied. Peaks at 1800 and 600 cm^{-1} represent carboxylate functional groups and C−F [51], C−C, and C−H stretching, respectively [50]. In particular, the vibrational peak at approximately 1102 cm^{-1} appeared in all samples with PFOA, representing the presence of C–F stretching bonds. This indicated the interaction between PFOA and the ZnO-NPs. The absorbance at 1645 cm^{-1} is due to H−O−H bending. The infrared band at 1010 cm^{-1} in all the examples except for ZnO-NPs and ZnO + PFOA is due to the stretching of N–H bonds [50] present in an organic compound, such as HA. The peaks between 1050 and 750 cm^{-1} could be due to K-potassium and P-phosphorous (from a buffer solution used to maintain pH 7) stretching with O and C. The absorbance at 3490 cm^{-1} is due to O-H stretching [50]. The peaks at approximately 933 and 871 cm^{-1} could be due to triatomic inorganic molecules (calcium chlorine), while 670 cm^{-1} could be due to CO_2 from the atmosphere [50].

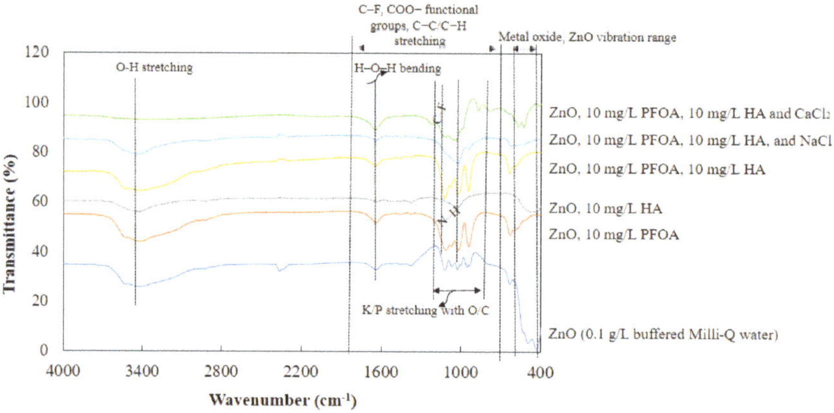

Figure 6. FTIR analysis of interactions between ZnO-NPs and PFOA in the presence of HA and electrolytes.

3.5. Interaction Scheme

A schematic diagram illustrating the interaction mechanisms is shown in Figure 7, depicting the proposed interactions between ZnO-NPs and PFOA and HA in the absence and presence of cations in simulated waters under controlled laboratory conditions. ZnO-NPs tend to aggregate in aqueous media (0.1 g/L) due to van der Waals and hydrophobic interactions surrounded by hydrogen bonding between water molecules (Figures 1 and 7). The hydrophobic molecules of PFOA could be comparatively easily adsorbed on the porous surface of the ZnO-NPs, which subsequently enhanced the size of the nanoparticles through electrostatic interactions, hydrogen bonding and van der Waals interactions. Specific surface area and surface morphology of the ZnO-NPs are key factors that impact the interaction mechanisms between ZnO-NPs and PFOA. The roughness, specific surface area, porosity, and shape and size of ZnO-NPs may significantly affect the interactions of PFOA molecules with ZnO-NPs. For instance, porous surfaces with more irregularly sized crystal structures, may provide more active sites for the sorption of PFOA molecules. The surface energy of ZnO-NPs is also associated with their surface morphology. The higher the surface energy, the stronger the interactions with PFOA molecules, leading to more adsorption. However, both the specific surface area and surface morphology of the ZnO-NPs influence the interaction mechanisms (such as adsorption) of PFOA onto ZnO NPs. Van der Waals and hydrophobic interactions may cause the nanoparticles to agglomerate. The coexistence of electrolytes in water systems can screen the charge on the surface of nanoparticles by counter ions and consequently make them unstable at high salt concentrations, in addition to agglomerating. The electrical potential of the ZnO-NPs was changed by varying the type of electrolyte used during the reaction. The type of electrolyte affects the alterations in the surface charge of the ZnO-NPs differently. The aggregation of ZnO-NPs in the presence of monovalent and divalent electrolytes is influenced by the surface charge via electrostatic interactions. Furthermore, cation bridging and ligand binding could also be considered.

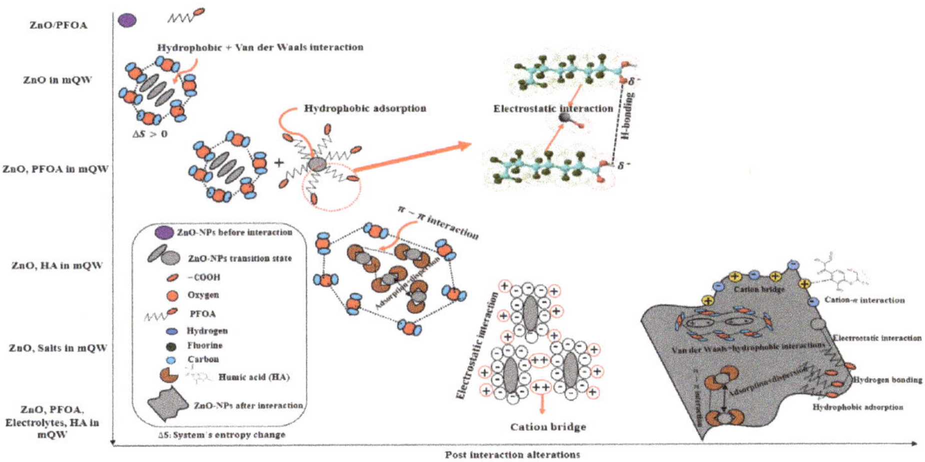

Figure 7. Possible potential interaction mechanisms between ZnO-NPs (size $\leq$ 100 nm) and PFOA in the presence of key factors (e.g., natural organic substances and electrolytes). The + and – symbols represent positive and negative charges respectively.

The addition of HA dispersed the ZnO-NPs by coating them (due to their large molecules), which meant that the ZnO-NPs had the least opportunity to attach to other contaminants. The presence of salts promoted electrostatic interactions and bridging effects as well. Moreover, the presence of PFOA, HA, and salts influenced the surface charge by adsorbing on the active surfaces, covering the surface made possible by the complex and large structure of HA molecules and dispersing nanoparticles and cation bridging, respectively. These factors could lead to alterations in the particle size and morphology of the ZnO-NPs. In summary, once ZnO-NPs are released in ecosystems containing various types of organic and inorganic pollutants, humic substances, and electrolytes, alterations in the parental nanoparticles can be expected. This is in terms of their agglomeration state, crystallinity morphology, purity, size, and surface chemistry. Comparison between this study with previous studies for the responsible interaction mechanisms are tabulated (Table S2).

4. Conclusions

This study revealed that the size, shape, crystallinity, surface charge, and morphology of the ZnO-NPs were altered after they interacted with PFOA, coexisting HA, and salts for various durations as aging factors under controlled pH (i.e., pH 7) in simulated water systems. Due to its large complex structure, humic acid behaves as a dispersant by covering ZnO-NPs and leaving the least opportunity for other compounds to adsorb. Variations in the size, shape, and surface charge due to aging of the ZnO-NPs could also be key factors in their fate and behaviour in combination with other environmental factors. PFOA-sorbed ZnO-NPs may slowly sink and reach sediments in the form of agglomerates in the presence of other substances in water media, such as altered/toxic substances. In addition to this, a toxicity study is recommended in future to determine the toxic impacts of PFOA-adsorbed ZnO-NPs on the living organisms.

Supplementary Materials: The following supporting information can be downloaded at: https://www.mdpi.com/article/10.3390/toxics12080602/s1, Figure S1: Particle size distribution of ZnO-NPs after 1 day (a) and 1 week (b) of interaction with PFOA; ZnO-NPs 0.1 g/L, pH 7; Figure S2a: TEM elemental analysis of ZnO-NPs with PFOA after 1 day of interaction; Figure S2b: TEM analysis (lattice spacing, crystallite size) of ZnO-NPs with PFOA after 1 day of interaction; Figure S3: Concentration of dissolved zinc at pH 7 after 1 day and 2 weeks of interaction; Figure S4: Particle size distribution of

ZnO-NPs after 1 day (a) and 1 week (b) of interaction with PFOA in the presence of electrolytes at pH 7 and room temperature (i.e., 20 °C); Figure S5: Particle size distribution of ZnO-NPs after 1 day (a) and 1 week (b) of interaction in the presence of various concentrations of PFOA and HA at pH 7 and room temperature (i.e., 20 °C); Figure S6: Adsorption of PFOA on the surface of ZnO-NPs after 2 weeks of interaction in the presence of various concentrations of PFOA and HA; Figure S7: Dissolved zinc (mg/L) at pH 7 after 1 day, 1 week, 2 weeks, and 3 weeks of interaction with various concentrations of PFOA and HA; Figure S8a: Elemental composition comparisons of the ZnO-NPs with contaminants; Figure S8b: XRD analysis of ZnO-NPs powder, HA powder, ZnO-NPs with and without interactions of 10 mg/L PFOA and 10 mg/L HA after 1 day and 1 weeks of interactions; Figure S9: Particle size distribution of ZnO-NPs after 1 day (a) and 1 week (b) of interaction with PFOA in the presence of electrolytes and HA at pH 7 and room temperature (i.e., 20 °C); Figure S10: TEM images of ZnO NP under various concentrations of $CaCl_2$ after 0 h and 24 h of interaction in solution (drops taken on a TEM grid from solution); ZnO, 10 mg/L PFOA, 10 mg/L HA, 5 mM $CaCl_2$ (a,b); ZnO, 10 mg/L PFOA, 10 mg/L HA, and 10 mM $CaCl_2$ (c,d); Table S1: Adsorption of PFOA on ZnO NPs (mg/g) in the presence of electrolytes and various concentrations of HA; Table S2: Interaction mechanisms of engineered nanoparticles with organic substances. References [15,22,24,25,68] are cited in the supplementary materials.

Author Contributions: A.U.H.K.: Methodology, experimental work and analysis, writing—initial draft writing. Y.L.: supervision, conceptualization, visualization, writing—review and editing. R.N.: supervision, funding acquisition, writing—review and editing. C.F.: supervision, writing—review and editing. H.K.S.: supervision, writing—review and editing; H.Z.: TEM imaging, EDS, mapping analysis and discussions; R.D.: supervision, writing—review and editing. All authors have read and agreed to the published version of the manuscript.

Funding: This research received no external funding.

Institutional Review Board Statement: Not applicable.

Informed Consent Statement: Not applicable.

Data Availability Statement: All relevant data are reported in main text or Supplementary Information.

Acknowledgments: The first author acknowledges The University of Newcastle (UON), Australia, for granting a fully funded PhD (ECRHDR UNRS Central and UNIPRS) scholarship. The first author acknowledges crcCARE for providing a scholarship. The authors are grateful to crcCARE and GCER, UON, for providing financial assistance and research facilities to complete this study. The authors also acknowledge the support of Graeme Jameson, Kitty Tang from the Centre for Multiphase Processes, UON, and Erica Wanless and Sonia Khandaker from the School of Environmental and Life Sciences for providing the facilities and training to conduct nanoparticle size and surface analysis. The authors are grateful to Abhishek Sharma, Francisca Munyeza, and Yunlong Luo from the GCER, UON for assisting in the ICP-OES, LC-MS, and Raman analysis, respectively.

Conflicts of Interest: There are no conflicts of interest among the authors regarding this research.

References

1. Stoller, M.; Ochando-Pulido, J.M. ZnO nano-particles production intensification by means of a spinning disk reactor. *Nanomaterials* **2020**, *10*, 1321. [CrossRef]
2. Bundschuh, M.; Filser, J.; Lüderwald, S.; Mckee, M.S.; Metreveli, G.; Schaumann, G.E.; Schulz, R.; Wagner, S. Nanoparticles in the environment: Where do we come from, where do we go to? *Environ. Sci. Eur.* **2018**, *30*, 6. [CrossRef] [PubMed]
3. Lead, J.R.; Batley, G.E.; Alvarez, P.J.J.; Croteau, M.N.; Handy, R.D.; McLaughlin, M.J.; Judy, J.D.; Schirmer, K. Nanomaterials in the environment: Behavior, fate, bioavailability, and effects—An updated review. *Environ. Toxicol. Chem.* **2018**, *37*, 2029–2063. [CrossRef]
4. Osmond, M.J.; McCall, M.J. Zinc oxide nanoparticles in modern sunscreens: An analysis of potential exposure and hazard. *Nanotoxicology* **2009**, *4*, 15–41. [CrossRef]
5. Ge, Y.; Schimel, J.P.; Holden, P.A. Evidence for negative effects of TiO2 and ZnO nanoparticles on soil bacterial communities. *Environ. Sci. Technol.* **2011**, *45*, 1659–1664. [CrossRef] [PubMed]
6. Jeng, H.A.; Swanson, J. Toxicity of metal oxide nanoparticles in mammalian cells. *J. Environ. Sci. Heal. Part A* **2006**, *41*, 2699–2711. [CrossRef] [PubMed]
7. Lin, D.; Drew Story, S.; Walker, S.L.; Huang, Q.; Cai, P. Influence of extracellular polymeric substances on the aggregation kinetics of TiO_2 nanoparticles. *Water Res.* **2016**, *104*, 381–388. [CrossRef]

8. Peng, Y.-H.; Tsai, Y.-C.; Hsiung, C.-E.; Lin, Y.-H.; Shih, Y.-H. Influence of water chemistry on the environmental behaviors of commercial ZnO nanoparticles in various water and wastewater samples. *J. Hazard. Mater.* **2017**, *322*, 348–356. [CrossRef] [PubMed]
9. Nowack, B.; Bucheli, T.D. Occurrence, behavior and effects of nanoparticles in the environment. *Environ. Pollut.* **2007**, *150*, 5–22. [CrossRef]
10. Park, C.M.; Chu, K.H.; Her, N.; Jang, M.; Baalousha, M.; Heo, J.; Yoon, Y. Occurrence and removal of engineered nanoparticles in drinking water treatment and wastewater treatment processes. *Sep. Purif. Rev.* **2017**, *46*, 255–272. [CrossRef]
11. Hou, J.; Miao, L.; Wang, C.; Wang, P.; Ao, Y.; Qian, J.; Dai, S. Inhibitory effects of zno nanoparticles on aerobic wastewater biofilms from oxygen concentration profiles determined by microelectrodes. *J. Hazard. Mater.* **2014**, *276*, 164–170. [CrossRef] [PubMed]
12. Liu, G.; Wang, D.; Wang, J.; Mendoza, C. Effect of ZnO particles on activated sludge: Role of particle dissolution. *Sci. Total Environ.* **2011**, *409*, 2852–2857. [CrossRef]
13. Liu, W.S.; Peng, Y.H.; Shiung, C.E.; Shih, Y.H. The effect of cations on the aggregation of commercial ZnO nanoparticle suspension. *J. Nanoparticle Res.* **2012**, *14*, 1259. [CrossRef]
14. Philippe, A.; Schaumann, G.E. Interactions of dissolved organic matter with natural and engineered inorganic colloids: A review. *Environ. Sci. Technol.* **2014**, *48*, 8946–8962. [CrossRef]
15. Khan, A.U.H.; Liu, Y.; Naidu, R.; Fang, C.; Dharmarajan, R.; Shon, H. Interactions between zinc oxide nanoparticles and hexabromocyclododecane in simulated waters. *Environ. Technol. Innov.* **2021**, *24*, 102078. [CrossRef]
16. Khan, R.; Inam, M.A.; Khan, S.; Park, D.R.; Yeom, I.T. Interaction between persistent organic pollutants and ZnO NPs in synthetic and natural waters. *Nanomaterials* **2019**, *9*, 472. [CrossRef]
17. Yang, K.; Lin, D.; Xing, B. Interactions of humic acid with nanosized inorganic oxides. *Langmuir* **2009**, *25*, 3571–3576. [CrossRef] [PubMed]
18. Bian, S.W.; Mudunkotuwa, I.A.; Rupasinghe, T.; Grassian, V.H. Aggregation and dissolution of 4 nm ZnO nanoparticles in aqueous environments: Influence of pH, ionic strength, size, and adsorption of humic acid. *Langmuir* **2011**, *27*, 6059–6068. [CrossRef]
19. Han, Y.; Kim, D.; Hwang, G.; Lee, B.; Eom, I.; Kim, J.P.; Tong, M.; Kim, H. Aggregation and dissolution of ZnO nanoparticles synthesized by different methods: Influence of ionic strength and humic acid. *Colloids Surf. A Physicochem. Eng. Asp.* **2014**, *451*, 7–15. [CrossRef]
20. Shrestha, S.; Wang, B.; Dutta, P. Nanoparticle processing: Understanding and controlling aggregation. *Adv. Colloid Interface Sci.* **2020**, *279*, 102162. [CrossRef]
21. Keller, A.A.; Wang, H.; Zhou, D.; Lenihan, H.S.; Cherr, G.; Cardinale, B.J.; Miller, R.; Zhaoxia, J.I. Stability and aggregation of metal oxide nanoparticles in natural aqueous matrices. *Environ. Sci. Technol.* **2010**, *44*, 1962–1967. [CrossRef]
22. Khan, A.U.H.; Liu, Y.; Fang, C.; Naidu, R.; Shon, H.K.; Rogers, Z.; Dharmarajan, R. A comprehensive physicochemical characterization of zinc oxide nanoparticles extracted from sunscreens and wastewaters. *Environ. Adv.* **2023**, *12*, 100381. [CrossRef]
23. Majedi, S.M.; Kelly, B.C.; Lee, H.K. Combined effects of water temperature and chemistry on the environmental fate and behavior of nanosized zinc oxide. *Sci. Total Environ.* **2014**, *496*, 585–593. [CrossRef]
24. Ding, N.; Chen, X.; Wu, C.M.L. Interactions between polybrominated diphenyl ethers and graphene surface: A DFT and MD investigation. *Environ. Sci. Nano* **2014**, *1*, 55–63. [CrossRef]
25. Wang, X.; Adeleye, A.S.; Wang, H.; Zhang, M.; Liu, M.; Wang, Y.; Li, Y.; Keller, A.A. Interactions between polybrominated diphenyl ethers (PBDEs) and TiO2 nanoparticle in artificial and natural waters. *Water Res.* **2018**, *146*, 98–108. [CrossRef]
26. Coperchini, F.; Croce, L.; Ricci, G.; Magri, F.; Rotondi, M.; Imbriani, M.; Chiovato, L. Thyroid Disrupting Effects of Old and New Generation PFAS. *Front. Endocrinol.* **2021**, *11*, 612320. [CrossRef]
27. Li, P.; Xiao, Z.; Xie, X.; Li, Z.; Yang, H.; Ma, X.; Sun, J.; Li, J. Perfluorooctanoic acid (PFOA) changes nutritional compositions in lettuce (Lactuca sativa) leaves by activating oxidative stress. *Environ. Pollut.* **2021**, *285*, 117246. [CrossRef]
28. Lang, J.R.; Allred, B.M.; Field, J.A.; Levis, J.W.; Barlaz, M.A. National Estimate of Per- and Polyfluoroalkyl Substance (PFAS) Release to U.S. Municipal Landfill Leachate. *Environ. Sci. Technol.* **2017**, *51*, 2197–2205. [CrossRef] [PubMed]
29. Gallen, C.; Eaglesham, G.; Drage, D.; Nguyen, T.H.; Mueller, J.F. A mass estimate of perfluoroalkyl substance (PFAS) release from Australian wastewater treatment plants. *Chemosphere* **2018**, *208*, 975–983. [CrossRef]
30. Cordner, A.; De La Rosa, V.Y.; Schaider, L.A.; Rudel, R.A.; Richter, L.; Brown, P. Guideline levels for PFOA and PFOS in drinking water: The role of scientific uncertainty, risk assessment decisions, and social factors. *J. Expo. Sci. Environ. Epidemiol.* **2019**, *29*, 157–171. [CrossRef]
31. Schultz, M.M.; Higgins, C.P.; Huset, C.A.; Luthy, R.G.; Barofsky, D.F.; Field, J.A. Fluorochemical mass flows in a municipal wastewater treatment facility. *Environ. Sci. Technol.* **2006**, *40*, 7350–7357. [CrossRef] [PubMed]
32. Sinclair, G.M.; Long, S.M.; Jones, O.A.H. What are the effects of PFAS exposure at environmentally relevant concentrations? *Chemosphere* **2020**, *258*, 127340. [CrossRef]
33. Zhou, Y.; Wang, T.; Jiang, Z.; Kong, X.; Li, Q.; Sun, Y.; Wang, P.; Liu, Z. Ecological effect and risk towards aquatic plants induced by perfluoroalkyl substances: Bridging natural to culturing flora. *Chemosphere* **2017**, *167*, 98–106. [CrossRef] [PubMed]
34. Arvaniti, O.S.; Stasinakis, A.S. Review on the occurrence, fate and removal of perfluorinated compounds during wastewater treatment. *Sci. Total Environ.* **2015**, *524–525*, 81–92. [CrossRef]

35. Savoca, D.; Melfi, R.; Palumbo Piccionello, A.; Barreca, S.; Buscemi, S.; Arizza, V.; Arculeo, M.; Pace, A. Presence and biodistribution of perfluorooctanoic acid (PFOA) in Paracentrotus lividus highlight its potential application for environmental biomonitoring. *Sci. Rep.* **2021**, *11*, 18763. [CrossRef]
36. Khan, A.U.H.; Naidu, R.; Dharmarajan, R.; Fang, C.; Shon, H.; Dong, Z.; Liu, Y. The interaction mechanisms of co-existing polybrominated diphenyl ethers and engineered nanoparticles in environmental waters: A critical review. *J. Environ. Sci.* **2023**, *124*, 227–252. [CrossRef]
37. Domingos, R.F.; Rafiei, Z.; Monteiro, C.E.; Khan, M.A.K.; Wilkinson, K.J. Agglomeration and dissolution of zinc oxide nanoparticles: Role of pH, ionic strength and fulvic acid. *Environ. Chem.* **2013**, *10*, 306–312. [CrossRef]
38. Yu, S.; Liu, J.; Yin, Y.; Shen, M. Interactions between engineered nanoparticles and dissolved organic matter: A review on mechanisms and environmental effects. *J. Environ. Sci.* **2018**, *63*, 198–217. [CrossRef]
39. Blok, L.; Bruyn, P.L.D. The ionic double layer at the ZnO solution interface, I. The experimental point of zero charge. *J. Colloid Interface Sci.* **1970**, *32*, 518–526. [CrossRef]
40. Liu, L.; Liu, Y.; Li, C.; Ji, R.; Tian, X. Improved sorption of perfluorooctanoic acid on carbon nanotubes hybridized by metal oxide nanoparticles. *Environ. Sci. Pollut. Res.* **2018**, *25*, 15507–15517. [CrossRef]
41. Mosharraf, M.; Nyström, C. The effect of particle size and shape on the surface specific dissolution rate of microsized practically insoluble drugs. *Int. J. Pharm.* **1995**, *122*, 35–47. [CrossRef]
42. Cardoso, D.; Narcy, A.; Durosoy, S.; Bordes, C.; Chevalier, Y. Dissolution kinetics of zinc oxide and its relationship with physicochemical characteristics. *Powder Technol.* **2021**, *378*, 746–759. [CrossRef]
43. Wöll, C. The chemistry and physics of zinc oxide surfaces. *Prog. Surf. Sci.* **2007**, *82*, 55–120. [CrossRef]
44. Talam, S.; Karumuri, S.R.; Gunnam, N. Synthesis, characterization, and spectroscopic properties of ZnO nanoparticles. *ISRN Nanotechnol.* **2012**, *2012*, 372505. [CrossRef]
45. Zak, A.K.; Razali, R.; Majid, W.H.A.; Darroudi, M. Synthesis and characterization of a narrow size distribution of zinc oxide nanoparticles. *Int. J. Nanomed.* **2011**, *6*, 1399–1403. [CrossRef]
46. Gordeeva, A.; Hsu, Y.J.; Jenei, I.Z.; Brant Carvalho, P.H.B.; Simak, S.I.; Andersson, O.; Haüssermann, U. Layered Zinc Hydroxide Dihydrate, $Zn_5(OH)_{10}\cdot 2H_2O$, from Hydrothermal Conversion of ϵ-$Zn(OH)_2$ at Gigapascal Pressures and its Transformation to Nanocrystalline ZnO. *ACS Omega* **2020**, *5*, 17617–17627. [CrossRef]
47. Rao, M.S.; Satyavathi, K.; Naga Bhaskararao, Y.; Cole, S. Structural and spectral investigations of undoped and Mn^{2+} ion doped $Zn_3(PO_4)_2ZnO$ nanocrystalline phosphor materials. *J. Alloys Compd.* **2016**, *682*, 7–13. [CrossRef]
48. Srivastava, V.; Gusain, D.; Sharma, Y.C. Synthesis, characterization and application of zinc oxide nanoparticles (n-ZnO). *Ceram. Int.* **2013**, *39*, 9803–9808. [CrossRef]
49. Gharagozlou, M.; Naghibi, S. Sensitization of ZnO nanoparticle by vitamin B12: Investigation of microstructure, FTIR and optical properties. *Mater. Res. Bull.* **2016**, *84*, 71–78. [CrossRef]
50. Stuart, B.H. *Infrared Spectroscopy: Fundamentals and Applications*; John Wiley & Sons Ltd.: Hoboken, NJ, USA, 2004; pp. 1–224. [CrossRef]
51. Lin, H.; Wang, Y.; Niu, J.; Yue, Z.; Huang, Q. Efficient Sorption and Removal of Perfluoroalkyl Acids (PFAAs) from Aqueous Solution by Metal Hydroxides Generated in Situ by Electrocoagulation. *Environ. Sci. Technol.* **2015**, *49*, 10562–10569. [CrossRef]
52. Lu, M.; Su, M.; Liu, N.; Zhang, J. Effects of environmental salinity on the immune response of the coastal fish Scatophagus argus during bacterial infection. *Fish Shellfish. Immunol.* **2022**, *124*, 401–410. [CrossRef]
53. Javed, R.; Zia, M.; Naz, S.; Aisida, S.O.; Ain, N.U.; Ao, Q. Role of capping agents in the application of nanoparticles in biomedicine and environmental remediation: Recent trends and future prospects. *J. Nanobiotechnol.* **2020**, *18*, 172. [CrossRef]
54. Zhang, X.; Servos, M.R.; Liu, J. Ultrahigh nanoparticle stability against salt, pH and solvent with retained surface accessibility via depletion stabilization. *J. Am. Chem. Soc.* **2012**, *134*, 9910–9913. [CrossRef] [PubMed]
55. Enustun, B.; Turkevich, J. Coagulation of colloidal gold. *J. Am. Chem. Soc.* **1963**, *85*, 3317–3328. [CrossRef]
56. Hyun, S.; Lee, L.S. Quantifying the contribution of different sorption mechanisms for 2,4-dichlorophenoxyacetic acid sorption by several variable-charge soils. *Environ. Sci. Technol.* **2005**, *39*, 2522–2528. [CrossRef]
57. Wang, F.; Shih, K. Adsorption of perfluorooctanesulfonate (PFOS) and perfluorooctanoate (PFOA) on alumina: Influence of solution pH and cations. *Water Res.* **2011**, *45*, 2925–2930. [CrossRef]
58. You, C.; Jia, C.; Pan, G. Effect of salinity and sediment characteristics on the sorption and desorption of perfluorooctane sulfonate at sediment-water interface. *Environ. Pollut.* **2010**, *158*, 1343–1347. [CrossRef] [PubMed]
59. Wang, F.; Liu, C.; Shih, K. Adsorption behavior of perfluorooctanesulfonate (PFOS) and perfluorooctanoate (PFOA) on boehmite. *Chemosphere* **2012**, *89*, 1009–1014. [CrossRef]
60. Jahan, S.; Bin Yusoff, I.; Alias, Y.B.; Bin Abu Bakar, A.F. Reviews of the toxicity behavior of five potential engineered nanomaterials (ENMs) into the aquatic ecosystem. *Toxicol. Rep.* **2017**, *4*, 211–220. [CrossRef]
61. Kilduff, J.E.; Karanfil, T.; Chin, Y.P.; Weber, W.J. Adsorption of natural organic polyelectrolytes by activated carbon: A size-exclusion chromatography study. *Environ. Sci. Technol.* **1996**, *30*, 1336–1343. [CrossRef]
62. Wang, B.; Lee, L.S.; Wei, C.; Fu, H.; Zheng, S.; Xu, Z.; Zhu, D. Covalent triazine-based framework: A promising adsorbent for removal of perfluoroalkyl acids from aqueous solution. *Environ. Pollut.* **2016**, *216*, 884–892. [CrossRef] [PubMed]
63. Ouyang, K.; Yu, X.Y.; Zhu, Y.; Gao, C.; Huang, Q.; Cai, P. Effects of humic acid on the interactions between zinc oxide nanoparticles and bacterial biofilms. *Environ. Pollut.* **2017**, *231*, 1104–1111. [CrossRef] [PubMed]

64. Xia, T.; Kovochich, M.; Liong, M.; Mädler, L.; Gilbert, B.; Shi, H.; Yeh, J.I.; Zink, J.I.; Nel, A.E. Comparison of the mechanism of toxicity of zinc oxide and cerium oxide nanoparticles based on dissolution and oxidative stress properties. *ACS Nano* **2008**, *2*, 2121–2134. [CrossRef] [PubMed]
65. Li, M.; Lin, D.; Zhu, L. Effects of water chemistry on the dissolution of ZnO nanoparticles and their toxicity to Escherichia coli. *Environ. Pollut.* **2013**, *173*, 97–102. [CrossRef] [PubMed]
66. Wu, C.; Zhang, K.; Huang, X.; Liu, J. Sorption of pharmaceuticals and personal care products to polyethylene debris. *Environ. Sci. Pollut. Res.* **2016**, *23*, 8819–8826. [CrossRef]
67. Chandrasekar, M.; Panimalar, S.; Uthrakumar, R.; Kumar, M.; Saravanan, M.E.R.; Gobi, G.; Matheswaran, P.; Inmozhi, C.; Kaviyarasu, K. Preparation and characterization studies of pure and Li+ doped ZnO nanoparticles for optoelectronic applications. *Mater. Today Proc.* **2021**, *36 Pt 2*, 228–231. [CrossRef]
68. Nuerla, A.; Qiao, X.; Li, J.; Zhao, D.; Yang, X.; Xie, Q.; Chen, J. Effects of substituent position on the interactions between PBDEs/PCBs and DOM. *Chin. Sci. Bull.* **2013**, *58*, 884–889. [CrossRef]

toxics

MDPI

Article

Multiphase Partitioning of Estrogens in a River Impacted by Feedlot Wastewater Discharge

Kuo-Hui Yang [1], Hao-Shen Hung [1], Wei-Hsiang Huang [1,2], Chi-Ying Hsieh [1,*] and Ting-Chien Chen [1,2,*]

[1] Department of Environmental Science and Engineering, National Pingtung University of Science and Technology, Pingtung 91201, Taiwan; pepea@mail.npust.edu.tw (K.-H.Y.); newhon16@yahoo.com.tw (H.-S.H.); carl7510@mail.npust.edu.tw (W.-H.H.)

[2] Disaster Prevention and Mitigation Technology Research Center, General Research Service Center, National Pingtung University of Science and Technology, Pingtung 91201, Taiwan

* Correspondence: chiying@mail.npust.edu.tw (C.-Y.H.); chen5637@mail.npust.edu.tw (T.-C.C.); Tel.: +886-8-7703202 (ext. 7081) (T.-C.C.); Fax: +886-8-7740256 (T.-C.C.)

Abstract: Estrogens in river systems can significantly impact aquatic ecosystems. This study aimed to investigate the multiphase partitioning of estrogens in Wulo Creek, Taiwan, which receives animal feedlot wastewater, to understand their distribution and potential environmental implications. Water samples were separated into suspended particulate matter (SPM), colloidal, and soluble phases using centrifugation and cross-flow ultrafiltration. Concentrations of estrone (E1), 17β-estradiol (E2), and estriol (E3) in each phase were analyzed using LC/MS/MS. Partition coefficients were calculated to assess estrogen distribution among phases. Estrogens were predominantly found in the soluble phase (85.8–87.3%). The risk assessment of estrogen equivalent (EEQ) values suggests that estrogen concentration in water poses a higher risk compared to SPM, with a majority of the samples indicating a high risk to aquatic organisms. The colloidal phase contained 12.7–14.2% of estrogens. The log K_{COC} values (4.72–4.77 L/kg-C) were significantly higher than the log K_{OC} and log K_{POC} values (2.02–3.40 L/kg-C) for all estrogens. Colloids play a critical role in estrogen distribution in river systems, potentially influencing their fate, transport, and biotoxicity. This finding highlights the importance of considering colloidal interactions in assessing estrogen behavior in aquatic environments.

Keywords: multiphase partitioning; estrogens; suspended particulate matter (SPM); colloid; soluble

1. Introduction

With the rapid development of industrialization and urbanization, the presence and distribution of endocrine-disrupting substances in aquatic environments have become a global environmental issue [1–4]. Studies have shown that natural estrogens, such as estrone (E1), 17β-estradiol (E2), and estriol (E3), even at extremely low concentrations (in the ng/L range) [1,5–7], can have serious effects on the reproductive systems of aquatic organisms, such as gender ratio imbalance and reduced reproductive capacity [5,6,8,9].

The main sources of natural estrogens are human and animal excreta [1,2,5,10,11], sewage discharges from livestock facilities, and agricultural fertilization, which are the primary pathways for estrogen entry into water bodies [12–17]. The distribution of estrogens in water bodies poses a potential threat to ecological safety. Animal excreta may account for 50% to 90% of natural estrogens in the environment [1,18–21].

The distribution of natural estrogens in aquatic environments is not limited to the dissolved phase; they also interact with suspended particulate matter (SPM), colloids, and sediments [6,22–26]. The distribution between these different phases affects the migration, transformation, and fate of estrogens, further impacting the quality of benthic organisms and sediments [22,27–30]. For example, colloidal particles have a strong adsorption capacity for estrogens due to their high specific surface area and abundant organic carbon content, affecting their bioavailability and toxicity [6,22,27,30,31]. Colloids can transport estrogens over long distances, thus affecting the quality of water bodies far from pollution sources [29,32]. Additionally, estrogens in the dissolved phase are more readily absorbed by aquatic organisms, increasing their toxicity risk [6,31].

Although studies have investigated the concentration and distribution of estrogens in water bodies, most have focused on the dissolved and particulate phases such as sediment and SPM, with relatively fewer studies on the colloid phase [6,28,30,33]. Furthermore, colloids of different origins and properties exhibit significant differences in their adsorption capacity for estrogens, making a comprehensive assessment of the multiphase distribution of estrogens in aquatic environments more complex [6,23,28,29,34].

Studies indicate that colloids are pivotal in the distribution of estrogens in aquatic environments. According to Huang et al. [31], the proportion of estrogens (E1, E2, and E3) in SPM fluctuates between 0 and 19.8%, 0 and 18.9%, and 0 and 6.9%, respectively. Conversely, in the colloid phase, estrogen percentages range from 14.6% to 35.5% for E1, 7.1% to 39.5% for E2, and 19.3% to 46.4% for E3. These findings underscore a notable presence of estrogens in colloids compared to SPM, illustrating the crucial role of colloids in the aquatic transport and behavior of estrogens.

Additionally, research reveals significant variations in the binding or partitioning coefficients of estrogens between distinct phases, namely SPM and the liquid phase (K_{OC}), SPM and the dissolved phase (K_{POC}), and colloid and the dissolved phase (K_{COC}). Huang et al. [31] documented average log K_{OC}, log K_{POC}, and log K_{COC} values for E1, E2, and E3, demonstrating the differential affinities estrogens exhibit towards diverse phases. Similarly, Nie et al. [29] provided comparative log K_{OC}, log K_{POC}, and log K_{COC} values for these estrogens, highlighting distinct partitioning behaviors in aquatic environments.

Crucially, K_{COC} values were found to be one to three orders of magnitude higher than K_{OC} and K_{POC}, suggesting that colloids possess a considerably higher affinity for estrogens compared to suspended particulate matter. These partition coefficients' variability is significantly influenced by the source of emissions and the physicochemical characteristics of both colloids and SPM in aquatic systems. These insights point to the necessity of incorporating colloids into assessments of estrogen distribution and environmental behavior in water bodies.

Despite existing research indicating the multiphase distribution characteristics of natural estrogens in aquatic environments [29,31,34], there are still challenges and shortcomings in current research. The multiphase distribution characteristics of estrogens under different environmental conditions are not fully understood [6,29]. A thorough investigation of the multiphase distribution of natural estrogens in aquatic environments will help reveal their environmental behavior and fate, understand the distribution and transformation patterns of estrogens between different phases, and provide a scientific basis for pollution control and remediation [6,28,34].

This study seeks to examine the distribution of natural estrogens in a river that receives significant wastewater discharges from animal husbandry. The specific objectives are as follows: 1. Quantify the concentrations of estrone (E1), 17β-estradiol (E2), and estriol (E3) in SPM, filtrate, colloidal, and soluble phases. 2. Determine the partition coefficients (K_{OC}, K_{POC}, K_{COC}) of these estrogens among different phases. 3. Assess the risk of estrogen in water and SPM. This comprehensive analysis of estrogen distribution will contribute to a better understanding of its behavior in aquatic environments.

2. Materials and Methods

2.1. Chemical Reagents

Estrone (E1, 99%), 17β-estradiol (E2, 99%), and estriol (E3, 97%) were procured from Cerilliant (Oakville, ON, Canada) Methanol of HPLC grade (99%) was sourced from Tedia Company Inc., (Fairfield, OH, USA), Ammonium hydroxide of ACS grade (28–30%) was obtained from J.T. Baker (Phillipsburg, NJ, USA). Additionally, acetonitrile of HPLC grade

(99.9%) was acquired from (Charlotte, NC, USA). Sulfuric acid of EP grade (95%) was purchased from Union Chemical Works, Taiwan.

2.2. Sampling Site

The water and SPM samples were taken from two specific sites along Wulo Creek, which is a tributary of the Gaoping River in Pingtung County, Taiwan. The Wulo Creek Basin experiences heavy rainfall during the wet season, with an average annual precipitation of around 2500–3000 mm. The temperature in the area ranges from 15 °C to 35 °C throughout the year, with the highest temperatures usually recorded in July and August. The high flow occurs during the wet season, from May to September, while the low flow occurs during the dry season, from October to April. As a result, this study's sampling period is from January to April each year to avoid rainy days.

This particular area is known for its large-scale animal husbandry practices, supporting a significant population of around 4500 cattle, 500,000 pigs, 3 million laying hens, and 5 million broilers.

The selection of these sampling sites was based on previous research by Chen et al. [35] and Hung et al. [36], which identified these locations (WL-1 and WL-4) as areas with high estrogen concentrations. These two sites are about 5 km apart. Figure 1 shows the geological map and sampling sites, and # indicates the number of animals.

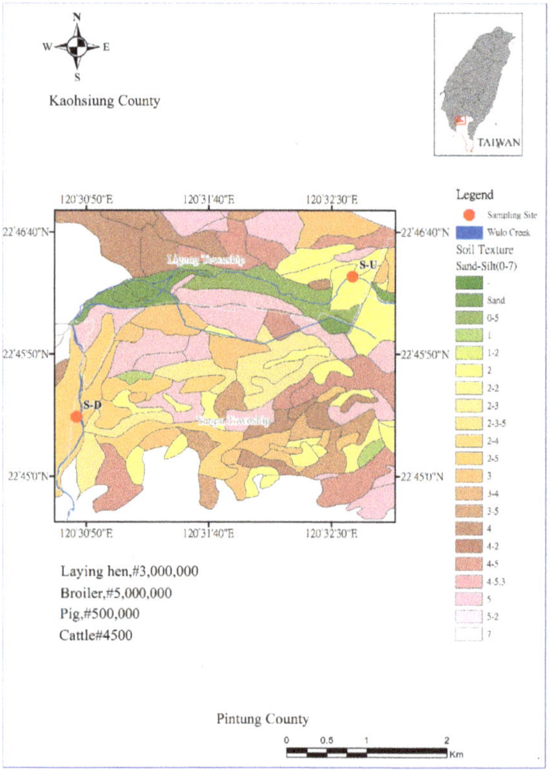

Figure 1. Geographical map of sampling site locations (adapted from Hung et al. [37], and Ministry of Agriculture, Republic of China (Taiwan)).

In our study, the upstream site (referred to as site-S-U) is located near the main discharge points of animal wastewater, while the downstream site (referred to as site-S-D) is further downstream. Both sampling sites can be accessed via bridges across the creek. The samples were collected from the central location of the creek to ensure a representative sampling of the water and SPM. This strategic approach was used to capture the overall environmental conditions and pollutant levels present in the water body. A total of 25 validated samples were collected for water and SPM. The detailed sampling description is provided in the Supplementary Information. The collected samples underwent basic water chemistry measurements, the results of which are listed in Table S1. The water quality, as per the River Pollution Index (RPI), was classified as "Moderately Polluted", reflecting the properties of animal wastewater.

2.3. Sampling Procedure and Treatment

The sampling of water and SPM involved using a peristaltic pump (Subaru, EY 20-3D-5.0 HP, Shibuya-ku, Tokyo, Japan) to extract river water onto the bridge and into a clean, pre-washed 20 L sampling bucket. Approximately 50–75 buckets, totaling around 1000–1500 L of water, were collected each time. To prevent estrogen degradation, 10 g of sodium azide was added to each full bucket of water. The suspended solid was separated from water samples using a continuous high-speed centrifuge (20,000 rpm) (Quantai Iron Works, LOS100LS, Taiwan). The water sample underwent solid–liquid separation by using a peristaltic pump (Cole-parmer MasterFlex, 7553-70, Gelsenkirchen, Germany) to feed the original water sample into the centrifuge tube of the continuous separation machine (Quantai Iron Works, LOS100LS, Taiwan), maintaining a flow rate of approximately 0.5 L/min. The high-speed centrifuge causes the water sample to adhere to the centrifuge tube wall, effectively collecting suspended solids; for each sample collection, the separation process took over 1 day each time. After solid–liquid centrifugal separation, the collected suspended solids were scraped off the centrifuge tube wall and placed into brown wide-mouth bottles, removing non-SPM samples such as small stones. These SPM samples were then placed in a freezer, and after 24 h, approximately 50 g of frozen SPM samples were further processed using a freeze dryer (FDU-1200, EYELA, Tokyo, Japan). The water-containing SPM sample was drained at −50 °C, under 5 Pa vacuum, and the freeze-drying time was about 3–5 days (depending on the solids' moisture content). The freeze-dried SPM samples were crushed and passed through a #20 mesh sieve (0.814 mm) to obtain the SPM sample.

After separating the solid and liquid components, the liquid samples were first filtered through a 0.7 μm coarse filter and then through a 0.22 μm fine filter using cellulose acetate membrane materials to remove suspended solids. Following the filtration, 12,000 mL of the filtered liquid was stored in a refrigerator at 4 °C. This was carried out in preparation for subsequent separation of the colloidal and dissolved phases and then solid-phase extraction (SPE).

2.4. The Separation Process of Water Samples

The liquid samples were filtered and then separated into colloidal and dissolved samples using a feed flow rate of 1.7–2.0 L/min. Throughout the separation process, the concentration volume factor was maintained at 10, and the volume ratio of dissolved sample ($V_{dissolved}$) to colloidal sample ($V_{colloid}$) was kept at 1:9. The mass recovery ratios (MB, %) for estrogens and DOC were calculated using Equation (1), while the mass percentages (MF_i %) of the colloidal and dissolved phases for estrogens and DOC were calculated using Equation (2) [38].

$$MB\ (\%) = \frac{\sum \left((C_{dissolved} \times V_{dissolved}) + (C_{colloid} \times V_{colloid}) \right)}{C_{filtrate} \times V_{filtrate}} \times 100 \quad (1)$$

$$MF_i(\%) = \frac{C_i \times V_i}{\sum \left((C_{dissolved} \times V_{dissolved}) + (C_{colloid} \times V_{colloid}) \right)} \times 100 \quad (2)$$

where C_i and V_i represent the concentration and volume of dissolved organic carbon (DOC) and estrogens in both colloidal and dissolved samples. $C_{filtrate}$ and $V_{filtrate}$ denote the concentration and volume of DOC and estrogens in the filtrate solution. $C_{colloid}$ and $V_{colloid}$ refer to the concentration and volume of DOC and estrogens in the colloidal solution. Similarly, $C_{dissolved}$ and $V_{dissolved}$ indicate the concentration and volume of DOC and estrogens in the dissolved solution.

During the separation process, the ultrafiltration membrane must be carefully cleaned. This process involved using reverse osmosis water, neutral detergent, 0.5 N sodium hydroxide solution, and deionized water (DDW) in a specific sequence.

2.5. Pretreatment of Liquid and SPM Samples for Estrogen Analysis

A volume of 1000 mL of the filtered liquid was subjected to solid-phase extraction (SPE). First, the filtered liquid samples were acidified using sulfuric acid to achieve a pH of 3.0. The target compound was then concentrated by passing through SPE cartridges using Oasis HLB (200 mg/6 mL, 30 μm particle size, Waters, Milford, Massachusetts, USA), which were pre-conditioned with 6 mL of methanol and 6 mL of deionized water. The acidified liquid sample was passed through the SPE column at a flow rate of 3–6 mL/min. Subsequently, 5 mL of 5% methanol was used to elute the cartridges, followed by drying under vacuum for 30 min. The column was then eluted with 3 mL of 5% methanol/double-distilled water (DDW) and further with 6 mL of methanol. Finally, the cartridge was eluted with 6 mL of 2% acetonitrile/methanol (ACN/MeOH) containing 0.5% NH_4OH.

The eluted sample was then evaporated using slowly blown nitrogen until it reached a near-dry state and subsequently redissolved with a 2 mL solution of ACN/DDW (v/v = 10/90). Finally, it underwent filtration through a 0.22 μm polytetrafluoroethylene syringe filter for LC/MS/MS analysis. It is important to note that this method was modified from previous studies to optimize the parameters [36,38–40].

To analyze the estrogen content of SPM, a modified method was used based on previous studies [41–43]. Initially, 2 g of freeze-dried SPM (#20 mesh screening) was mixed with a 10 mL solution of acetonitrile and distilled deionized water (ACN/DDI = 9:1). The mixture was then sonicated using a sonicator at 180–240 W and a frequency of 20 kHz for 5 s, followed by a 5 s pause, and then resumed for 12 min. The sample was subjected to sonic extraction twice. Afterward, the solution was centrifuged at 2000 rpm for 20 min to obtain the supernatant, resulting in a total extract of 20 mL. The extract was then concentrated to 2 mL by purging with nitrogen. This concentrated solution was then dissolved in 98 mL of distilled deionized water (DDW). Finally, the solution was processed and analyzed following the pretreatment steps outlined for estrogen-containing liquid samples.

2.6. LC/MS/MS Analysis Conditions

The analytes were separated using an Agilent 1200 HPLC system (Agilent Technologies, Palo Alto, CA, USA) with a Phenomenex Gemini C-18 Column (100 × 2.0 mm, dp = 3 μm) at a constant temperature of 40 °C. Injections of 25 μL were analyzed in a mobile phase consisting of ultrapure water with 0.05% ammonium hydroxide (A) and acetonitrile with 0.05% ammonium hydroxide (B). The chromatographic gradient started at a flow rate of 0.2 mL/min, maintaining 20% of component B for 1 min, increasing to 35% within 1.5 min, reaching 58% over another 1.5 min, and then held for 0.3 min. At 6.7 min, component B was increased to 85%, and the total analysis duration was 11 min. Retention time (Rt), parent ion production, fragmentation voltage, and collision voltage adhered to the parameters optimized for standard products as per Hung et al. [36], using an Agilent 6410 LC/MS/MS system. Negative-ion electrospray ionization (ESI) was employed. The procedure utilized dynamic multiple reaction monitoring (MRM) with a drying gas temperature of 325 °C, drying gas flow of 8 L/min, nebulizer pressure of 30 psi, and capillary voltage of 4000 V.

2.7. Quality Assurance and Control

The calibration curves for estrogen-containing samples spiked with varying concentrations of three target estrogens (0.5 to 200 μg/L) demonstrated strong linearity (R^2 >

0.995). To ensure the absence of contamination and proper instrument performance, procedural blanks and sample replicates were analyzed for each batch. Recovery rates were determined using a standard addition method with 10 ng/L spiking concentration and ranged from 79% to 91%. The method's detection limits were 0.3, 0.5, and 0.4 ng/L for E1, E2, and E3, respectively, in water (Table S2). The internal quality assay was maintained below 5%, while the intermediate assay was below 8% at the quantitation limit, meeting the acceptable criteria for environmental analysis. Detailed quality assurance data for each compound in DI water and the liquid samples can be found in Hung et al. [36]

2.8. Partition Coefficient Calculation

In this study, we evaluated three types of partition coefficients: K_{OC}, K_{POC}, and K_{COC}. The calculation of K_{OC}, K_{POC}, and K_{COC} is performed according to Equations (3)–(5), respectively [26,30,31].

$$K_{OC} = C_{SPM}/C_{filtrate}/f_{OC} \tag{3}$$

$$K_{POC} = C_{SPM}/C_{dissolved}/f_{OC} \tag{4}$$

$$K_{COC} = C_{colloid}/C_{dissolved}/DOC_{colloid} \tag{5}$$

where C_{SPM}, $C_{filtrate}$, $C_{colloid}$, and $C_{dissolved}$ indicate estrogen concentrations of SPM and the estrogen concentrations of filtrate, colloid, and dissolved solutions. f_{oc} and $DOC_{colloid}$ indicate the TOC fraction of SPM and the DOC concentration of colloid.

2.9. 17β-Estradiol Equivalent (EEQ) Calculation

To assess the potential risks of estrogenic activity to aquatic organisms, estradiol equivalents (EEQs) in the filtrate and SPM collected from Wulo Creek were calculated.

The EEQ in filtrate water of the analyzed estrogens was calculated following Equation (6):

$$EEQ = \sum EEQ_i = \sum (C_i \times EEF_i) \tag{6}$$

The risk assessment for the SPM was carried out by converting the estrogenic activities of estrogens into their corresponding EEQs in water, which were expressed as the bioavailable fractions of the estrogens in the SPM. The EEQ_{water} in the SPM was calculated following Equation (7) [44,45]:

$$EEQ_{water}\ (\text{ng/L}) = \sum EEQ_i = \sum (1000 \times C_{s,i}\ (\text{ng/g}) \times EEF_i)/K_{oc,i}\ (\text{L/kg}) \tag{7}$$

where EEQ_i is the EEQ value of the selected estrogen i, and C_i and $C_{s,i}$ are the concentration of the selected estrogen i in filtrate and SPM. EEF_i is the estrogenic equivalent factor relative to E2 and the EEF values of E1, E2, and E3 were 0.25, 1, and 5.9×10^{-3}, respectively [46]. $K_{oc,i}$ is the organic carbon standardized partitioning coefficient of chemical i between SPM and filtrate.

2.10. Statistical Analysis

In this study, various tests were conducted using S-Plus software (version 6.2) at a significance level of $p < 0.05$. The t-test method was used to compare estrogen differences between two groups, and the ANOVA test method was used for comparing differences between three groups, followed by Tukey's post hoc test.

3. Results

3.1. Dissolved Organic Carbon Concentrations in Liquid Samples

In this study, 25 water samples were collected from sampling sites. Thirteen and twelve samples were taken from site-U and site-D, respectively. Table 1 lists the concentrations of dissolved organic carbon (DOC) and estrogens in the liquid samples. The samples include the filtrate (<0.45 μm), colloidal (1 kDa-0.45 μm), and soluble (<1 kDa) phases. The average DOC concentration in the total filtrate was 17.1 ± 9.7 mg/L at site-U, which was higher than the concentration of 8.2 ± 6.8 mg/L at site-D (p = 0.014). Site-U is located near animal feedlot wastewater discharge sources. This indicates that the river carries wastewater from agricultural areas with high concentrations of DOM, resulting in a high DOC concentration and poor water quality. Similar findings have been reported in other studies [35]. The lower DOC concentration at site-D was due to the aggregation and deposition of large particulate dissolved organic matter (DOM) as the river water flowed downstream.

Table 1. Concentrations of estrogens and DOC of filtrate (<0.45 μm), colloidal (1 kDa-0.45 μm), and soluble (<1 kDa) phases.

Site	Phase	E1, ng/L	E2, ng/L	E3, ng/L	DOC, mg/L
	Filtrate	22.8 ± 11.9	4.1 ± 2.6	5.0 ± 1.6	27.8 ± 9.4
S-U_{13}	Colloid	25.6 ± 14.2	5.8 ± 4.8	6.8 ± 3.6	49.2 ± 15.2
	Soluble	16.4 ± 6.7	5.3 ± 4.4	3.8 ± 1.9	20.5 ± 9.7
	Filtrate	46.8 ± 32.5	4.2 ± 3.8	4.0 ± 3.8	11.6 ± 8.6
S-D_{13}	Colloid	53.0 ± 47.6	7.9 ± 4.4	6.0 ± 4.6	31.4 ± 13.6
	Soluble	36.7 ± 26.0	5.9 ± 2.7	4.4 ± 3.0	9.0 ± 9.1
	Filtrate	164.7 ± 108.6	6.09 ± 4.50	3.53 ± 1.01	12.4 ± 4.9
S-U_{14}	Colloid	239 ± 103	7.8 ± 2.5	5.5 ± 4.3	28.9 ± 19.9
	Soluble	201 ± 93	7.0 ± 5.6	4.0 ± 2.3	12.8 ± 5.0
	Filtrate	328.3 ± 171.0	6.48 ± 2.77	4.63 ± 1.05	4.84 ± 0.6
S-D_{14}	Colloid	377.1 ± 219.1	7.9 ± 3.6	6.5 ± 1.6	18.5 ± 3.9
	Soluble	275.3 ± 120.5	4.4 ± 1.4	5.0 ± 0.4	3.0 ± 0.4

S-U_{13}, S-U_{14}: site-U sampling in 2013 and 2014, respectively; S-D_{13}, S-D_{14}: site-D sampling in 2013 and 2014, respectively.

The mass recovery ratios of filtrate separation into colloidal and soluble phase for organic carbon (OC) and estrogens were calculated using Equation (1). The OC recovery ratios ranged from 73% to 157%, with an average of 97% across all samples. Most of the OC recovery ratios fell between 80% and 120% for 20 out of 25 samples. These average OC recovery ratios were considered acceptable. For instance, in a study by Nie et al. [29] on water from the Huangpu River in China, the OC recovery ratios ranged from 88% to 128%, with an average of 108%. Similarly, Chuang et al. [38] found an average OC recovery ratio of 105% when separating river water into five types of size-fractionated DOM.

Figure 2 displays the OC mass percentages of colloidal and soluble phases using Equation (2) for two years and two sites. Despite the colloid having a higher DOC concentration compared to the soluble phase (Table 1), the volume fractions were 0.1 and 0.9 for colloidal and soluble phases, respectively. Consequently, the colloidal mass percentages were lower than those of the soluble phase. On average, the colloidal OC percentage was 31.5% with a standard deviation of 14.8%. At site-D, the colloidal percentage was significantly higher at 39.4% compared to site-U, which had a percentage of 22.0% (p = 0.003). The colloidal OC mass percentages in our study were slightly lower than the 32–55% range reported in freshwater DOM studies [38,47,48]. However, they are comparable to the findings of Chen et al. [30], whose study showed an average colloidal OC mass percentage of 30.0% with a standard deviation of 11.5%.

In our study, the low colloidal OC percentages indicate a low level of humification in the dissolved organic matter, suggesting a lack of synthesis of humic and fulvic acid- like organic matter due to the low humification extent of the water receiving animal wastewater discharge DOM. Additionally, the soluble phase of dissolved organic matter may aggregate to form a larger colloidal phase from site-U to site-D, increasing the percentage of the colloidal phase at site-D.

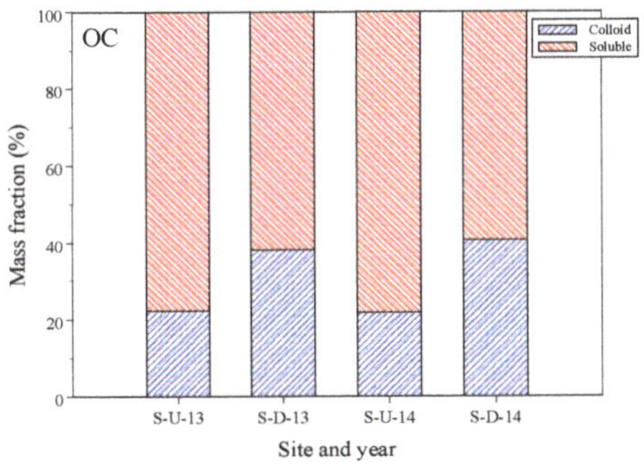

Figure 2. The mass percentages of organic carbon (OC) distributed on colloidal and soluble phases.

3.2. Estrogen Concentrations in Water Samples

The concentrations of three natural estrogens, E1, E2, and E3, were detected in all liquid phases. Table 1 lists the average estrogen concentrations of the filtrate samples. The order of total average concentrations of the filtrate was 153.0 ± 153.7 ng/L (E1) > 5.4 ± 3.7 ng/L (E2) > 4.2 ± 2.1 ng/L (E3), and the concentrations showed considerable variation. The sampling sites affected by the wastewater of animal feedlot discharge resulted in high estrogen concentrations [5,13,29,35,49,50]. The order and concentrations of estrogen were similar to animal-polluted river water [35,49], but the concentrations were higher than river water receiving effluent of STP wastewater [1,29,31]. The filtrate's total average E1 concentration of 121.0 ± 112.0 ng/L at site-U was lower than the concentration of 187.6 ± 112.0 ng/L at site-D without significant difference ($p = 0.302$). Additionally, the E2 and E3 concentrations were insignificantly different at the two sites. The average E2 concentrations were 5.5 ± 4.0 ng/L and 5.3 ± 3.4 ng/L at site-U and site-D, respectively ($p = 0.93$). The average E3 concentrations were 4.0 ± 1.4 ng/L and 4.3 ± 2.7 ng/L at site-U and site-D, respectively ($p = 0.68$). According to Liu et al. [49] and Chen et al. [35], high estrogen concentrations were found upstream of Wulo Creek, with an E1 concentration exceeding 1000 ng/L, suggesting that animal wastewater could carry high estrogen concentrations to receiving water.

Table 1 also presents the estrogen concentrations in colloidal and soluble phases. The estrogen concentrations in these phases were measured directly and did not account for volume factors, leading to higher concentrations in the colloidal phase compared to the filtrate and soluble phases. After separation, the estrogen concentration of individual species in the colloidal and soluble phases was comparable to that in the filtrate.

The average estrogen mass recovery ratios were 88.8%, 92.0%, and 103.8% for E1, E2, and E3, respectively, in the filtrate separated into colloidal and soluble phases, which were within an acceptable range [27,29,30]. For example, in the partition study of estrogens between colloidal and soluble phases, the average mass recovery ratios were 131%, 99%, and 144% for E1, E2, and E3, respectively, during the separation processes [29]. Yan et al. [27] reported mass recovery ratios ranging from 64.2% to 117.8% for 27 emerging organic compounds in a separation test.

Figure 3a–c illustrate the estrogen mass percentages of E1, E2, and E3, respectively, at site-U and site-D during the years 2013 and 2014. The overall average mass fractions of the colloidal phase were 12.7 ± 2.6%, 14.2 ± 5.5%, and 13.5 ± 5.5% for E1, E2, and E3, respectively. Additionally, the colloidal mass percentages of the estrogens did not show significant differences ($p = 0.46$); the mass percentage for each estrogen was not significantly different between 2013 and 2014 (p = 0.07–0.35) and between the two sampling sites (p = 0.32–0.97).

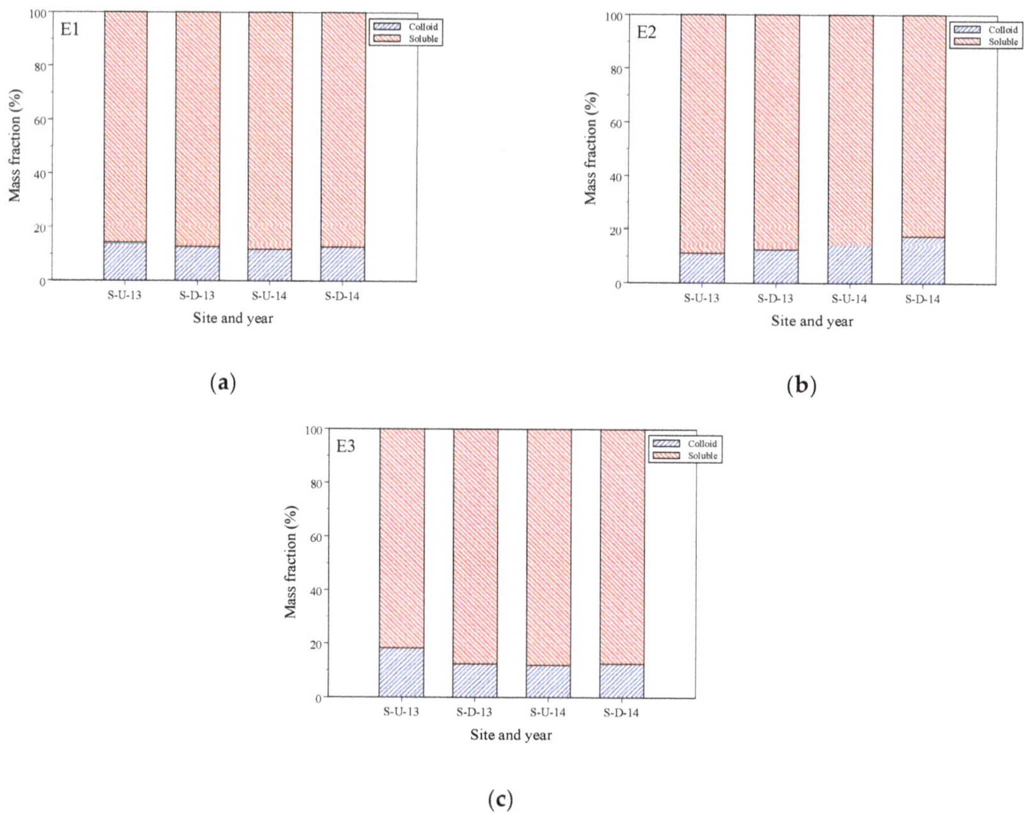

Figure 3. The mass percentages of estrogens distributed on colloidal and soluble phases: (**a**) (E1), (**b**) (E2), and (**c**) (E3).

Previous studies on the multiphase partitioning of estrogens in in situ river water have demonstrated a wide range of percentages for colloid-bound estrogens. Nie et al. [29] reported ranges of 2.0–58.4%, 8.4–72.0%, and 3.4–62.7% for E1, E2, and E3, respectively. These percentages significantly differed from the present study due to variations in colloid sources. Huang et al. [31] found average colloid-bound phases of 26.3%, 15.1%, and 19.5% for E1, E2, and E3, respectively, in the Shaying River, China; these percentages were slightly higher than those observed in the present study. In a constructed wetland, Chen et al. [30] reported colloid-bound estrogen percentages ranging from 7.3% to 8.5% for E1, E2, and E3, which were lower than the percentages found in the present study. In four advanced wastewater treatment processes, Huang et al. [28] reported the mass percentages of three natural estrogens in different phases (suspended particulate matter, colloidal, and soluble phases); the colloid-bound fractions were 3.9–19.4%, 15.1–31.7%, and 54.0–77.8% for E1, E2, and E3, respectively. The percentages for E1 and E2 were similar to those in the present study, but the percentage for E3 was higher. The distribution of estrogen in the colloidal phase is influenced by sources, biochemical processes, and other factors [23,27,29–31,51,52].

3.3. Estrogen Concentrations in SPM

In this study, a total of 23 suspended particulate matter (SPM) samples were analyzed for estrogen concentrations. Eleven samples were taken from site-U and twelve from site-D. Table 2 presents the average SPM concentrations and estrogen concentrations extracted from SPM across the two sampling years and both sites. The overall average SPM concentration at site-U was 126.8 ± 33.8 mg/L, significantly higher than the 37.4 ± 19.2 mg/L observed at site-D ($p < 0.001$). As river water flows downstream from site-U to site-D, larger suspended particulates aggregate and settle into the sediment, thereby reducing the SPM concentration downstream.

Table 2. The SPM concentrations of water samples and TOC and estrogen concentrations in SPM samples.

Site	SPM (mg/L)	TOC (%)	E1 (μg/kg)	E2 (μg/kg)	E3 (μg/kg)
S-U_{13}	98 ± 14	33.5 ± 0.1	1.05 ± 1.21	0.30 ± 0.35	0.70 ± 0.81
S-D_{13}	35 ± 18	22.3 ± 11.3	2.27 ± 1.34	1.27 ± 1.04	2.13 ± 1.80
S-U_{14}	140 ± 32	16.4 ± 1.2	0.72 ± 1.11	ND	0.64 ± 1.28
S-D_{14}	40 ± 22	23.1 ± 5.9	1.65 ± 0.44	ND	1.40 ± 0.86

ND: not detected.

Among the 23 SPM samples analyzed for estrogen concentrations, E2 was detected in 2013 but not in 2014. Some SPM samples had estrogen concentrations below the detection limit, with the number of detectable samples being 17 for E1, 6 for E2, and 13 for E3. Estrogen concentrations below the detection limit were considered zero in statistical calculations. The total average estrogen concentrations in SPM were 1.37 ± 1.19, 0.35 ± 0.73, and 1.19 ± 1.36 μg/kg for E1, E2, and E3, respectively, with E1 concentrations being significantly higher than E2 ($p = 0.004$). The E1 concentrations were 0.82 ± 1.10 μg/kg at site-U and 1.96 ± 1.01 μg/kg at site-D, while the E3 concentrations were 0.66 ± 1.12 μg/kg at site-U and 1.77 ± 1.40 μg/kg at site-D. Both E1 and E3 concentrations were higher at site-D than at site-U ($p = 0.013$ for E1; $p = 0.042$ for E3). This higher estrogen concentration at site-D could be attributed to the deposition of large SPM containing low estrogen concentrations.

The order of estrogen concentrations in SPM differed from that in filtrated water samples. In the water samples, E1 concentrations were significantly higher than those of E2 and E3, whereas in SPM samples, E1 and E3 concentrations were higher than E2 concentrations. E2, originally excreted by animals, is metabolized to E3 and eventually to E1 [53]. This metabolic pathway typically results in higher E1 concentrations in river water [35,49]. However, in the present study, similar concentrations of E1 and E3 in SPM suggest that E3 had not yet fully metabolized to E1. Therefore, the high E3 concentration in SPM indicates that E3 was initially discharged from animal waste and wastewater.

Huang et al. [31] reported mean estrogen concentrations in SPM of 4.56, 7.95, and 16.5 μg/kg for E1, E2, and E3, respectively, which were higher and followed a different order than those in the present study. In the Huangpu River, China, average concentrations were 81.2, 44.9, and 72.9 μg/kg for E1, E2, and E3 in SPM, respectively, in river water affected by animal feeding operation wastewater [29]. In their study, average estrogen concentrations were less than 20 μg/kg for E1, E2, and E3 in mainstream and tributary mouth waters. Studies by Huang et al. [31] and Nie et al. [29] demonstrated that SPM had high estrogen concentrations primarily due to source factors. In Nie et al. [29], E1 and E3 concentrations were higher than E2, similarly to the present study, whereas in Huang et al. [31], E3 concentrations were higher than both E1 and E2. Sources, SPM characteristics, hydrolysis, and oxidation mechanisms may affect estrogen concentrations in SPM [29,31,53–56].

Previous studies typically analyzed estrogen concentrations in SPM residues on filter membranes [29,31,34,55], involving small amounts of SPM mass. In the present study, estrogen concentrations were analyzed using a consistent SPM mass of two grams dry weight per analysis. This study collected SPM samples from large water volumes (1000–1500 L) using a high-speed (20,000 rpm) continuous-flow centrifuge. The SPM collection period lasted more than one day, during which estrogen metabolism and decomposition might have occurred, resulting in low estrogen concentrations in SPM. However, analyzing large SPM masses enhances the precision of measurements for very low concentrations of organic compounds such as estrogens in SPM. For instance, Gong et al. [34] used a continuous-flow centrifuge to collect SPM from 250 to 400 L of water. They detected E1 in two of eight SPM samples, with concentrations of 4.7 and 6.0 μg/kg, whereas E2 and E3 were below the detection limit in all samples.

Estrogen concentrations in SPM were converted to the liquid phase. Since colloidal- and soluble-phase estrogen mass fractions were discussed in Section 3.2, this section investigates mass fractions in SPM and liquid phases. In SPM, the average estrogen fractions were 0.16%, 0.21%, and 1.36% for E1, E2, and E3, respectively, suggesting that most estrogens were present in the liquid phase. These fractions were lower than those reported by Huang et al. [31] for the Shaying River, China, where estrogen mass fractions in SPM were 0–19.8%, 0–18.9%, and 0–6.88% for E1, E2, and E3, respectively. In the present study, the difference between total solid and suspended solid was 487.7 mg/L, implying a large amount of dissolved organic matter in the liquid phase, likely binding a high fraction of estrogens. Additionally, SPM discharged from animal wastewater is characterized by low humification and a low adsorption capacity for estrogens.

3.4. Multiphase Partition Coefficient of Estrogens

The multiphase partition coefficient is essential for understanding the fate of estrogens in different phases within river systems [6,27–29,31,34,55]. This study investigated the organic carbon-normalized estrogen partition coefficients for SPM-filtrate (K_{OC}), SPM-soluble (K_{POC}), and colloid-soluble (K_{COC}) phases, calculated according to Equations (3)–(5) [30,31,34]. Table 3 provides the partition coefficients for the three estrogens. The log K_{OC} values ranged from 1.08 to 2.96, 2.43 to 3.05, and 2.80 to 4.07 L/kg-C for E1, E2, and E3, respectively. The log K_{POC} values ranged from 1.23 to 2.97, 2.26 to 3.09, and 2.85 to 4.19 for E1, E2, and E3, respectively. The log K_{COC} values ranged from 4.23 to 5.26, 4.17 to 5.59, and 4.08 to 5.45 for E1, E2, and E3, respectively. Table 4 outlines the ranges and average estrogen partition coefficients of log K_{OC}, log K_{POC}, and log K_{COC} in this study and as reported in previous studies. Given the limited research on colloid-soluble partition coefficients, log K_{COC} values from in situ river water and laboratory batch experiments were adopted for comparison.

Table 3. The organic carbon-normalized partition coefficients of log K_{OC} (SPM-filtrate), log K_{POC} (SPM-soluble), and log K_{COC} (colloid-soluble) for estrogens.

Site	Coefficients	E1	E2	E3
S-U$_{13}$	log K_{OC}	2.41 ± 0.02	2.58 ± 0.03	2.95 ± 0.03
	log K_{POC}	2.58 ± 0.02	2.42 ± 0.05	3.09 ± 0.11
	log K_{COC}	4.48 ± 0.19	4.36 ± 0.19	4.57 ± 0.14
S-D$_{13}$	log K_{OC}	2.54 ± 0.57	3.46 ± 0.05	4.00 ± 0.07
	log K_{POC}	2.64 ± 0.58	3.52 ± 0.01	4.02 ± 0.11
	log K_{COC}	4.70 ± 0.13	4.68 ± 0.08	4.70 ± 0.14
S-U$_{14}$	log K_{OC}	2.30 ± 0.12	NA	3.60 ± 0.07
	log K_{POC}	NA	NA	3.53 ± 0.11
	log K_{COC}	4.75 ± 0.41	4.81 ± 0.54	4.94 ± 0.47
S-D$_{14}$	log K_{OC}	1.39 ± 0.31	NA	3.25 ± 0.21
	log K_{POC}	1.46 ± 0.29	NA	3.18 ± 0.23
	log K_{COC}	4.85 ± 0.19	4.99 ± 0.29	4.85 ± 0.17

NA: not available.

Table 4. The average and ranges of estrogen partition coefficients of log K_{OC}, log K_{POC}, and log K_{COC} in this study and reported in previous studies.

	E1	E2	E3	Matrix and Method	Ref.
log K_{OW}	3.43	3.94	2.81		[50]
log K_{OC}	1.08–2.96 (2.02, 17) *	2.43–3.05 (2.74, 4) *	2.80–4.07 (3.38, 11) *	In situ river water (CFUF)	PS
log K_{OC}	3.21	3.38	2.84	In situ river water (CFUF)	[31]
log K_{OC}	1.67–3.46	2.27–2.46	3.53–4.07	In situ river water (CFUF)	[29]
log K_{POC}	1.23–2.97 (2.06, 14) *	2.26–3.09 (2.69, 4) *	2.85–4.19 (3.40, 13) *	In situ river water (CFUF)	PS
log K_{POC}	1.75–3.60	2.48–2.60	3.79–4.49	In situ river water (CFUF)	[29]
log K_{POC}	3.35	3.45	2.94	In situ river water (CFUF)	[31]
log K_{COC}	4.23–5.26 (4.72, 22) *	4.17–5.59 (4.75, 20) *	4.08–5.45 (4.77, 20) *	In situ river water (CFUF)	PS
log K_{COC}	4.18–4.85	3.96–4.2		In situ river water (CFUF)	[24]
log K_{COC}	4.18–4.23	3.96–4.20		River water (CFUF)	[23]
log K_{COC}	6.60–6.81	6.42–6.78	7.09–7.85	In situ river water (CFUF)	[22]
log K_{COC}	5.03	4.76	4.25	In situ river water (CFUF)	[31]
log K_{COC}	5.4		6.11	In situ river water (CFUF)	[22]
log K_{COC}	7.09	7.58	7.8	In situ river water (CFUF)	[27]
log K_{COC}	4.08	4.04	4.11	In situ river water (CFUF)	[30]
log K_{COC}		4.57–4.94		Commercial DOM (FQ)	[48]
log K_{COC}		4.08–4.68		Biological wastewater (FQ)	[54]
log K_{COC}		<3–5.25		Biological wastewater (FQ)	[49]
log K_{COC}	3.98	3.93–4.12		Commercial DOM (FQ)	[55]
log K_{COC}	4.52–6.02	3.42–5.11		Commercial DOM (SPME)	[33]

PS: present study, FQ: fluorescence quenching; CFUF: cross-flow ultrafiltration; SPME: solid-phase microextraction. * (mean value, sample numbers).

The order of total average coefficients for log K_{OC} was 3.38 ± 0.43 (E3) > 2.74 ± 0.33 (E2) > 2.02 ± 0.58 (E1), and for log K_{POC} it was 3.40 ± 0.41 (E3) > 2.69 ± 0.45 (E2) > 2.06 ± 0.64 (E1). The partition coefficients for log K_{OC} and log K_{POC} for E3 were significantly higher than those for E1 ($p < 0.001$). However, the log K_{OC} and log K_{POC} values for individual estrogens were not significantly different (p = 0.87–0.91). These values were comparable to the ranges reported in previous studies [29,31]. The octanol–water partition coefficient (log K_{OW}) typically indicates the capacity of estrogens to bind to organic matter [6,53]. In this study, the log K_{OC} and log K_{POC} values were independent of the log K_{OW} values, which are 3.94, 3.13, and 2.85 for E2, E1, and E3, respectively (Table 4). This suggests that the hydrophobicity of estrogens is not the dominant factor controlling their adsorption onto SPM [31,34].

Conventionally, the partition coefficient is the ratio of estrogen concentrations between SPM and liquid phases [31]. In the present study, E1 had significantly higher concentrations than E2 and E3 in the liquid phase, whereas E1 and E3 had higher concentrations than E2 in SPM. This concentration difference resulted in E3 having the highest log K_{OC} and log K_{POC} values, and E1 having the lowest. Therefore, the partition coefficients of organic compounds are influenced by their sources and metabolic conditions in both SPM and liquid phases [29,30].

In this study, the total average log K_{COC} coefficients were 4.72 ± 0.27, 4.75 ± 0.40, and 4.75 ± 0.30 kg/L-C for E1, E2, and E3, respectively. The log K_{COC} values were comparable to or higher than those studied in the in situ river water and laboratory batch experiments listed in Table 4 [23,24,30,31,33,48,49,57,58]. However, some studies reported higher log K_{COC} values in in situ river water [22,27,29]. The high log K_{COC} values may be due to the characteristics of colloids, which have a strong estrogen sorption ability. Log K_{COC} values obtained in laboratory batch experiments were generally lower than those in in situ river water because laboratory conditions typically involve shorter durations and higher sorbate concentrations [34,59]. Different experimental conditions lead to variations in the partition coefficients of estrogens bound to colloids. Moreover, the log K_{COC} values were not significantly different among the three estrogens (p = 0.88) and were one to two orders of magnitude higher than the log K_{OC}, log K_{POC}, and log K_{OW} values in this study. This is consistent with findings that the estrogen partition coefficients of log K_{COC} in in situ river water are much higher than log K_{OC} and log K_{POC} [29,31]. The high log K_{COC} values suggest that colloids play a critical role in influencing estrogen distribution, which could further affect estrogen transport, toxicity, and fate in rivers [22,27–29,31]. Colloids have a strong ability to absorb estrogen from the liquid phase due to their high specific surface area, sorption site density, organic carbon content, and strong absorption capacity [29,31].

Highly hydrophobic compounds such as PAHs, PCBs, and pesticides have shown a high correlation between log K_{COC} and log K_{OW}, suggesting that hydrophobic interactions are the major contributors to the sorption of these compounds to organic matter [6]. However, estrogens, being moderately hydrophobic compounds, might have different sorption mechanisms, forming estrogen–DOM complexes [6,24,51,52,57]. In this study, log K_{COC} values were not significantly correlated with the hydrophobicity of estrogens (log K_{OW}) (r = 0.09). Weak correlations between log K_{COC} and log K_{OW} for estrogens have been observed in previous studies, suggesting that partition mechanisms other than nonspecific hydrophobic interactions play a key role in estrogen sorption [24,51,52,57]. Previous studies have found that estrogen log K_{COC} has a strong correlation with the aromaticity and phenolic functional groups of DOM [51,52,57], suggesting that π-π electron interactions, hydrogen bonding, phenolic groups, and ligand exchange play important roles in the binding of estrogens to colloids [6,51,52,59].

3.5. Risk Assessment of Estrogens in Filtrate Water and SPM

The $\sum$EEQ values in the filtrate of water and SPMs are summarized in Table 5. In total, the $\sum$EEQ values in the filtrate ranged from 3.6 to 142.1 ng/L, with an average value of 43.7 ng/L. In the SPMs, the $\sum$EEQ values ranged from 0.0 to 40.6 ng/L, with an average value of 6.8 ng/L. The average $\sum$EEQ values of filtrate were 13.5 and 63.8 ng/L for the years 2013 and 2014 and were 28.5 and 52.3 ng/L for site-U and site-D, respectively. The average $\sum$EEQ values of SPM were 3.7 and 8.9 ng/L for the years 2013 and 2014, respectively, and were 2.4 and 12.2 ng/L for site-U and site-D, respectively. The $\sum$EEQ value in site-D was significantly higher than site-U and was significantly different for the years 2013 and 2014, indicating variations in temporal and spatial estrogen risk in river water receiving animal feedlot wastewater. In total (filtrate and SPM), the $\sum$EEQ values ranged from 4.0 to 182.7 ng/L with an average of 50.5 ng/L. The contribution of $\sum$EEQ values was 87.4%, and 12.6% was attributed to filtrate water and SPM, respectively.

Table 5. EEQ values (ng/L) of the filtrate and SPM in Wulo Creek.

Species	EEQ of Filtrate		EEQ of SPM	
	Mean ± SD	Ranges	Mean ± SD	Ranges
E1	38.2 ± 38.4	3.1–132.3	9.1 ± 12.4	0.4–40.6
E2	5.4 ± 3.6	0.0–16.7	3.6 ± 1.5	2.2–5.1
E3	0.0 ± 0.0	0.0–0.1	0.0 ± 0.0	0.0–0.0
Sum EEQ	43.7 ± 40.6	3.6–142.1	6.8 ± 11.0	0.0–40.6

The estrogenic risk level of $\sum$EEQ values to organisms is categorized as follows: $\sum$EEQ < 1 ng/L is a low estrogenic risk, while 1 < $\sum$EEQ < 10 attributes moderate risk and $\sum$EEQ > 10 attributes high estrogenic risk. In the present study, the $\sum$EEQ values of filtrate water were 16% and 84%, respectively, which represent median and high risk for aquatic organisms. The $\sum$EEQ values of the SPM samples were 40%, 44%, and 16%, which are attributed to low, medium, and high estrogenic risk to aquatic organisms. Furthermore, in water, the contributions of $\sum$EEQ values were 81.7%, 18.2%, and 0.1% by E1, E2, and E3, respectively. In SPM, the contributions of $\sum$EEQ values were 78.4%, 11.0%, and 10.6%

by E1, E2, and E3, respectively. E1 was the major contributor to the ∑EEQs in the surface water and SPM.

4. Conclusions

This study provides important insights into how estrogens behave in river systems affected by animal wastewater. Our findings show that estrogen concentrations vary significantly. E1 is the most dominant in the liquid portion, and higher concentrations are found downstream than upstream. Most estrogens (98.64–99.84%) were discovered in the liquid phase, mainly in the soluble fraction (85.8–87.3%). Our results confirm the original hypothesis that colloids play a crucial role in the distribution of estrogens. The considerably higher log K_{COC} values than log K_{OC} and log K_{POC} emphasize the importance of colloidal interactions in influencing estrogen behavior in water environments. We found weak correlations between partition coefficients and log K_{OW}, indicating that factors beyond hydrophobicity impact estrogen sorption onto suspended particulate matter and colloids. The risk assessment of E2 estrogen equivalent (EEQ) values suggests that estrogen concentration in water poses a higher risk compared to SPM, with a majority of the samples indicating a high risk to aquatic organisms. The significant role of colloids in estrogen transport suggests that traditional water quality models may underestimate the mobility of these compounds. This study emphasizes the complex interactions between estrogens and different water phases, highlighting the often overlooked role of colloids. It is important to note that our study has limitations, including potential seasonal variations not covered in our sampling period. Further research should investigate temporal dynamics and encompass various aquatic environments to enhance the generalizability of our findings.

Supplementary Materials: The following supporting information can be downloaded at https://www.mdpi.com/article/10.3390/toxics12090671/s1: Table S1: Basic water chemistry of Wulo Creek. Table S2: Quality assurance data for the analysis of each target compound in DI water and river water.

Author Contributions: Conceptualization, C.-Y.H. and T.-C.C.; methodology, K.-H.Y., H.-S.H., and W.-H.H.; formal analysis, T.-C.C., H.-S.H., and W.-H.H.; investigation, H.-S.H., K.-H.Y., and W.-H.H.; writing—original draft preparation, K.-H.Y., H.-S.H., and T.-C.C.; writing—review and editing, C.-Y.H. and T.-C.C.; visualization, K.-H.Y., H.-S.H., W.-H.H., and T.-C.C. All authors have read and agreed to the published version of the manuscript.

Funding: This research was funded by the National Science and Technology Council, R.O.C. grant number MOST 103-2221-E-020-008.

Institutional Review Board Statement: Not applicable.

Informed Consent Statement: Not applicable.

Data Availability Statement: Data are available through request to the corresponding author.

Acknowledgments: The authors would like to thank the Center for Agricultural and Aquacultural Product Inspection and Certification of National Pingtung University of Science and Technology, Taiwan, for assisting in the development of analytical methods. The LC-MS/MS used for estrogen analyses was financially supported by the National Pingtung University of Science and Technology.

Conflicts of Interest: The authors declare no conflicts of interest.

References

1. Adeel, M.; Song, X.; Wang, Y.; Francis, D.; Yang, Y. Environmental impact of estrogens on human, animal and plant life: A critical review. *Environ. Int.* **2017**, *99*, 107–119. [CrossRef] [PubMed]
2. Ciślak, M.; Kruszelnicka, I.; Zembrzuska, J.; Ginter-Kramarczyk, D. Estrogen pollution of the European aquatic environment: A critical review. *Water Res.* **2023**, *229*, 119413. [CrossRef] [PubMed]
3. Du, B.; Fan, G.; Yu, W.; Yang, S.; Zhou, J.; Luo, J. Occurrence and risk assessment of steroid estrogens in environmental water samples: A five-year worldwide perspective. *Environ. Pollut.* **2020**, *267*, 115405. [CrossRef] [PubMed]
4. Ojoghoro, J.; Scrimshaw, M.; Sumpter, J. Steroid hormones in the aquatic environment. *Sci. Total Environ.* **2021**, *792*, 148306. [CrossRef] [PubMed]

5. Grzegorzek, M.; Wartalska, K.; Kowalik, R. Occurrence and sources of hormones in water resources—Environmental and health impact. *Environ. Sci. Pollut. Res.* **2024**, *31*, 37907–37922. [CrossRef] [PubMed]
6. Ma, L.; Yates, S.R. Dissolved organic matter and estrogen interactions regulate estrogen removal in the aqueous environment: A review. *Sci. Total Environ.* **2018**, *640*, 529–542. [CrossRef]
7. Zhong, R.; Zou, H.; Gao, J.; Wang, T.; Bu, Q.; Wang, Z.-L.; Hu, M.; Wang, Z. A critical review on the distribution and ecological risk assessment of steroid hormones in the environment in China. *Sci. Total Environ.* **2021**, *786*, 147452. [CrossRef]
8. Arnon, S.; Dahan, O.; Elhanany, S.; Cohen, K.; Pankratov, I.; Gross, A.; Ronen, Z.; Baram, S.; Shore, L.S. Transport of testosterone and estrogen from dairy-farm waste lagoons to groundwater. *Environ. Sci. Technol.* **2008**, *42*, 5521–5526. [CrossRef]
9. Jobling, S.; Williams, R.; Johnson, A.; Taylor, A.; Gross-Sorokin, M.; Nolan, M.; Tyler, C.R.; van Aerle, R.; Santos, E.; Brighty, G. Predicted exposures to steroid estrogens in UK rivers correlate with widespread sexual disruption in wild fish populations. *Environ. Health Perspect.* **2006**, *114*, 32–39. [CrossRef]
10. Johnson, A.; Williams, R.; Matthiessen, P. The potential steroid hormone contribution of farm animals to freshwaters, the United Kingdom as a case study. *Sci. Total Environ.* **2006**, *362*, 166–178. [CrossRef]
11. Hutchins, S.R.; White, M.V.; Hudson, F.M.; Fine, D.D. Analysis of lagoon samples from different concentrated animal feeding operations for estrogens and estrogen conjugates. *Environ. Sci. Technol.* **2007**, *41*, 738–744. [CrossRef] [PubMed]
12. Lafrance, P.; Caron, E. Impact of recent manure applications on natural estrogen concentrations in streams near agricultural fields. *Environ. Res.* **2013**, *126*, 208–210. [CrossRef] [PubMed]
13. Rechsteiner, D.; Wettstein, F.E.; Pfeiffer, N.; Hollender, J.; Bucheli, T.D. Natural estrogen emissions to subsurface tile drains from experimental grassland fields in Switzerland after application of livestock slurries and free compounds. *Sci. Total Environ.* **2021**, *779*, 146351. [CrossRef] [PubMed]
14. Casey, F.X.; Hakk, H.; DeSutter, T.M. Free and conjugated estrogens detections in drainage tiles and wells beneath fields receiving swine manure slurry. *Environ. Pollut.* **2020**, *256*, 113384. [CrossRef] [PubMed]
15. Havens, S.M.; Hedman, C.J.; Hemming, J.D.; Mieritz, M.G.; Shafer, M.M.; Schauer, J.J. Occurrence of estrogens, androgens and progestogens and estrogenic activity in surface water runoff from beef and dairy manure amended crop fields. *Sci. Total Environ.* **2020**, *710*, 136247. [CrossRef]
16. Xu, P.; Zhou, X.; Xu, D.; Xiang, Y.; Ling, W.; Chen, M. Contamination and risk assessment of estrogens in livestock manure: A case study in Jiangsu Province, China. *Int. J. Environ. Res. Public Health* **2018**, *15*, 125. [CrossRef]
17. Gomes, F.B.R.; Fernandes, P.A.A.; Bottrel, S.E.C.; Brandt, E.M.F.; Pereira, R.d.O. Fate, occurrence, and removal of estrogens in livestock wastewaters. *Water Sci. Technol.* **2022**, *86*, 814–833. [CrossRef]
18. Tang, Z.; Wan, Y.-p.; Liu, Z.-h.; Wang, H.; Dang, Z.; Liu, Y. Twelve natural estrogens in urines of swine and cattle: Concentration profiles and importance of eight less-studied. *Sci. Total Environ.* **2022**, *803*, 150042. [CrossRef]
19. Combalbert, S.; Hernandez-Raquet, G. Occurrence, fate, and biodegradation of estrogens in sewage and manure. *Appl. Microbiol. Biot.* **2010**, *86*, 1671–1692. [CrossRef]
20. Hanselman, T.A.; Graetz, D.A.; Wilkie, A.C. Manure-borne estrogens as potential environmental contaminants: A review. *Environ. Sci. Technol.* **2003**, *37*, 5471–5478. [CrossRef]
21. Khanal, S.K.; Xie, B.; Thompson, M.L.; Sung, S.; Ong, S.-K.; Van Leeuwen, J. Fate, transport, and biodegradation of natural estrogens in the environment and engineered systems. *Environ. Sci. Technol.* **2006**, *40*, 6537–6546. [CrossRef] [PubMed]
22. Yan, C.; Yang, Y.; Zhou, J.; Nie, M.; Liu, M.; Hochella Jr, M.F. Selected emerging organic contaminants in the Yangtze Estuary, China: A comprehensive treatment of their association with aquatic colloids. *J. Hazard. Mater.* **2015**, *283*, 14–23. [CrossRef] [PubMed]
23. Zhou, J.; Liu, R.; Wilding, A.; Hibberd, A. Sorption of selected endocrine disrupting chemicals to different aquatic colloids. *Environ. Sci. Technol.* **2007**, *41*, 206–213. [CrossRef] [PubMed]
24. Liu, R.; Wilding, A.; Hibberd, A.; Zhou, J.L. Partition of endocrine-disrupting chemicals between colloids and dissolved phase as determined by cross-flow ultrafiltration. *Environ. Sci. Technol.* **2005**, *39*, 2753–2761. [CrossRef]
25. Carballa, M.; Fink, G.; Omil, F.; Lema, J.M.; Ternes, T. Determination of the solid–water distribution coefficient (Kd) for pharmaceuticals, estrogens and musk fragrances in digested sludge. *Water Res.* **2008**, *42*, 287–295. [CrossRef]
26. Lee, J.; Cho, J.; Kim, S.H.; Kim, S.D. Influence of 17β-estradiol binding by dissolved organic matter isolated from wastewater effluent on estrogenic activity. *Ecotoxicol. Environ. Saf.* **2011**, *74*, 1280–1287. [CrossRef]
27. Yan, C.; Nie, M.; Yang, Y.; Zhou, J.; Liu, M.; Baalousha, M.; Lead, J.R. Effect of colloids on the occurrence, distribution and photolysis of emerging organic contaminants in wastewaters. *J. Hazard. Mater.* **2015**, *299*, 241–248. [CrossRef]
28. Huang, Y.; Xie, X.; Zhou, L.J.; Ji, X.; Gao, B.; Xu, G.Z.; Li, A. Multi-phase distribution and risk assessment of endocrine disrupting chemicals in the surface water of the Shaying River,-Huai River Basin, China. *Ecotoxicol. Environ. Saf.* **2019**, *173*, 45–53. [CrossRef]
29. Nie, M.; Yang, Y.; Liu, M.; Yan, C.; Shi, H.; Dong, W.; Zhou, J.L. Environmental estrogens in a drinking water reservoir area in Shanghai: Occurrence, colloidal contribution and risk assessment. *Sci. Total Environ.* **2014**, *487*, 785–791. [CrossRef]
30. Chen, T.-C.; Yeh, K.-J.C.; Kuo, W.-C.; Chao, H.-R.; Sheu, S.-C. Estrogen degradation and sorption onto colloids in a constructed wetland with different hydraulic retention times. *J. Hazard. Mater.* **2014**, *277*, 62–68. [CrossRef]
31. Huang, Y.; Li, W.; Qin, L.; Xie, X.; Gao, B.; Sun, J.; Li, A. Distribution of endocrine-disrupting chemicals in colloidal and soluble phases in municipal secondary effluents and their removal by different advanced treatment processes. *Chemosphere* **2019**, *219*, 730–739. [CrossRef] [PubMed]

32. Gudda, F.O.; Ateia, M.; Waigi, M.G.; Wang, J.; Gao, Y. Ecological and human health risks of manure-borne steroid estrogens: A 20-year global synthesis study. *J. Environ. Manag.* **2022**, *301*, 113708. [CrossRef] [PubMed]
33. Neale, P.A.; Escher, B.I.; Schäfer, A.I. pH dependence of steroid hormone—Organic matter interactions at environmental concentrations. *Sci. Total Environ.* **2009**, *407*, 1164–1173. [CrossRef] [PubMed]
34. Gong, J.; Huang, Y.; Huang, W.; Ran, Y.; Chen, D. Multiphase partitioning and risk assessment of endocrine-disrupting chemicals in the Pearl River, China. *Environ. Toxicol. Chem.* **2016**, *35*, 2474–2482. [CrossRef] [PubMed]
35. Chen, T.-S.; Chen, T.-C.; Yeh, K.-J.C.; Chao, H.-R.; Liaw, E.-T.; Hsieh, C.-Y.; Chen, K.-C.; Hsieh, L.-T.; Yeh, Y.-L. High estrogen concentrations in receiving river discharge from a concentrated livestock feedlot. *Sci. Total Environ.* **2010**, *408*, 3223–3230. [CrossRef] [PubMed]
36. Hung, H.-S.; Yeh, K.-J.C.; Chen, T.-C. Investigation of free and conjugated estrogen fate and emission coefficients in three duck farms. *Environ. Sci. Pollut. Res.* **2023**, *30*, 9874–9885. [CrossRef]
37. Hung, H.-S.; Yeh, K.-J.C.; Hsieh, C.-Y.; Chen, T.-C. Occurrence and Degradation of Free and Conjugated Estrogens in a River Receiving Feedlot Animal Discharge. *Appl. Sci.* **2022**, *12*, 11961. [CrossRef]
38. Chuang, C.-W.; Hsu, L.-F.; Tsai, H.-C.; Liu, Y.-Y.; Huang, W.-S.; Chen, T.-C. Nickel Binding Affinity with Size-Fractioned Sediment Dissolved and Particulate Organic Matter and Correlation with Optical Indicators. *Appl. Sci.* **2020**, *10*, 8995. [CrossRef]
39. Isobe, T.; Shiraishi, H.; Yasuda, M.; Shinoda, A.; Suzuki, H.; Morita, M. Determination of estrogens and their conjugates in water using solid-phase extraction followed by liquid chromatography–tandem mass spectrometry. *J. Chromatogr. A* **2003**, *984*, 195–202. [CrossRef]
40. Kumar, V.; Nakada, N.; Yasojima, M.; Yamashita, N.; Johnson, A.C.; Tanaka, H. Rapid determination of free and conjugated estrogen in different water matrices by liquid chromatography–tandem mass spectrometry. *Chemosphere* **2009**, *77*, 1440–1446. [CrossRef]
41. Bartelt-Hunt, S.L.; Snow, D.D.; Kranz, W.L.; Mader, T.L.; Shapiro, C.A.; Donk, S.J.v.; Shelton, D.P.; Tarkalson, D.D.; Zhang, T.C. Effect of growth promotants on the occurrence of endogenous and synthetic steroid hormones on feedlot soils and in runoff from beef cattle feeding operations. *Environ. Sci. Technol.* **2012**, *46*, 1352–1360. [CrossRef]
42. Bevacqua, C.E.; Rice, C.P.; Torrents, A.; Ramirez, M. Steroid hormones in biosolids and poultry litter: A comparison of potential environmental inputs. *Sci. Total Environ.* **2011**, *409*, 2120–2126. [CrossRef] [PubMed]
43. Zheng, W.; Yates, S.R.; Bradford, S.A. Analysis of steroid hormones in a typical dairy waste disposal system. *Environ. Sci. Technol.* **2008**, *42*, 530–535. [CrossRef] [PubMed]
44. Zhao, J.; Ying, G.; Chen, F.; Liu, Y.; Wang, L.; Yang, B.; Liu, S.; Tao, R. Estrogenic activity profiles and risks in surface waters and sediments of the Pearl River system in south China assessed by chemical analysis and in vitro bioassay. *J. Environ. Monit.* **2011**, *13*, 813–821. [CrossRef] [PubMed]
45. Jin, S.; Yang, F.; Xu, Y.; Dai, H.; Liu, W. Risk assessment of xenoestrogens in a typical sewage-holding lake in China. *Chemosphere* **2013**, *93*, 892–898. [CrossRef] [PubMed]
46. Beck, I.-C.; Bruhn, R.; Gandrass, J. Analysis of estrogenic activity in coastal surface waters of the Baltic Sea using the yeast estrogen screen. *Chemosphere* **2006**, *63*, 1870–1878. [CrossRef] [PubMed]
47. Chuang, C.-W.; Huang, W.-S.; Chen, H.-S.; Hsu, L.-F.; Liu, Y.-Y.; Chen, T.-C. Sorption Constant of Bisphenol A and Octylphenol Onto Size-Fractioned Dissolved Organic Matter Using a Fluorescence Method. *Int. J. Environ. Res. Public Health* **2021**, *18*, 1102. [CrossRef]
48. Ilina, S.M.; Lapitskiy, S.A.; Alekhin, Y.V.; Viers, J.; Benedetti, M.; Pokrovsky, O.S. Speciation, size fractionation and transport of trace elements in the continuum soil water–mire–humic lake–river–large oligotrophic lake of a Subarctic watershed. *Aquat. Geochem.* **2016**, *22*, 65–95. [CrossRef]
49. Liu, Y.-Y.; Lin, Y.-S.; Yen, C.-H.; Miaw, C.-L.; Chen, T.-C.; Wu, M.-C.; Hsieh, C.-Y. Identification, contribution, and estrogenic activity of potential EDCs in a river receiving concentrated livestock effluent in Southern Taiwan. *Sci. Total Environ.* **2018**, *636*, 464–476. [CrossRef]
50. Sutaswiriya, N.; Homklin, S.; Kreetachat, T.; Vaithanomsat, P.; Kreetachat, N. Monitoring estrogen and androgen residues from livestock farms in Phayao Lake, Thailand. *Environ. Monit. Assess.* **2021**, *193*, 812. [CrossRef]
51. Yamamoto, H.; Liljestrand, H.M.; Shimizu, Y.; Morita, M. Effects of physical– chemical characteristics on the sorption of selected endocrine disruptors by dissolved organic matter surrogates. *Environ. Sci. Technol.* **2003**, *37*, 2646–2657. [CrossRef] [PubMed]
52. Holbrook, R.D.; Love, N.G.; Novak, J.T. Sorption of 17β-estradiol and 17α-ethinylestradiol by colloidal organic carbon derived from biological wastewater treatment systems. *Environ. Sci. Technol.* **2004**, *38*, 3322–3329. [CrossRef] [PubMed]
53. Ying, G.-G.; Williams, B.; Kookana, R. Environmental fate of alkylphenols and alkylphenol ethoxylates—A review. *Environ. Int.* **2002**, *28*, 215–226. [CrossRef] [PubMed]
54. Holthaus, K.I.; Johnson, A.C.; Jürgens, M.D.; Williams, R.J.; Smith, J.J.; Carter, J.E. The potential for estradiol and ethinylestradiol to sorb to suspended and bed sediments in some English rivers. *Environ. Toxicol. Chem. Int. J.* **2002**, *21*, 2526–2535. [CrossRef]
55. Nie, M.; Yan, C.; Dong, W.; Liu, M.; Zhou, J.; Yang, Y. Occurrence, distribution and risk assessment of estrogens in surface water, suspended particulate matter, and sediments of the Yangtze Estuary. *Chemosphere* **2015**, *127*, 109–116. [CrossRef]
56. Liu, Y.-H.; Zhang, S.-H.; Ji, G.-X.; Wu, S.-M.; Guo, R.-X.; Cheng, J.; Yan, Z.-Y.; Chen, J.-Q. Occurrence, distribution and risk assessment of suspected endocrine-disrupting chemicals in surface water and suspended particulate matter of Yangtze River (Nanjing section). *Ecotoxicol. Environ. Saf.* **2017**, *135*, 90–97. [CrossRef]

57. Holbrook, R.D.; Love, N.G.; Novak, J.T. Biological wastewater treatment and estrogenic endocrine disrupting compounds: Importance of colloid organic carbon. *Pract. Period. Hazard. Toxic. Radioact. Waste Manag.* **2003**, *7*, 289–296. [CrossRef]
58. Qiao, X.; Carmosini, N.; Li, F.; Lee, L.S. Probing the primary mechanisms affecting the environmental distribution of estrogen and androgen isomers. *Environ. Sci. Technol.* **2011**, *45*, 3989–3995. [CrossRef]
59. Hu, Y.; Yan, X.; Shen, Y.; Di, M.; Wang, J. Occurrence, behavior and risk assessment of estrogens in surface water and sediments from Hanjiang River, Central China. *Ecotoxicology* **2019**, *28*, 143–153. [CrossRef]

toxics

MDPI

Article

Enhanced Adsorption of Cadmium by a Covalent Organic Framework-Modified Biochar in Aqueous Solution

Yanwei Hou [1,2], Shanna Lin [2], Jiajun Fan [2], Youchi Zhang [2], Guohua Jing [1,*] and Chao Cai [2,*]

[1] College of Chemical Engineering, Huaqiao University, Xiamen 361021, China; houyw@hqu.edu.cn
[2] Key Lab of Urban Environment and Health, Institute of Urban Environment, Chinese Academy of Sciences, Xiamen 361021, China; snlin@iue.ac.cn (S.L.); jjfan@iue.ac.cn (J.F.); yczhang@iue.ac.cn (Y.Z.)
* Correspondence: zhoujing@hqu.edu.cn (G.J.); ccai@iue.ac.cn (C.C.)

Abstract: In the environmental field, the advancement of new high-efficiency heavy metal adsorption materials remains a continuous research focus. A novel composite, covalent organic framework-modified biochar (RH-COF), was fabricated via an in-situ polymerization approach in this study. The COF-modified biochar was characterized by elemental analysis, BET analysis, SEM, FT-IR, and XPS. The nitrogen and oxygen content in the modified material increased significantly from 0.96% and 15.50% to 5.40% and 24.08%, respectively, indicating the addition of a substantial number of nitrogen- and oxygen-containing functional groups to the RH-COF surface, thereby enhancing its adsorption capacity for Cd from 4.20 mg g^{-1} to 58.62 mg g^{-1}, representing an approximately fourteen-fold increase. Both the pseudo-second-order model and the Langmuir model were suitable for describing the kinetics and isotherms of Cd^{2+} adsorption onto RH-COF. The adsorption performance of Cd^{2+} by RH-COF showed minimal sensitivity to pH values between 4.0 and 8.0, but could be slightly influenced by ionic strength. Mechanistic analysis showed that the Cd^{2+} adsorption on RH-COF was dominated by surface complexation and chelation, alongside electrostatic adsorption, surface precipitation, and Cπ–cation interactions. Overall, these findings suggest that the synthesis of COF-biochar composite may serve as a promising remediation strategy while providing scientific support for applying COF in environmental materials.

Keywords: COF-modified biochar; rice husk; cadmium; adsorption

1. Introduction

Biochar has been an extremely hot scientific subject over the past two decades, and is regarded as a green environmental material due to its wide availability, facile preparation, low risk of secondary pollution, and low cost [1–3]. The application of biochar in the environment can bring numerous benefits, such as enhancing crop yield and soil health, reducing greenhouse gas emissions, reutilizing waste, and alleviating climate change. Owing to its favorable pore structure and abundant surface functional group structure, biochar, as an excellent alternative adsorbent, has been extensively studied in terms of its effect and mechanism for removing heavy metals from both soil and water [4,5]. Technology based on biochar has emerged as one of the highly potential and promising tools for environmental remediation.

Biochar can be derived from numerous agricultural wastes, such as manure, corn stalk, rice straw, rice husk, and bean stalk. Despite the similar appearance of different biochars, their behaviors vary significantly due to the diverse physicochemical properties of biochars, which are profoundly influenced by the type of feed stock [6]. Consequently, some biochars exhibit excellent adsorption performance, while others do not. Generally speaking, biochars produced from poultry manure possess a higher metal adsorption capacity compared to those derived from plants [7–9]. Xu et al. reported that biochar originated from dairy manure demonstrated superior performance to rice husk biochar, which has a low surface area and limited functional groups in removing Cu^{2+}, Cd^{2+}, Zn^{2+}, and Pb^{2+} from water [10]. However, the application of these biochars in heavy metal removal is largely hindered by their poor adsorption performance. Cd is a toxic heavy metal element that poses a significant risk to human health. Epidemiological evidence suggests that occupational and environmental cadmium exposure may be associated with many types of cancer, including breast, lung, prostate, nasopharyngeal, pancreatic, and kidney cancers [11]. Therefore, it is essential for these biochars to be modified to enhance their metal adsorption ability.

Consequently, diverse modification approaches of biochars, such as acidification, alkalization, amination, magnetic modification, metal ion incorporation, oxidation using oxidizing agents and surfactant modification, have been developed to obtain specific advantages for numerous particular applications. These approaches may dramatically alter the physicochemical properties of biochars and significantly improve their capacity for adsorbing heavy metals [12]. Chen et al. fabricated sulfur modified wheat straw biochar, which exhibited a superior adsorption capacity for Cd^{2+} from aqueous solutions compared to the original biochar [13]. Fan et al. prepared thiol-modified rice straw biochar with maximum adsorption capacities of 45.1 mg g^{-1} for Cd^{2+} and 61.4 mg g^{-1} for Pb^{2+} [14]. Lee and Shin modified three distinct types of biochar (rice husk, wood chip, and mixture) through five different methods (acid, alkaline, oxidic, manganese oxide, and iron oxide). The results indicated that MnOx-modified biochar achieved the highest adsorption capacities, which increased the adsorption capacity of Cd^{2+} from 3 mg g^{-1} to 10 mg g^{-1} [15]. To date, producing modified biochar material has emerged as an important practice for expanding the environmental applications of biochar.

Covalent organic frameworks (COFs) represent an emerging category of porous substances with pre-designable architectures [16]. COF materials possess numerous specific properties, such as elevated thermal/chemical stability, substantial porosity and surface area, as well as uniform and adjustable pore sizes, endowing them with potential as remediation substances [17,18]. Introducing a covalent organic framework (COF) to the outermost layer of the adsorbent material would generate more oxygen- and nitrogen-containing organic groups, which might enhance the material's adsorption capacity for target metal ions [19]. The COF-modified material would offer superior porosity, high stability, novel synergistic properties, large specific surface area, high density of nitrogen- and oxygen-containing functional groups, high crystallinity, and novel forms for environmental applications [20–22], facilitating the adsorption of metal ions. Many COF-modified materials have demonstrated significant potential in removing heavy metal species. For example, Sun et al. fabricated a COF-S-SH material through post-synthesis modification for Hg^{2+} removal [23]. Jiang et al. prepared an EDTA@COF for the removal of heavy metal ions [24]. Yang et al. made an EB-COF:Br material as an adsorbent for As (V) [25]. Li et al. modified COF on the absorbent material and augmented the adsorption capacity of Cu^{2+} compared with the raw material [26]. Consequently, it is anticipated to enhance the adsorption capacity of COF-modified biochar composites, yet few studies have explored the adsorption performance of this composite.

In this study, a novel composite, COF-modified biochar derived from rice husk, was fabricated. The properties of COF-modified biochar were characterized using modern chemical analysis approaches, and a series of adsorption experiments was carried out to investigate the adsorption performance and mechanism of COF-modified biochar on Cd^{2+}, which was selected as a representative heavy metal in environmental media. The outcomes of this study will offer scientific support for the development of efficient adsorbents for the remediation of cadmium-contaminated water.

2. Materials and Methods

2.1. Chemicals and Reagents

All reagents used in this study were analytical grade and used as received. Ethylene glycol, ammonium hydroxide, N, N-dimethylacetamide (DMAC), dioxane, acetic acid solution, methanol, acetone, tetrahydrofuran, sodium nitrate, disodium hydrogen phosphate, lead nitrate, copper nitrate hydrate, calcium nitrate, magnesium nitrate, and sodium bicarbonate were purchased from Sinopharm Reagent Co., Ltd. (Shanghai, China). Cadmium nitrate and sodium humate were purchased from Aladdin Reagent Co., Ltd. (Shanghai, China). The 2,4,6-trihydroxy-1,3,5-benzenetricarboxaldehyde (Tp) and 4,4′-diamino-[1,1′-biphenyl]-2,2′-dicarboxylic acid (DBd) were purchased from Chinese Academy of Sciences-Yanshen Technology Co., Ltd. (Changchun, China).

2.2. Preparation of Materials

Rice husk biochar (RH) was prepared from air-dried rice husk. Rice husk was pyrolyzed at 500 °C for 2 h under oxygen-limited conditions in a muffle furnace (KSL, Kejing Inc., Hefei, China). The RH was ground to pass through a 0.15 mm sieve and then thoroughly mixed for further application.

COF was solvothermally synthesized by the co-condensation of 2,4,6-trihydroxy-1,3,5-benzenetricarboxaldehyde (Tp) and 4,4′-diamino-[1,1′-biphenyl]-2,2′-dicarboxylic acid (DBd), with acetic acid serving as a catalyst. Briefly, 1.60 mmol of Tp and 2.40 mmol of DBd in 120 mL N, N-dimethylacetamide (DMAC)/dioxane (1:11) and 10 mL acetic acid solution (6 mol L^{-1}) were mixed in a 200 mL Teflon-tube under ultrasonication. The mixture was subsequently transferred into a 200 mL Teflon-lined autoclave for solvothermal reaction at 120 °C for 3 days. After the reaction, the precipitate was filtered, washed several times with dioxane, methanol, acetone and tetrahydrofuran (THF), and dried at 40 °C for 12 h in a vacuum. The COF was ground to pass through a 0.15 mm sieve and then mixed thoroughly for further experiments.

RH (2.00 g), serving as a raw material, was added to 80 mL ethylene glycol under ultrasonication. Subsequently, an additional 6 mL of ammonium hydroxide was introduced, and the resulting mixed solution was transferred to a Teflon-lined autoclave for solvothermal reaction at 180 °C for 10 h. After the reaction was concluded, the precipitate was filtered, washed repeatedly with distilled water, and dried in a vacuum at 40 °C for 12 h. After pretreatment, the RH was ground to pass through a 0.15 mm sieve and then mixed thoroughly for further application.

The synthesis process of the COF-modified rice husk biochar composite (RH-COF) was identical to that of COF, except that 0.80 g of pretreated RH was added to the mixed solution prior to ultrasonication.

2.3. Characterization of Materials

The pH value of the material was determined by weighing a 0.3 g sample and placing it in a 15 mL centrifuge tube, adding 6 mL ultra-pure water, mixing it well, and oscillating for 1 h at 180 $r{\cdot}min^{-1}$ at 25 °C. Subsequently, the pH value of the sample solution was determined using a pH meter (STARTER 3100/F, Ohaus Inc., Changzhou, China) [14]. The contents of C, N, O, and H in the samples were quantified using a Vario Max CHNS-O-Cl Analyser (Elementar Analysysteme GmbH, Hanau, Germany) [27]. The structural characteristics and elemental composition of the samples were examined via field emission scanning electron microscopy (FE-SEM) (S-4800, Hitachi, Tokyo, Japan) [28,29]. The specific surface area and pore size distribution of the samples were determined with an ASAP 2020 M+C specific surface analyser (Micromeritics, GA, Atlanta, USA) and analyzed by the Brunauer–Emmett–Teller (BET) and Barret–Joyner–Halenda (BJH) methods with N_2 adsorption isotherms, respectively. The functional groups of the samples were identified utilizing Fourier transform infrared spectroscopy (FT-IR, Nicolet iS10, Thermo, Waltham, MA, USA). The surface element analysis was performed by X-ray photoelectron spectroscopy (Axis Supra, Kratos, Tokyo, Japan) with an Al Kα source (hv = 1486.6 eV) [14]. The surface charge properties of the samples were investigated by zeta potential measurements at varying equilibrium pH values using Zeta PALS (Malvern, Malvern City, UK) [30].

2.4. Batch Adsorption Experiments

The adsorption performance of Cd^{2+} by RH and RH-COF was investigated by batch adsorption experiments in a 15 mL polyethylene centrifuge tube. The $Cd(NO_3)_2$ solution in an electrolyte background of 0.01 mol L^{-1} $NaNO_3$ was mixed with either RH or RH-COF at a dosage of 2 g L^{-1}. The pH of the $Cd(NO_3)_2$ solution was adjusted to the desired values using 1 mol L^{-1} HNO_3 and 1 mol L^{-1} NaOH.

The Cd^{2+} adsorption kinetics experiments were carried out with 0.20 g of either RH or RH-COF in 100 mL solution containing either 100 or 250 mg L^{-1} Cd^{2+} (pH = 5) in a series of 250 mL glass conical flasks. The samples were periodically withdrawn (0–48 h) and filtered through a 0.45 μm PES water filtration membrane (Keyilong Lab Equipment Co., Ltd., Tianjin, China). The concentration of Cd^{2+} in the filtrate was determined by an inductively coupled plasma optical emission spectrometer (ICP-OES, Optima 7000DV, PerkinElmer, Waltham, MA, USA) after dilution with 2% HNO_3 solution. Both pseudo-first-order and pseudo-second-order kinetic models were employed to analyze the kinetic parameters.

The Cd^{2+} adsorption isotherms were established by introducing 0.01 g of RH or RH-COF into 5 mL of Cd^{2+} solutions with varying concentrations in 15 mL polyethylene centrifuge tubes. Cd^{2+} concentrations (pH = 5) ranged from 5 to 700 mg L^{-1}. The mixture was agitated at a speed of 150 rpm at 25 $\pm$ 0.5 °C for 48 h, and then the suspension was filtered through a 0.45 μm PES water filtration membrane to analyze the residual Cd^{2+} concentrations. The adsorption isotherm parameters were derived using both Langmuir and Freundlich models.

The influence of pH on the adsorption of Cd^{2+} was assessed across a range from pH 2.0 to 8.0 with initial pH adjustments using either 1 mol L^{-1} HNO_3 or NaOH solutions. Additionally, the effect of ionic strength on the adsorption of Cd^{2+} was evaluated by varying $NaNO_3$ concentrations between 0.001 and 0.100 mol L^{-1}. Furthermore, the influence of $H_2PO_4{}^{2-}$, $HCO_3{}^{-}$, Cu^{2+}, Pb^{2+}, Ca^{2+}, Mg^{2+}, and HA on the adsorption of Cd^{2+} were investigated by supplementary additions of $H_2PO_4{}^{-}$ (10^{-4} mol L^{-1}), $HCO_3{}^{-}$ (10^{-3} mol L^{-1}), Cu^{2+} (100 mg L^{-1}), Pb^{2+} (100 mg L^{-1}), Ca^{2+} (10^{-2} mol L^{-1}), Mg^{2+} (10^{-2} mol L^{-1}), and HA (10^{-3} mol L^{-1}) with 250 mg L^{-1} Cd^{2+} solution (pH = 5), respectively. All adsorption experiments were conducted in triplicate.

2.5. Data Processing and Statistical Analysis

All data were presented as the means plus or minus standard deviation. Differences among treatments were examined using one-way analysis of variance (ANOVA). Data processing was performed using Origin2017 (OriginLab Corporation, Northampton, MA, USA), while the statistical analysis was conducted using IBM SPSS Statistics 23 software (SPSS Inc., IBM, Armonk, New York, NY, USA).

3. Results and Discussion

3.1. Physiochemical Properties of Materials

Field emission scanning electron microscopy (FE-SEM) images of RH, COF, and RH-COF are presented in Figure 1. The RH sample exhibited a rough surface with an irregular porous structure and some attached particles, while a spherical COF structure was observed. The SEM image of the RH-COF composite demonstrated that COF had successfully adhered to both the surface and pores of the composite.

The physicochemical properties of RH, COF, and RH-COF are summarized in Table 1. The elemental analysis indicated that a significant alteration in the composition of RH following modification. The nitrogen content in RH-COF was elevated compared to that in RH, increasing from 0.96% to 5.40%, thereby suggesting the successful modification of RH with COF.

Figure 1. SEM micrographs of RH (**a**) at 1000× magnification, COF (**b**), and RH-COF (**c**) at 5000× magnification.

Table 1. Physicochemical properties of materials.

	Sample		
	RH	COF	RH-COF
C (%)	53.34 ± 2.24 a	60.18 ± 1.16 b	36.31 ± 1.20 c
N (%)	0.96 ± 0.03 a	8.88 ± 0.12 b	5.40 ± 0.05 c
O (%)	15.50 ± 2.72 a	22.37 ± 2.88 a	24.08 ± 6.91 a
H (%)	4.01 ± 0.27 a	5.61 ± 0.11 b	4.49 ± 0.09 c
Molar H/C (%)	0.08	0.09	0.12
Molar O/C (%)	0.29	0.37	0.66
Molar (O + N)/C (%)	0.31	0.52	0.81
pH	9.67 ± 0.04 a	7.95 ± 0.04 b	7.90 ± 0.01 b
Bet surface area (m^2 g^{-1})	2.70	37.37	33.41
Average pore width (nm)	9.24	6.12	6.30
Pore volume (cm^3 g^{-1})	0.01	0.06	0.05

Different lowercase letters in the table indicate significant differences between treatments at $p < 0.05$.

Furthermore, the pH of RH-COF was lower than that of RH (pH 9.67) as shown in Table 1. The pH level of biochar is primarily influenced by its surface functional groups [31]. The observed decrease in pH for RH-COF can be partially attributed to the protonation or deprotonation reaction occurring on the surface of the composite, which might affect its adsorption performance during the application [27].

The nitrogen adsorption/desorption isotherms and corresponding pore size distribution curves for RH, COF, and RH-COF are presented in Figure 2. The nitrogen adsorption/desorption isotherms of RH, COF, and RH-COF exhibited a characteristic IV-type isotherm with an H3 hysteresis loop, indicating the presence of mesopores in these materials (Figure 2) [32]. Moreover, the nitrogen adsorption isotherm demonstrated that following modification RH-COF possessed greater abundance of mesopores compared to RH (Figure 2a,c).

In terms of the surface properties, post-modification analysis demonstrated an increase in both the surface area and average pore volume of RH-COF. Specifically, the surface area of RH-COF increased from 2.70 m^2 g^{-1} to 33.41 m^2 g^{-1}, however, the pore diameter of RH-COF decreased from 9.24 nm to 6.30 nm due to infilling by COF with a large specific surface area and pore volume (Table 1). The abundant pores facilitate not only the adsorption of solute onto the outer surface of RH-COF but also their diffusion into the pore channels where they can bind to the activated adsorption sites on the inner surface. Notably, since the pore diameter of RH-COF exceeds that of the cadmium at approximately 0.19 nm, it exhibits distinctive adsorption performance towards Cd^{2+}.

The analysis of Fourier transform infrared (FT-IR) spectroscopy was carried out to elucidate the chemical changes of the functional groups on the adsorbent surface. The FT-IR spectrum of RH (Figure 3) presented distinct peaks corresponding to O-H (3416 cm^{-1} and 1438 cm^{-1}), C=O/C-C (1580 cm^{-1}), C-O (1099 cm^{-1}), and Si-O (803 cm^{-1} and 463 cm^{-1}) [33–37], indicating the presence of carboxyl and hydroxyl functional groups on the surface of RH [28,38]. In contrast, the FT-IR spectrum of COF presented obvious peaks for N-H/O-H (3418 cm^{-1}), CH=O (2850 cm^{-1}), C=O (1717 cm^{-1}), C=C (1455 cm^{-1}

and 1593 cm^{-1}), as well as C-N (1280 cm^{-1}) [33,36,39,40]. Notably, the FT-IR spectrum of RH-COF presented minimal variation in the composition of the functional groups compared to either COF or RH. Specifically, the peaks for O-H/N-H (3416 cm^{-1} and 3418 cm^{-1}) and CH=O (2850 cm^{-1}) were absent, while a significant reduction in the peak intensity for C=O (1717 cm^{-1}) was observed along with a shift to a new position at 1709 cm^{-1}. Additionally, the new peaks corresponding to C=N (1615 cm^{-1}), C=C (1452 cm^{-1} and 1593 cm^{-1}), and C-N (1292 cm^{-1}) were identified in the FT-IR spectrum of RH-COF [39–43]. Therefore, the introduction of nitrogen-containing functional groups (C-N/C=N) and carbon-containing functional groups (C=C/C=O) in RH-COF demonstrated that successful loading of COF on RH had occurred. These functional groups are intrinsically linked to the materials' physical and chemical properties, and they can influence adsorption capacity.

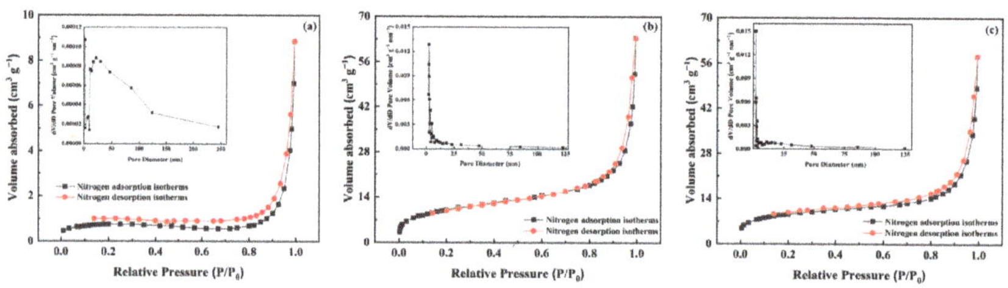

Figure 2. Nitrogen adsorption/desorption isotherms and pore size distributions of RH (**a**), COF (**b**), and RH-COF (**c**).

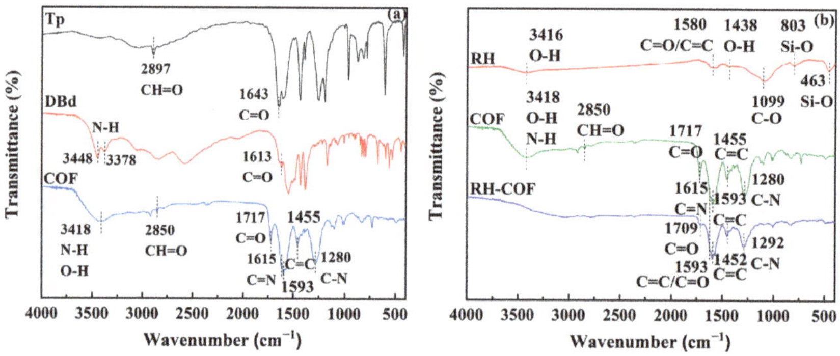

Figure 3. Comparison of the FT-IR spectra of Tp, DBd and COF (**a**), Comparison of the FT-IR spectra of RH, COF and RH-COF (**b**).

3.2. *Adsorption Behaviour of Materials for Cd^{2+}*

3.2.1. Adsorption Kinetics

Adsorption kinetics are crucial for estimating the rate of adsorption and offering valuable insights into the adsorption mechanism. The adsorption kinetics of Cd^{2+} on RH and RH-COF were examined, and the results are shown in Figure 4. The adsorption amount of Cd^{2+} on RH increased gradually, reaching adsorption equilibrium after approximately 2 h. In contrast, the initial adsorption of Cd^{2+} on RH-COF was rapid, achieving equilibrium within 0.5 h, indicating a significant enhancement in the adsorption rate following modification.

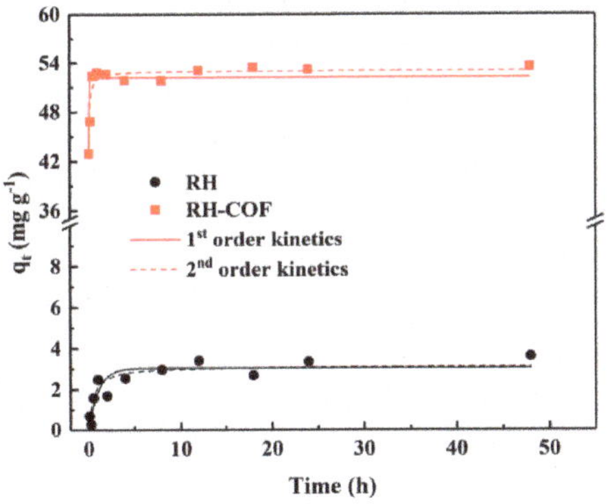

Figure 4. Adsorption kinetics of Cd^{2+} on RH and RH-COF (initial concentration of Cd^{2+}: 100 mg L^{-1} (RH) and 250 mg L^{-1} (RH-COF), adsorbent dosage: 2.0 g L^{-1}, pH 5).

To elucidate the adsorption mechanisms, the adsorption kinetics data were fitted to both pseudo-first-order and pseudo-second-order kinetic models. Detailed information on these adsorption kinetics and fitting parameters can be found in the Supplementary Information (Table S1). For both RH and RH-COF, the coefficient of determination (R^2) for the pseudo-second-order kinetic model (0.92 and 0.90) surpassed that for the pseudo-first-order kinetic model (0.77 and 0.69). This suggests that the adsorption of Cd^{2+} by both materials is more accurately described by the pseudo-second-order kinetic model than by the pseudo-first-order kinetic model, implying that the adsorption predominately governs this process. The process of the pseudo-second-order model can be divided into two distinct kinetic phases. In the initial stage of adsorption, due to the existence of a large number of adsorption sites on the surface of the adsorbent, the adsorption rate was fast. The K_2 value of RH-COF for Cd^{2+} exceeded that of RH (Table S1), suggesting a faster adsorption of Cd^{2+} on RH-COF compared to its counterpart. However, as time progressed and available adsorption sites were gradually filled, there was a corresponding decline in adsorption rate until the adsorption equilibrium was reached where the difference between their adsorption rates diminished.

3.2.2. Adsorption Isotherms

The influence of varying initial Cd^{2+} concentrations on the adsorption isotherms of Cd^{2+} by RH and RH-COF was investigated, with results presented in Figure 5. Comprehensive details regarding the adsorption isotherms and the model parameters fitted using the Langmuir and Freundlich models are available in the Supplementary Information (Table S2). The maximum adsorption capacity of RH-COF for Cd^{2+} surpassed that of RH. As indicated in Table S2, the Langmuir adsorption capacity of RH-COF for Cd^{2+} reached 58.62 mg g^{-1}, nearly 14 times greater than that of RH (4.20 mg g^{-1}). This finding indicates a significant enhancement in the adsorption capacity of RH for Cd^{2+} following COF modification. Moreover, when compared to previously reported modified biochar adsorbents and other biocarbon sources, RH-COF exhibited a competitive adsorption capacity for Cd^{2+} (Table 2), highlighting distinct advantages associated with COF-modified biochar. According to Table S2, the experimental data for both RH and RH-COF were more accurate by the Langmuir model (R^2 = 0.92 and 0.90, respectively) than by the Freundlich model (R^2 = 0.83 and 0.84, respectively). The results imply that metal ion adsorption occurred at a homogeneous surface by monolayer sorption [44].

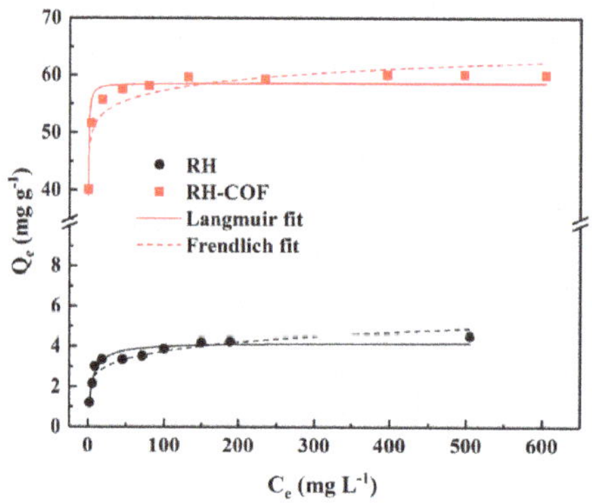

Figure 5. Adsorption isotherms of Cd^{2+}on RH and RH-COF (adsorbent dosage: 2.0 g L^{-1}, pH 5, contact time: 48 h).

Table 2. The adsorption of Cd^{2+} on other modified biochars.

Adsorbent	pH	Q_m (mg g^{-1})	Reference
RH-COF	5.0	58.62	This study
Magnesium oxide–rice husk biochar composite (MgO-BCR)	5.0	18.1	[45]
Calcium-based magnetic biochar (Ca-MBC)	6.0	10.1	[46]
Chitosan–pyromellitic dianhydride modified biochar(CPMB)	5.0	30.12–38.24	[12]
Rice husk	-	7.8	[10]
Modified rice husk	-	8.58–20.24	[47]
RHB/MgAl-layered double hydroxide-coated rice husk(MgAl–LDH@RHB)	6.0	27.46/113.99	[48]
RHB 300, RHB 500, RHB 700/RHB 300-Si, RHB 500-Si, RHB 700-Si	-	52.65, 58.62, 76.55/44.75, 47.83, 60.37	[49]

3.2.3. Effect of Environmental Factors on the Adsorption of Cd^{2+}

The pH plays a crucial role in the adsorption process, influencing not only the protonation degree of the functional groups on the adsorbent surface but also the chemical speciation of Cd in the solution [50,51]. This study investigated the effect of an initial pH range from 2.0 to 8.0 on the adsorption of Cd^{2+} by RH and RH-COF, with results illustrated in Figure 6a. As pH increased, the amount of Cd^{2+} adsorption on RH gradually increased at a slow rate. In contrast, the adsorption capacity of RH-COF for Cd^{2+} exhibited a sharp increase as pH transitioned from 2.0 to 4.0, followed by a gradual rise that stabilized beyond pH 5.0. At a pH of 2.0, both RH-COF and RH displayed similar and notably low adsorption capacities for Cd^{2+}. However, from pH values ranging between 3.0 and 8.0, the adsorption capacity of RH-COF was significantly higher than that of RH in this study. The influence of pH on the adsorption process may be attributed to the electrostatic interactions and competition between H^+ and Cd^{2+}. The point-of-zero charge (pHpzc) values for RH and RH-COF were found to be 2.49 and 2.84, respectively (Figure S1). When the pH falls below its respective pHpzc, the surface becomes positively charged with an abundance of available H^+ occupying potential adsorption sites; consequently, these H^+ hinder the binding interactions between Cd^{2+} and adsorbents. As pH increases, the deprotonation on the adsorbent surface becomes more pronounced, while the competition for binding sites between Cd^{2+} and H^+ decreases, resulting in more available binding sites which facilitate electrostatic interactions leading to enhanced coupling with both RH and RH-COF by Cd^{2+}. When the pH value exceeds 3.0, the adsorption capacity of Cd^{2+} by RH-COF significantly surpasses that of RH, a phenomenon attributed to successful introduction of COF with a high specific surface area and pores, which are favorable for Cd^{2+} adsorption.

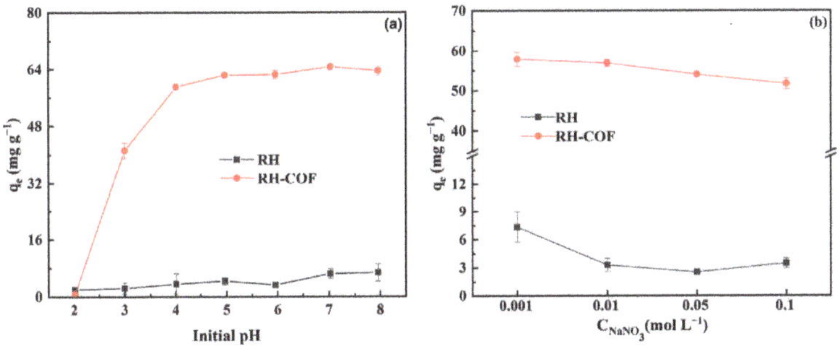

Figure 6. Effect of pH (**a**) and ionic strength (**b**) on the adsorption of Cd^{2+} by RH and RH-COF (initial concentration of Cd^{2+}: 250 mg L^{-1}, adsorbent dosage: 2.0 g L^{-1}, contact time: 48 h).

The influence of ionic strength on adsorption can be used to determine the adsorption type for Cd^{2+}. The impact of ionic strength on the adsorption of Cd^{2+} by RH and RH-COF was investigated and the results are presented in Figure 6b. The effects of ionic strength varied between the two materials. As ionic strength increased, the Cd^{2+} adsorption capacity of RH was drastically decreased by 52.80–65.62%, indicating that Cd^{2+} was primarily adsorbed on the RH surface as outer-sphere complexes, which was nonspecific adsorption. However, the reduction in adsorption of Cd^{2+} by RH-COF was relatively modest, ranging from 1.68% to10.66% with rising Na^+ concentration. This indicates that Na^+ competes with Cd^{2+} for available adsorption sites in this context. The diverse and abundant functional groups on RH-COF facilitated heavy metal complexation or coordination; consequently, the influence of Na^+ on Cd^{2+} adsorption of RH-COF was weaker than that of RH.

The influences of coexisting ions, including $H_2PO_4^-$, HCO_3^-, Cu^{2+}, Pb^{2+}, Ca^{2+}, Mg^{2+}, and HA, on the adsorption of Cd^{2+} by RH and RH-COF were investigated, with results illustrated in Figure 7. The effects of coexisting ions and HA on Cd^{2+} adsorption were basically in the order of Cu^{2+} > Pb^{2+} > Ca^{2+} > Mg^{2+} > HA, $H_2PO_4^-$, HCO_3^-. It was found that $H_2PO_4^-$, HCO_3^-, and HA exert a minimal effect on the adsorption of Cd^{2+} on RH-COF. Conversely, the adsorption capacity of Cd^{2+} was significantly decreased in the presence of Cu^{2+}, Pb^{2+}, Ca^{2+}, and Mg^{2+}. For example, the adsorption capacity of Cd^{2+} decreased to 13.44 mg g^{-1} and 34.15 mg g^{-1} in the presence of Cu^{2+} and Pb^{2+}, respectively, suggesting a stronger sorption affinity of Cu^{2+} and Pb^{2+} towards RH-COF. Previous studies demonstrated that certain heavy metals can efficiently inhibit the adsorption of Cd^{2+} due to competition for available binding sites [52,53].

3.3. Possible Adsorption Mechanism

In this study, a series of advanced characteristic techniques, such as SEM, EDS, FT-IR, XPS, etc., were used while combining the analysis of surface functional groups and heavy metal chemical forms to elucidate the possible mechanisms of Cd^{2+} adsorption on RH and new RH-COF material.

The SEM-EDS spectra of RH and RH-COF after the adsorption of Cd^{2+} were thoroughly analyzed. The results (Figure 8) showed that the characteristic peaks of Cd appeared in the EDS spectra, confirming successful adsorption of Cd^{2+} on the surface of the adsorbents. Furthermore, SEM mapping was conducted on RH and RH-COF following the adsorption of Cd^{2+} to investigate the final form of the resultant products. Figure 8 verifies that Cd was effectively adsorbed by RH and RH-COF. Notably, the distribution of Cd in the mapping diagram was relatively uniform and closely coincides with that of O. This observation suggests that the oxygen-containing functional groups present on both RH and RH-COF may significantly facilitate the adsorption of Cd^{2+}.

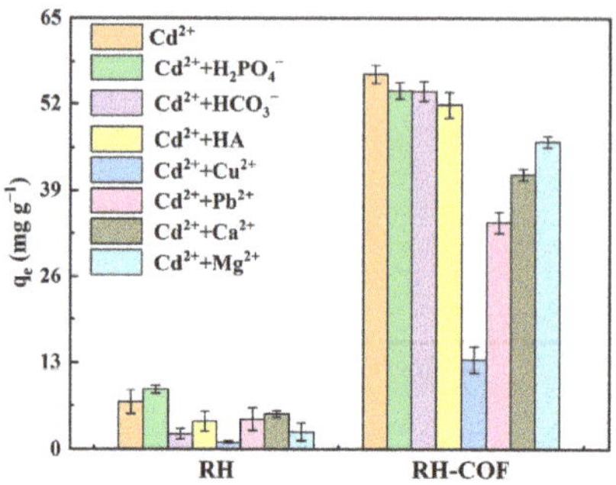

Figure 7. Effect of coexisting ions on the adsorption of Cd^{2+} by RH and RH-COF (initial concentration: 250 mg L^{-1} Cd^{2+}, 10^{-4} mol L^{-1} $H_2PO_4^-$, 10^{-3} mol L^{-1} HCO_3^-, 100 mg L^{-1} Cu^{2+}, 100 mg L^{-1} Pb^{2+}, 10^{-2} mol L^{-1} Ca^{2+}, 10^{-2} mol L^{-1} Mg^{2+}, and 10^{-3} mol L^{-1} HA, adsorbent dosage: 2.0 g L^{-1}, contact time: 48 h).

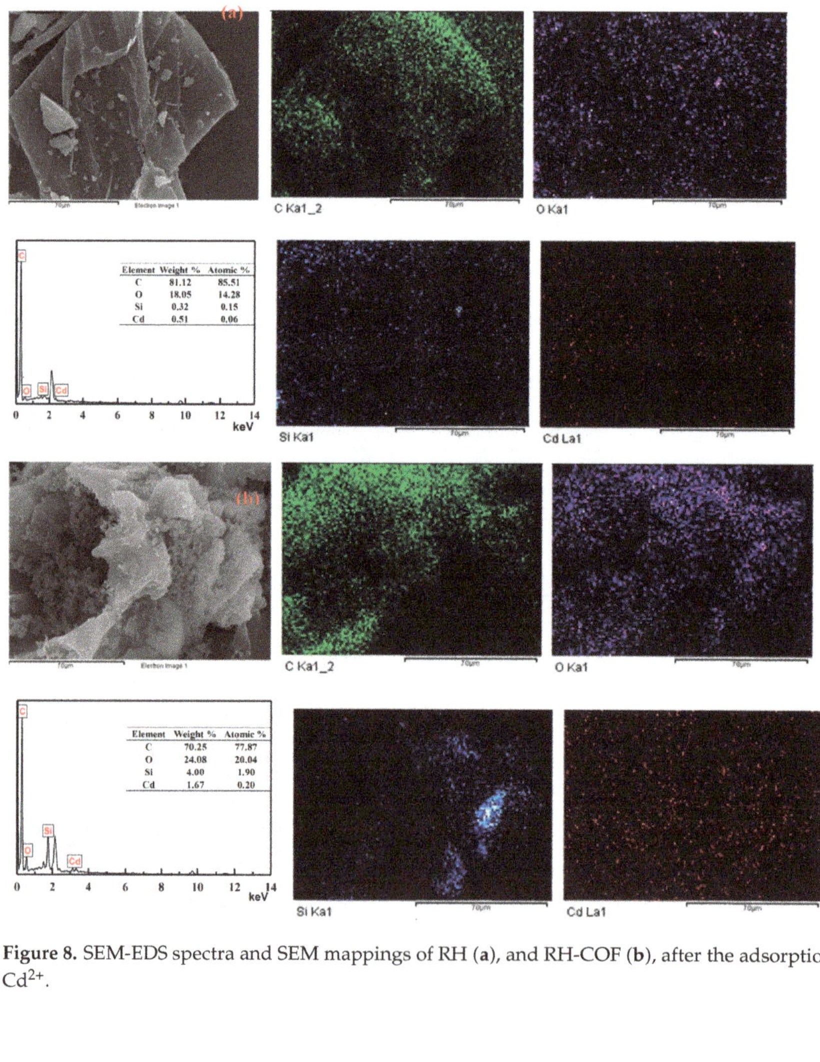

Figure 8. SEM-EDS spectra and SEM mappings of RH (**a**), and RH-COF (**b**), after the adsorption of Cd^{2+}.

The pHpzc values of RH and RH-COF were relatively low (Figure S1), resulting in the surface of both adsorbents being negatively charged in aqueous solutions, which facilitated the adsorption of Cd^{2+}. When the pH exceeded the pHpzc, deprotonation occurred on the surface of RH and RH-COF, allowing for electrostatic interactions that promote Cd^{2+} adsorption. No significant change in the zeta potential was observed for either adsorbent before or after adsorption, suggesting that electrostatic adsorption plays a role in the process, but its contribution is somewhat limited (Figure 9). Within a pH range of 3 to 8, the zeta potential of RH-COF was lower than that of RH, suggesting a significantly higher density of negative charges on the surface of RH-COF compared to RH, so the electrostatic interactions between RH-COF and Cd^{2+} were stronger. As illustrated in Figure 6a, when the pH exceeded 3, the capacity of Cd^{2+} adsorption by RH-COH surpassed that by RH. Furthermore, the equilibrium pH following Cd^{2+} adsorption on RH-COF is lower than that on RH, which indicates a higher concentration of H^+ in solution without occupying additional adsorption sites on the surface of RH-COF, consequently, more Cd^{2+} bonded with it by electrostatic interactions (Figure S2).

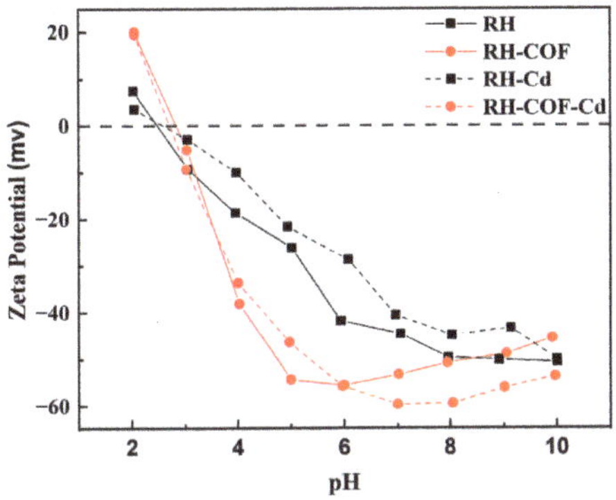

Figure 9. Zeta potential of RH and RH-COF before and after the adsorption of Cd^{2+}.

The FT-IR spectra of RH and RH-COF, both prior to and following the adsorption of Cd^{2+}, were recorded, with the results presented in Figure 10. After the adsorption of Cd^{2+}, the peak intensities of O-H (3416 cm^{-1}), C=O/C-C (1580 cm^{-1}), C-O (1099 cm^{-1}), and Si-O (803 cm^{-1} and 463 cm^{-1}) were markedly weaker and the peak of O-H (1438 cm^{-1}) disappeared compared with RH (Figure 10a). After the adsorption of Cd^{2+}, a reduction in the peak intensities for C=C/C=O (1593 cm^{-1}), C=C (1452 cm^{-1}), and C-N (1292 cm^{-1}) was observed along with a slight shift in their positions. The disappearance of the C=O peak at 1709 cm^{-1} was noted, along with new peaks of O-H (1384 cm^{-1} and 3422 cm^{-1}) emerging relative to RH-COF (Figure 10b). These findings indicated that carboxyl and hydroxyl groups might form complexes with Cd^{2+} on the surface of both RH and RH-COF. The absence of the peak corresponding to C=O indicated surface complexation of heavy metals through delocalized π electrons [51,54], which might be the reason for the disappearance of the peak of C=O (1709 cm^{-1}) on RH-COF. Following adsorption, it was observed that the peak intensities of Si-O (803 cm^{-1} and 463 cm^{-1}) were weaker than those of RH (Figure 10a), indicating that surface complexation between Si-O and Cd^{2+} occurred on the RH surface [45,55].

To elucidate the adsorption mechanism of Cd^{2+} by RH and RH-COF, XPS analysis was conducted on RH and RH-COF after Cd^{2+} adsorption, with results presented in Figure 11. The survey spectra of RH and RH-COF after the adsorption of Cd^{2+} showed the presence of new peaks corresponding to Cd 3d, indicating the successful adsorption of Cd^{2+} onto the surface of both materials (Figure 11a,b). Following adsorption, the binding energies for the $Cd3d_{5/2}$ and $Cd3d_{3/2}$ levels in RH were observed at 405.94 eV and 412.74 eV as well as at 406.52 eV and 413.32 eV, respectively (Figure 11c), indicating the formation of $CdCO_3$

and $Cd(OH)_2$ or Cd-O complexes [56–58]. This implied that Cd^{2+} is coordinated to oxygen functional groups (OFGs) on the surface of RH [57,58]. As shown in Figure 11d, the binding energies for $Cd3d_{5/2}$ (405.23 eV) and $Cd3d_{3/2}$ (412.03 eV) levels in RH-COF indicated a similar formation of Cd-O, $Cd(OH)_2$, and $CdCO_3$ [59,60], alongside the coordination interactions between amino and Cd^{2+} [58], or the chelation involving Cd ion and four 'O' [61,62].

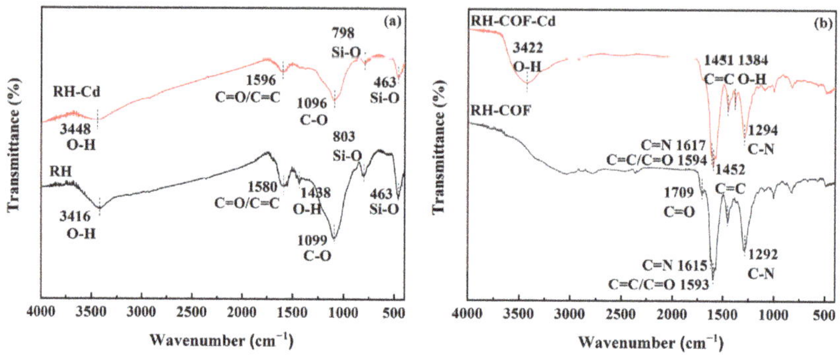

Figure 10. FT-IR spectra of RH (**a**) and RH-COF (**b**) before and after the adsorption of Cd^{2+}.

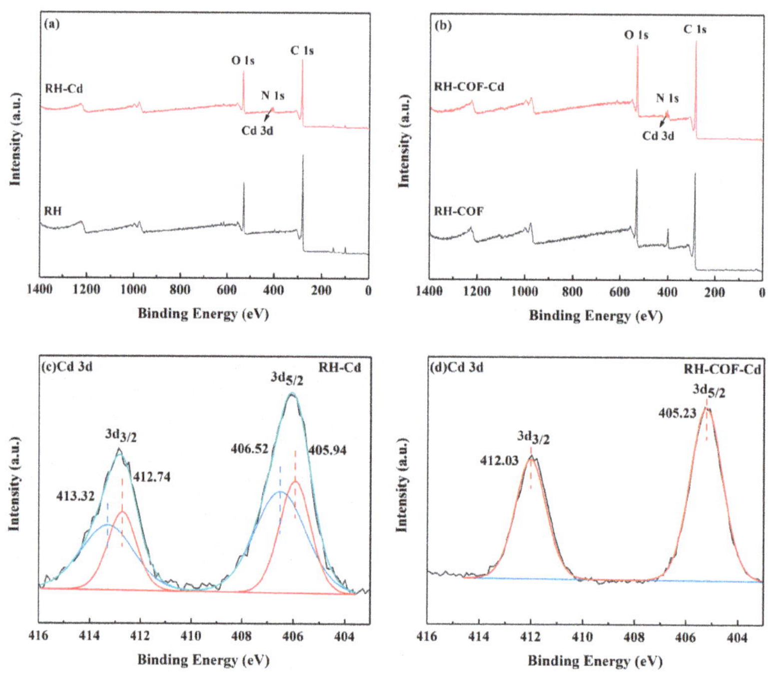

Figure 11. *Cont.*

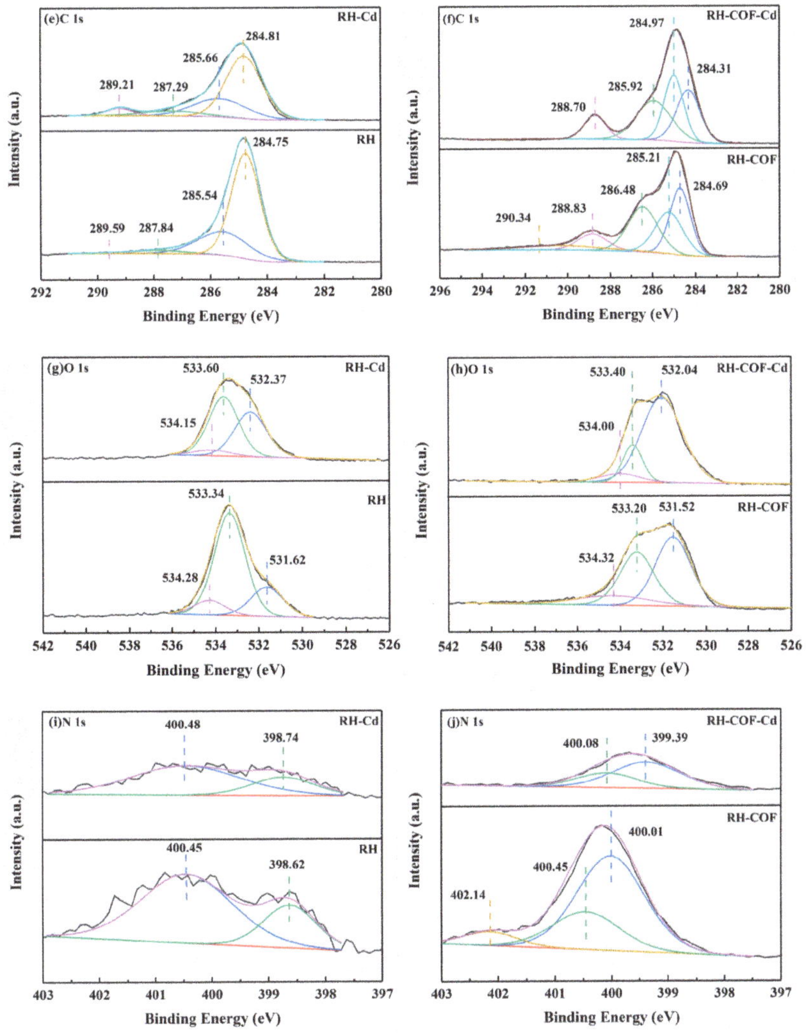

Figure 11. XPS spectra for RH (**a**) and RH-COF (**b**) before and after cadmium uptake, XPS spectra of Cd3d for RH (**c**) and RH-COF (**d**) after cadmium uptake, XPS spectra of C1s for RH (**e**) and RH-COF (**f**) before and after cadmium uptake, XPS spectra of O1s for RH (**g**) and RH-COF (**h**) before and after cadmium uptake, and XPS spectra of N1s for RH (**i**) and RH-COF (**j**) before and after cadmium uptake.

Following the adsorption of Cd^{2+}, a slight shift was observed in four characteristic peaks of RH in the C1s spectra (Figure 11e), including C-C/C=C (284.75 eV), C-O (285.54 eV), C=O (287.84 eV), and CO_3^{2-}/O=C-O (289.59 eV) [59,63–65]. The peak area corresponding to CO_3^{2-}/O=C-O increased from 0.64% to 5.25% (Table S3). This finding indicated that $CdCO_3$ precipitation may be formed after Cd^{2+} is adsorbed by RH. Three characteristic peaks of RH were identified in the O1s spectra, including C=O (531.62 eV), C-O/OH (533.34 eV), and O-C=O (534.28 eV) [63,66,67]. After the adsorption of Cd^{2+}, the binding energies of the peaks showed a slight shift, and the peak areas of C-O/OH and O-C=O decreased by 27.54% and 3.51%, respectively (Figure 11g) [63,66,67]. This indicated that the hydroxyl and carboxyl groups were involved in the adsorption of Cd^{2+}. After Cd^{2+}

adsorption of RH-COF, the binding energies of C=O (531.52 eV), C-O/OH (533.20 eV), and O-C=O (534.32 eV) exhibited minor shifts as well (Figure 11h) [63,66,67], concurrently, and the peak areas of C-O/OH and O-C=O decreased from 37.54% and 13.79% to 16.69% and 7.58%, respectively (Table S3). This further suggested that the hydroxyl and carboxyl groups contribute to the adsorption mechanism. At the same time, the peak of O-C=O in the C1s spectra disappeared (Figure 11f), indicating that carboxyl groups play a predominant role in the adsorption of Cd^{2+} by RH-COF.

Following adsorption, the binding energies and the peak area of N1s of RH exhibited no significant changes (Figure 11i), suggesting that the nitrogen-containing functional groups in RH are not involved in the adsorption process. As shown in Figure 11j, three characteristic peaks of RH-COF were identified in the N1s spectra, including $-NH_2$ (400.01 eV), C-N (400.45 eV), and C=N (402.14 eV) [67–70]. After the adsorption of Cd^{2+}, the binding energies of the peaks belonging to $-NH_2$ and C-N shifted to 399.39 eV and 400.08 eV, respectively, while the peak areas of $-NH_2$ and C-N increased by 0.73% and 6.83%, respectively. However, the peak of C=N disappeared. This indicated that the nitrogen-containing functional groups in RH-COF reacted with Cd^{2+}, and that amino functional groups may play an important role.

Based on the above results, it can be inferred that carboxyl, hydroxyl, and carbonate functional groups play an important role in the adsorption of Cd^{2+} on RH, primarily forming COO-Cd or Cd-O complexes and $CdCO_3$ precipitates. Furthermore, the oxygen-containing and nitrogen-containing functional groups present on the surface of RH-COF enhanced the adsorption capacity of Cd^{2+}.

In summary, the findings from SEM-EDS, zeta potential, FT-IR, and XPS analyses indicated that the potential adsorption mechanisms of Cd^{2+} by RH were dominated by surface complexation and surface precipitation, along with electrostatic adsorption. Conversely, the adsorption of RH-COF on Cd^{2+} was mainly caused by surface complexation and chelation, along with electrostatic adsorption, surface precipitation, and Cπ–cation interactions. The mechanism diagram is illustrated in Figure 12.

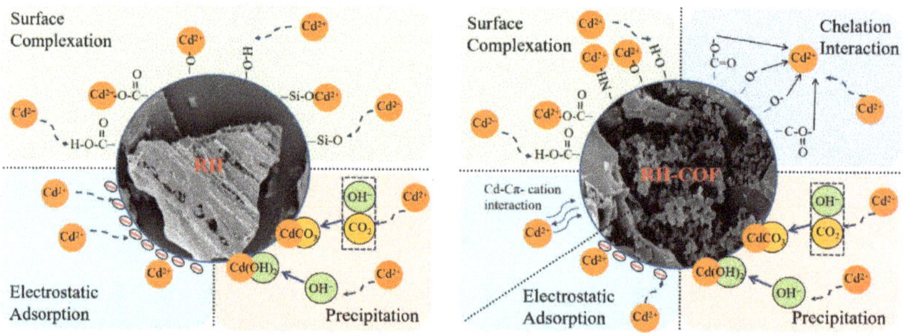

Figure 12. The mechanisms of Cd^{2+} adsorption onto RH and RH-COF.

4. Conclusions

In this work, rice husk biochar was used as the pristine material and a novel covalent organic framework-modified biochar was successfully synthesized via an in-situ polymerization method with Tp and DBd. The resultant COF-modified biochar composite was comprehensively characterized by elemental analysis, BET analysis, SEM, FT-IR, and XPS, confirming the successful modification of the COF on rice husk biochar. In comparison with RH, the pH decreased while the pHpzc value increased for RH-COF. Additionally, there was an enhancement in both the types and quantities of nitrogen- and oxygen-containing functional groups on the surface of RH-COF. The composite demonstrated efficient extraction and preconcentration of Cd^{2+} in aqueous solutions, with a maximum Cd^{2+} adsorption capacity of 58.62 mg g^{-1}, which was nearly 14 times greater than that of RH. Furthermore, RH-COF exhibited superior adsorption kinetics compared with RH, indicating that COF modification provided additional active adsorption sites for Cd^{2+}. The pseudo-second-order model and the Langmuir model effectively described both the kinetics and isotherms for Cd^{2+} adsorption onto RH-COF, and chemisorption on the monolayer surface played a dominant role in Cd^{2+} removal in this study. Notably, the adsorption performance of Cd^{2+} by RH-COF was hardly influenced by pH values ranging from 4.0 to 8.0. However, ionic strength could regulate the Cd^{2+} adsorption capacity of RH-COF slightly, while certain coexisting ions had more pronounced effects. Mechanistic analysis revealed that the adsorption of Cd^{2+} by RH-COF was dominated by surface complexation and chelation, along with electrostatic adsorption, surface precipitation, and Cπ–cation interactions. Overall, the findings from this study suggest that the synthesis of COF-biochar combined material could be a promising tool for remediation efforts targeting contaminated water.

Supplementary Materials: The following supporting information can be downloaded at https://www.mdpi.com/article/10.3390/toxics12100717/s1, Figure S1: Zeta Potential of RH and RH-COF; Figure S2: Equilibrium pH after adsorption of Cd^{2+} by RH and RH-COF at different initial solution pH; Table S1: Adsorption kinetic parameters of Cd^{2+} on RH and RH-COF (initial concentration of Cd^{2+}: 100 mg L^{-1} (RH) and 250 mg L^{-1} (RH-COF), adsorbent dosage: 2.0 g L^{-1}, pH 5); Table S2: Isotherm parameters of Langmuir and Freundlich for the adsorption of Cd^{2+} on RH and RH-COF (adsorbent dosage: 2.0 g L^{-1}, pH 5); Table S3: The Relative ratio of peaks in C1s, N1s and O1s from XPS.

Author Contributions: Investigation, Y.H. and S.L.; Formal analysis, Y.H., Y.Z. and C.C.; Resources, J.F.; Writing—original draft, Y.H.; Writing—review and editing, G.J. and C.C.; Methodology, C.C. All authors have read and agreed to the published version of the manuscript.

Funding: The research received funding from the Fujian Natural Science Foundation Project (2021J01314), National Key R&D Project (2023YFC3709700), Strategic Priority Research Program of the Chinese Academy of Sciences, China (XDA23030202).

Institutional Review Board Statement: Not applicable.

Informed Consent Statement: Not applicable.

Data Availability Statement: Dataset available on request from the authors.

Conflicts of Interest: The authors declare that they have no known competing financial interests or personal relationships that could have appeared to influence the work reported in this paper.

References

1. Khan, N.; Chowdhary, P.; Gnansounou, E.; Chaturvedi, P. Biochar and environmental sustainability: Emerging trends and techno-economic perspectives. *Bioresour. Technol.* **2021**, *332*, 125102. [CrossRef] [PubMed]
2. Tang, J.; Zhao, B.; Lyu, H.; Li, D. Development of a novel pyrite/biochar composite (BM-FeS_2@ BC) by ball milling for aqueous Cr (VI) removal and its mechanisms. *J. Hazard. Mater.* **2021**, *413*, 125415. [CrossRef] [PubMed]
3. Wang, T.; Li, G.; Yang, K.; Zhang, X.; Wang, K.; Cai, J.; Zheng, J. Enhanced ammonium removal on biochar from a new forestry waste by ultrasonic activation: Characteristics, mechanisms and evaluation. *Sci. Total Environ.* **2021**, *778*, 146295. [CrossRef]
4. Cho, E.-J.; Kang, J.-K.; Lee, C.-G.; Bae, S.; Park, S.-J. Use of thermally activated Fenton sludge for Cd removal in zinc smelter wastewater: Mechanism and feasibility of Cd removal. *Environ. Pollut.* **2023**, *334*, 122166. [CrossRef]
5. Yang, H.; Choi, G.-R.; Jeong, Y.J.; Cho, I.S.; Park, S.-J.; Lee, C.-G. Enhancing acetaminophen removal through persulfate activation with $ZnCl_2$-SPI biochar: A study on reactive oxygen species contribution according to acetaminophen concentration. *Chem. Eng. J.* **2024**, *496*, 154065. [CrossRef]
6. Higashikawa, F.S.; Conz, R.F.; Colzato, M.; Cerri, C.E.P.; Alleoni, L.R.F. Effects of feedstock type and slow pyrolysis temperature in the production of biochars on the removal of cadmium and nickel from water. *J. Clean. Prod.* **2016**, *137*, 965–972. [CrossRef]
7. Ahmad, M.; Rajapaksha, A.U.; Lim, J.E.; Zhang, M.; Bolan, N.; Mohan, D.; Vithanage, M.; Lee, S.S.; Ok, Y.S. Biochar as a sorbent for contaminant management in soil and water: A review. *Chemosphere* **2014**, *99*, 19–33. [CrossRef] [PubMed]
8. Mohan, D.; Sarswat, A.; Ok, Y.S.; Pittman, C.U., Jr. Organic and inorganic contaminants removal from water with biochar, a renewable, low cost and sustainable adsorbent—A critical review. *Bioresour. Technol.* **2014**, *160*, 191–202. [CrossRef] [PubMed]
9. Park, J.H.; Choppala, G.; Lee, S.J.; Bolan, N.; Chung, J.W.; Edraki, M. Comparative sorption of Pb and Cd by biochars and its implication for metal immobilization in soils. *Water Air Soil. Pollut.* **2013**, *224*, 1711. [CrossRef]
10. Xu, X.; Cao, X.; Zhao, L. Comparison of rice husk-and dairy manure-derived biochars for simultaneously removing heavy metals from aqueous solutions: Role of mineral components in biochars. *Chemosphere* **2013**, *92*, 955–961. [CrossRef]

11. Genchi, G.; Sinicropi, M.S.; Lauria, G.; Carocci, A.; Catalano, A. The effects of cadmium toxicity. *Int. J. Environ. Res. Public Health* **2020**, *17*, 3782. [CrossRef] [PubMed]
12. Deng, J.; Liu, Y.; Liu, S.; Zeng, G.; Tan, X.; Huang, B.; Tang, X.; Wang, S.; Hua, Q.; Yan, Z. Competitive adsorption of Pb (II), Cd (II) and Cu (II) onto chitosan-pyromellitic dianhydride modified biochar. *J. Colloid. Interface Sci.* **2017**, *506*, 355–364. [CrossRef]
13. Chen, D.; Wang, X.; Wang, X.; Feng, K.; Su, J.; Dong, J. The mechanism of cadmium sorption by sulphur-modified wheat straw biochar and its application cadmium-contaminated soil. *Sci. Total Environ.* **2020**, *714*, 136550. [CrossRef]
14. Fan, J.; Cai, C.; Chi, H.; Reid, B.J.; Coulon, F.; Zhang, Y.; Hou, Y. Remediation of cadmium and lead polluted soil using thiol-modified biochar. *J. Hazard. Mater.* **2020**, *388*, 122037. [CrossRef]
15. Lee, H.-S.; Shin, H.-S. Competitive adsorption of heavy metals onto modified biochars: Comparison of biochar properties and modification methods. *J. Environ. Manag.* **2021**, *299*, 113651. [CrossRef] [PubMed]
16. Lu, H.; Wang, C.; Chen, J.; Ge, R.; Leng, W.; Dong, B.; Huang, J.; Gao, Y. A novel 3D covalent organic framework membrane grown on a porous α-Al 2 O 3 substrate under solvothermal conditions. *Chem. Commun.* **2015**, *51*, 15562–15565. [CrossRef]
17. Zhang, N.; Ishag, A.; Li, Y.; Wang, H.; Guo, H.; Mei, P.; Meng, Q.; Sun, Y. Recent investigations and progress in environmental remediation by using covalent organic framework-based adsorption method: A review. *J. Clean. Prod.* **2020**, *277*, 123360. [CrossRef]
18. Ahmed, I.; Jhung, S.H. Covalent organic framework-based materials: Synthesis, modification, and application in environmental remediation. *Coord. Chem. Rev.* **2021**, *441*, 213989. [CrossRef]
19. Li, W.-T.; Shi, W.; Hu, Z.-J.; Yang, T.; Chen, M.-L.; Zhao, B.; Wang, J.-H. Fabrication of magnetic Fe_3O_4@ metal organic framework@ covalent organic framework composite and its selective separation of trace copper. *Appl. Surf. Sci.* **2020**, *530*, 147254. [CrossRef]
20. Zhu, R.; Zhang, P.; Zhang, X.; Yang, M.; Zhao, R.; Liu, W.; Li, Z. Fabrication of synergistic sites on an oxygen-rich covalent organic framework for efficient removal of Cd (II) and Pb (II) from water. *J. Hazard. Mater.* **2022**, *424*, 127301. [CrossRef]
21. Lu, X.-F.; Ji, W.-H.; Yuan, L.; Yu, S.; Guo, D.-S. Preparation of Carboxy-Functionalized Covalent Organic Framework for Efficient Removal of Hg^{2+} and Pb^{2+} from Water. *Ind. Eng. Chem. Res.* **2019**, *58*, 17660–17667. [CrossRef]
22. Wen, R.; Li, Y.; Zhang, M.; Guo, X.; Li, X.; Li, X.; Han, J.; Hu, S.; Tan, W.; Ma, L. Graphene-synergized 2D covalent organic framework for adsorption: A mutual promotion strategy to achieve stabilization and functionalization simultaneously. *J. Hazard. Mater.* **2018**, *358*, 273–285. [CrossRef]
23. Sun, Q.; Aguila, B.; Perman, J.; Earl, L.D.; Abney, C.W.; Cheng, Y.; Wei, H.; Nguyen, N.; Wojtas, L.; Ma, S. Postsynthetically modified covalent organic frameworks for efficient and effective mercury removal. *J. Am. Chem. Soc.* **2017**, *139*, 2786–2793. [CrossRef]
24. Jiang, Y.; Liu, C.; Huang, A. EDTA-functionalized covalent organic framework for the removal of heavy-metal ions. *ACS Appl. Mater. Interfaces* **2019**, *11*, 32186–32191. [CrossRef]
25. Yang, C.-H.; Chang, J.-S.; Lee, D.-J. Covalent organic framework EB-COF: Br as adsorbent for phosphorus (V) or arsenic (V) removal from nearly neutral waters. *Chemosphere* **2020**, *253*, 126736. [CrossRef] [PubMed]
26. Li, Y.; Wang, C.; Ma, S.; Zhang, H.; Ou, J.; Wei, Y.; Ye, M. Fabrication of hydrazone-linked covalent organic frameworks using alkyl amine as building block for high adsorption capacity of metal ions. *ACS Appl. Mater. Interfaces* **2019**, *11*, 11706–11714. [CrossRef]
27. Zhang, Y.; Fan, J.; Fu, M.; Ok, Y.S.; Hou, Y.; Cai, C. Adsorption antagonism and synergy of arsenate (V) and cadmium (II) onto Fe-modified rice straw biochars. *Environ. Geochem. Health* **2019**, *41*, 1755–1766. [CrossRef] [PubMed]
28. Zhang, J.; Shao, J.; Jin, Q.; Li, Z.; Zhang, X.; Chen, Y.; Zhang, S.; Chen, H. Sludge-based biochar activation to enhance Pb (II) adsorption. *Fuel* **2019**, *252*, 101–108. [CrossRef]
29. Ramola, S.; Belwal, T.; Li, C.J.; Liu, Y.X.; Wang, Y.Y.; Yang, S.M.; Zhou, C.H. Preparation and application of novel rice husk biochar–calcite composites for phosphate removal from aqueous medium. *J. Clean. Prod.* **2021**, *299*, 126802. [CrossRef]
30. Gao, R.; Fu, Q.; Hu, H.; Wang, Q.; Liu, Y.; Zhu, J. Highly-effective removal of Pb by co-pyrolysis biochar derived from rape straw and orthophosphate. *J. Hazard. Mater.* **2019**, *371*, 191–197. [CrossRef]
31. Yakout, S.M. Monitoring the changes of chemical properties of rice straw–derived biochars modified by different oxidizing agents and their adsorptive performance for organics. *Bioremediation J.* **2015**, *19*, 171–182. [CrossRef]
32. Fu, D.; He, Z.; Su, S.; Xu, B.; Liu, Y.; Zhao, Y. Fabrication of α-FeOOH decorated graphene oxide-carbon nanotubes aerogel and its application in adsorption of arsenic species. *J. Colloid. Interface Sci.* **2017**, *505*, 105–114. [CrossRef] [PubMed]
33. Hu, S.; Xiang, J.; Sun, L.; Xu, M.; Qiu, J.; Fu, P. Characterization of char from rapid pyrolysis of rice husk. *Fuel Process. Technol.* **2008**, *89*, 1096–1105. [CrossRef]
34. Liou, T.-H. Evolution of chemistry and morphology during the carbonization and combustion of rice husk. *Carbon* **2004**, *42*, 785–794. [CrossRef]
35. Yang, K.; Wang, X.; Cheng, H.; Tao, S. Enhanced immobilization of cadmium and lead adsorbed on crop straw biochars by simulated aging processes. *Environ. Pollut.* **2022**, *302*, 119064. [CrossRef]
36. Wang, Y.; Wang, L.; Li, Z.; Yang, D.; Xu, J.; Liu, X. MgO-laden biochar enhances the immobilization of Cd/Pb in aqueous solution and contaminated soil. *Biochar* **2021**, *3*, 175–188. [CrossRef]
37. Fan, S.; Wang, Y.; Wang, Z.; Tang, J.; Tang, J.; Li, X. Removal of methylene blue from aqueous solution by sewage sludge-derived biochar: Adsorption kinetics, equilibrium, thermodynamics and mechanism. *J. Environ. Chem. Eng.* **2017**, *5*, 601–611. [CrossRef]
38. Lyu, H.; Gao, B.; He, F.; Zimmerman, A.R.; Ding, C.; Huang, H.; Tang, J. Effects of ball milling on the physicochemical and sorptive properties of biochar: Experimental observations and governing mechanisms. *Environ. Pollut.* **2018**, *233*, 54–63. [CrossRef]

39. Wang, R.; Shi, X.; Xiao, A.; Zhou, W.; Wang, Y. Interfacial polymerization of covalent organic frameworks (COFs) on polymeric substrates for molecular separations. *J. Membr. Sci.* **2018**, *566*, 197–204. [CrossRef]
40. Yue, J.-Y.; Wang, L.; Ma, Y.; Yang, P.; Zhang, Y.-Q.; Jiang, Y.; Tang, B. Metal ion-assisted carboxyl-containing covalent organic frameworks for the efficient removal of Congo red. *Dalton Trans.* **2019**, *48*, 17763–17769. [CrossRef]
41. Lai, L.; Chen, L.; Zhan, D.; Sun, L.; Liu, J.; Lim, S.H.; Poh, C.K.; Shen, Z.; Lin, J. One-step synthesis of NH2-graphene from in situ graphene-oxide reduction and its improved electrochemical properties. *Carbon* **2011**, *49*, 3250–3257. [CrossRef]
42. Dinari, M.; Hatami, M. Novel N-riched crystalline covalent organic framework as a highly porous adsorbent for effective cadmium removal. *J. Environ. Chem. Eng.* **2019**, *7*, 102907. [CrossRef]
43. Li, Y.; Yang, C.-X.; Qian, H.-L.; Zhao, X.; Yan, X.-P. Carboxyl-Functionalized Covalent Organic Frameworks for the Adsorption and Removal of Triphenylmethane Dyes. *ACS Appl. Nano Mater.* **2019**, *2*, 7290–7298. [CrossRef]
44. Febrianto, J.; Kosasih, A.N.; Sunarso, J.; Ju, Y.-H.; Indraswati, N.; Ismadji, S. Equilibrium and kinetic studies in adsorption of heavy metals using biosorbent: A summary of recent studies. *J. Hazard. Mater.* **2009**, *162*, 616–645. [CrossRef]
45. Xiang, J.; Lin, Q.; Cheng, S.; Guo, J.; Yao, X.; Liu, Q.; Yin, G.; Liu, D. Enhanced adsorption of Cd (II) from aqueous solution by a magnesium oxide–rice husk biochar composite. *Environ. Sci. Pollut. Res.* **2018**, *25*, 14032–14042. [CrossRef]
46. Wu, J.; Huang, D.; Liu, X.; Meng, J.; Tang, C.; Xu, J. Remediation of As (III) and Cd (II) co-contamination and its mechanism in aqueous systems by a novel calcium-based magnetic biochar. *J. Hazard. Mater.* **2018**, *348*, 10–19. [CrossRef] [PubMed]
47. Kumar, U.; Bandyopadhyay, M. Sorption of cadmium from aqueous solution using pretreated rice husk. *Bioresour. Technol.* **2006**, *97*, 104–109. [CrossRef]
48. Li, A.; Zhang, Y.; Ge, W.; Zhang, Y.; Liu, L.; Qiu, G. Removal of heavy metals from wastewaters with biochar pyrolyzed from MgAl-layered double hydroxide-coated rice husk: Mechanism and application. *Bioresour. Technol.* **2022**, *347*, 126425. [CrossRef]
49. Huang, F.; Gao, L.-Y.; Wu, R.-R.; Wang, H.; Xiao, R.-B. Qualitative and quantitative characterization of adsorption mechanisms for Cd2+ by silicon-rich biochar. *Sci. Total Environ.* **2020**, *731*, 139163. [CrossRef]
50. Pehlivan, E.; Yanık, B.; Ahmetli, G.; Pehlivan, M. Equilibrium isotherm studies for the uptake of cadmium and lead ions onto sugar beet pulp. *Bioresour. Technol.* **2008**, *99*, 3520–3527. [CrossRef]
51. Tran, H.N.; You, S.-J.; Chao, H.-P. Effect of pyrolysis temperatures and times on the adsorption of cadmium onto orange peel derived biochar. *Waste Manag. Res.* **2016**, *34*, 129–138. [CrossRef] [PubMed]
52. Selim, H.; Zhang, H. Modeling approaches of competitive sorption and transport of trace metals and metalloids in soils: A review. *J. Environ. Qual.* **2013**, *42*, 640–653. [CrossRef]
53. Qi, X.; Yin, H.; Zhu, M.; Yu, X.; Shao, P.; Dang, Z. MgO-loaded nitrogen and phosphorus self-doped biochar: High-efficient adsorption of aquatic Cu2+, Cd2+, and Pb2+ and its remediation efficiency on heavy metal contaminated soil. *Chemosphere* **2022**, *294*, 133733. [CrossRef]
54. Xu, X.; Cao, X.; Zhao, L.; Wang, H.; Yu, H.; Gao, B. Removal of Cu, Zn, and Cd from aqueous solutions by the dairy manure-derived biochar. *Environ. Sci. Pollut. Res.* **2013**, *20*, 358–368. [CrossRef] [PubMed]
55. Wang, H.; Wang, X.; Ma, J.; Xia, P.; Zhao, J. Removal of cadmium (II) from aqueous solution: A comparative study of raw attapulgite clay and a reusable waste–struvite/attapulgite obtained from nutrient-rich wastewater. *J. Hazard. Mater.* **2017**, *329*, 66–76. [CrossRef] [PubMed]
56. Chi, T.; Zuo, J.; Liu, F. Performance and mechanism for cadmium and lead adsorption from water and soil by corn straw biochar. *Front. Environ. Sci. Eng.* **2017**, *11*, 15. [CrossRef]
57. Liu, M.; Liu, Y.; Shen, J.; Zhang, S.; Liu, X.; Chen, X.; Ma, Y.; Ren, S.; Fang, G.; Li, S. Simultaneous removal of Pb^{2+}, Cu^{2+} and Cd^{2+} ions from wastewater using hierarchical porous polyacrylic acid grafted with lignin. *J. Hazard. Mater.* **2020**, *392*, 122208. [CrossRef]
58. Liang, X.; Han, J.; Xu, Y.; Wang, L.; Sun, Y.; Tan, X. Sorption of Cd^{2+} on mercapto and amino functionalized palygorskite. *Appl. Surf. Sci.* **2014**, *322*, 194–201. [CrossRef]
59. Liang, X.; Han, J.; Xu, Y.; Sun, Y.; Wang, L.; Tan, X. In situ field-scale remediation of Cd polluted paddy soil using sepiolite and palygorskite. *Geoderma* **2014**, *235*, 9–18. [CrossRef]
60. Ramrakhiani, L.; Ghosh, S.; Mandal, A.K.; Majumdar, S. Utilization of multi-metal laden spent biosorbent for removal of glyphosate herbicide from aqueous solution and its mechanism elucidation. *Chem. Eng. J.* **2019**, *361*, 1063–1077. [CrossRef]
61. Shi, M.; Zhao, Z.; Song, Y.; Xu, M.; Li, J.; Yao, L. A novel heat-treated humic acid/MgAl-layered double hydroxide composite for efficient removal of cadmium: Fabrication, performance and mechanisms. *Appl. Clay Sci.* **2020**, *187*, 105482. [CrossRef]
62. Cui, S.; Zhang, R.; Peng, Y.; Gao, X.; Li, Z.; Fan, B.; Guan, C.-Y.; Beiyuan, J.; Zhou, Y.; Liu, J. New insights into ball milling effects on MgAl-LDHs exfoliation on biochar support: A case study for cadmium adsorption. *J. Hazard. Mater.* **2021**, *416*, 126258. [CrossRef] [PubMed]
63. Zhang, F.; Wang, X.; Yin, D.; Peng, B.; Tan, C.; Liu, Y.; Tan, X.; Wu, S. Efficiency and mechanisms of Cd removal from aqueous solution by biochar derived from water hyacinth (*Eichornia crassipes*). *J. Environ. Manag.* **2015**, *153*, 68–73. [CrossRef]
64. Wang, R.-Z.; Huang, D.-L.; Liu, Y.-G.; Zhang, C.; Lai, C.; Zeng, G.-M.; Cheng, M.; Gong, X.-M.; Wan, J.; Luo, H. Investigating the adsorption behavior and the relative distribution of Cd^{2+} sorption mechanisms on biochars by different feedstock. *Bioresour. Technol.* **2018**, *261*, 265–271. [CrossRef]

65. Cai, W.; Wei, J.; Li, Z.; Liu, Y.; Zhou, J.; Han, B. Preparation of amino-functionalized magnetic biochar with excellent adsorption performance for Cr (VI) by a mild one-step hydrothermal method from peanut hull. *Colloids Surf. Physicochem. Eng. Asp.* **2019**, *563*, 102–111. [CrossRef]
66. Wu, J.; Wang, T.; Zhang, Y.; Pan, W.-P. The distribution of Pb (II)/Cd (II) adsorption mechanisms on biochars from aqueous solution: Considering the increased oxygen functional groups by HCl treatment. *Bioresour. Technol.* **2019**, *291*, 121859. [CrossRef]
67. Zhu, L.; Tong, L.; Zhao, N.; Wang, X.; Yang, X.; Lv, Y. Key factors and microscopic mechanisms controlling adsorption of cadmium by surface oxidized and aminated biochars. *J. Hazard. Mater.* **2020**, *382*, 121002. [CrossRef]
68. Jevremović, A.; Bober, P.; Mičušík, M.; Kuliček, J.; Acharya, U.; Pfleger, J.; Milojević-Rakić, M.; Krajišnik, D.; Trchova, M.; Stejskal, J. Synthesis and characterization of polyaniline/BEA zeolite composites and their application in nicosulfuron adsorption. *Microporous Mesoporous Mater.* **2019**, *287*, 234–245. [CrossRef]
69. Jiang, C.; Sun, R.; Du, Z.; Singh, V.; Chen, S. A cationic Zr-based metal organic framework with enhanced acidic resistance for selective and efficient removal of CrO_4^{2-}. *New J. Chem.* **2020**, *44*, 12646–12653. [CrossRef]
70. Oh, Y.; Le, V.-D.; Maiti, U.N.; Hwang, J.O.; Park, W.J.; Lim, J.; Lee, K.E.; Bae, Y.-S.; Kim, Y.-H.; Kim, S.O. Selective and regenerative carbon dioxide capture by highly polarizing porous carbon nitride. *ACS Nano* **2015**, *9*, 9148–9157. [CrossRef]

toxics

MDPI

Article

Influence of Tetrabromobisphenol-A on the Fate and Behavior of Zinc Oxide Nanoparticles Affected by Salts, Humic Acid, and Bovine Serum Albumin in Water Systems

Anwar Ul Haq Khan [1,2], Yanju Liu [1,2,*], Ravi Naidu [1,2], Cheng Fang [1,2] and Ho Kyong Shon [3]

1 Global Centre for Environmental Remediation (GCER), College of Engineering Science and Environment, The University of Newcastle, Callaghan, NSW 2308, Australia; anwar.khan@uon.edu.au (A.U.H.K.); ravi.naidu@newcastle.edu.au (R.N.); cheng.fang@newcastle.edu.au (C.F.)
2 crc for Contamination Assessment and Remediation of the Environment (crcCARE), ATC Building, The University of Newcastle, Callaghan, NSW 2308, Australia
3 School of Civil and Environmental Engineering, University of Technology Sydney (UTS), City Campus, Broadway, NSW 2007, Australia; hokyong.shon-1@uts.edu.au
* Correspondence: yanju.liu@newcastle.edu.au; Tel.: +61-(02)-4913-8738

Abstract: The environmental release of zinc oxide nanoparticles (ZnO-NPs) may have consequences for ecosystems. The behavior and environmental effects of ZnO-NPs could change due to their interactions with other existing substances. This research explored how the presence of coexisting organic pollutants (like tetrabromobisphenol-A (TBBPA)), electrolytes (such as NaCl and $CaCl_2$), natural organic materials (including humic acid (HA)), and bovine serum albumin (BSA) in simulated water affected the behavior of ZnO-NPs. Various characterization techniques were used to analyze the size, shape, purity, crystallinity, and surface charge of ZnO-NPs following interactions (after one day, one week, two weeks, and three weeks) at pH 7. The findings demonstrated changes in both the size and zeta potential of the ZnO-NPs in isolation and when TBBPA and electrolytes were included in the suspension. The size and surface charge exhibited different variations across fixed concentrations (5 mM) of various electrolytes. HA and BSA contributed to the dispersion of ZnO-NPs by affecting the zeta potential. These dispersion effects were also observed in the presence of TBBPA and salts, attributed to their substantial aliphatic carbon content and complex structures. Potential interaction forces that could explain the adsorption of TBBPA include cation bridging, hydrophobic interactions, hydrogen bonding, electrostatic interactions, and van der Waals forces. The co-occurrence of organic pollutants (TBBPA) and natural organic compounds (HA and BSA) can alter the surface properties and behavior of ZnO-NPs in natural and seawater, aiding in the understanding of the fate and impact of engineered nanoparticles (such as ZnO-NPs) in the environment.

Keywords: zinc oxide nanoparticles; tetrabromobisphenol-A; salts; adsorption; zeta potential; aggregation

1. Introduction

Zinc oxide nanoparticles (ZnO-NPs), a type of inorganic mineral filter, are widely incorporated into various consumer products, such as cosmetics, paints, electronics, and textiles. Their popularity stems from their ability to effectively absorb and reflect ultraviolet (UV) radiation, providing protection and enhancing product durability [1–3]. ZnO-NPs are commonly produced inorganic substances, having around 10 million tons of total worldwide output [4,5]. ZnO-NPs rank as the third most widely produced metal-based

https://doi.org/10.3390/toxics13030148

ENPs worldwide, following silicon dioxide (SiO_2) and titanium dioxide (TiO_2), with annual production estimates ranging from 550 to 33,400 tons. Their release into water, soil, and sediments raises concerns about potential environmental risks, as their nanoscale size, high surface-area-to-volume ratio, and toxic properties can negatively impact various organisms, including plants, animals, microbes, and humans [6–8]. Adverse effects on marine life, such as sea urchins, have been documented [9], and daphnia [10], mammals [11], earthworms [12], marine diatoms (*Thalassiosira pseudonana*) [13], and plants [14] have also been reported. Likewise, the accumulation and harmful effects of ZnO-NPs from sunscreens in seawater have been documented [15]. Studies show that exposure to these nanoparticles can lead to growth inhibition, DNA damage, and oxidative stress in marine algae [15].

Various modeling studies based on material flow suggest that ZnO-NPs are released into surface water at concentrations ranging from 0.008 to 0.055 μg/L, while wastewater treatment plant effluents contain between 0.34 and 1.42 μg/L [16,17]. Estimated concentrations of ZnO-NPs in different natural environments have also been reported, including sediment (1.8–5.7 μg/kg/y), soil (6.8–22.3 μg/kg/y), and sludge (136–647 μg/kg/y) [18]. In the United States, wastewater treatment influent has been found to contain ZnO-NPs at levels between 20 and 212 μg/L [19]. Additionally, studies have detected the release of engineered nanoparticles (ENPs), including ZnO-NPs, from sunscreens into swimming pool water [20]. It is estimated that around 10–25% of manufactured ZnO-NPs may enter the environment and accumulate in freshwater systems [21]. The release of ZnO-NPs up to 0.05–10 μg/L (estimated based on model studies) in surface waters in the United States has been investigated [22].

The physical, chemical, and eco-toxicological behaviors of ZnO-NPs are critical for risk assessment upon their release into natural environments (e.g., recreational and swimming waters, wastewater and seawater bodies). Environmental factors, such as electrolytes, pH, organic and inorganic compounds, polymers, light, and heat, can significantly influence the behavior and toxicity of ZnO-NPs [23–25]. These interactions can have a substantial impact on their fate and behavior. The presence of proteins, such as BSA, humic substances, ultraviolet radiation, and salinity, can affect their interaction mechanisms [25–27]. For instance, organic pollutants, like brominated flame retardants, including polybrominated diphenyl ethers and hexabromocyclododecane, influence the physicochemical properties of ZnO-NPs in aqueous environments under certain conditions [28,29]. However, the potential formation of complex compounds through interactions with metal oxides and polymeric substances has not been accounted for. Studies indicate that upon interacting with HA, ZnO-NPs undergo dispersion and exhibit alterations in surface charge [30]. Electrolytes influence the stability of ZnO-NPs, leading to agglomeration at high salt concentrations due to electrical double-layer compression and reduced energy barriers [31–33]. The electrostatic attraction between BSA and the ZnO-NPs surface promotes adsorption, leading to reduced agglomeration and flocculation [34]. The hydrophobic nature of organic pollutants assists them in interacting with and sorbing onto the active sites of ZnO-NPs in aquatic systems. This process is driven by electrostatic and hydrophobic interactions, π–π stacking, van der Waals forces, ligand exchange, hydrogen bonding, and molecular bridging effects [35,36].

TBBPA remains one of the most widely used brominated flame retardants. In 2016, around 241,352 tons were produced, primarily in China, the USA, and the Middle East [37]. China produces approximately 180,000 tons of TBBPA annually [37]. TBBPA is primarily used in epoxy resins and polycarbonate, which are found in products such as electronics, furniture, keyboards, and other items [38]. TBBPA has been detected in indoor air and dust [39,40], sediments [41], soils [42], water [43], and sewage sludge, leading to its presence in the food chain [44]. TBBPA levels have been reported in various water bodies worldwide. In the River Skerne, a tributary of the River Tees in England, concentrations ranged from undetectable to 9800 ng/g dw. An unidentified river in the Netherlands had concentrations between 0.1 and 130 ng/g dw [45], while levels in South Korea's Nakdong River varied from 0.05 to 150 ng/g dw [46]. In eastern and southern China, TBBPA was found at concentrations ranging from below detection limits to 4870 ng/L [47]. Similarly, in the USA, measurements in the Detroit River and other industrial areas revealed concentrations between 600 and 1840 ng/L [45]. In China, TBBPA levels of <0.4 to 259 ng/g dry weight were detected in 52 sludge samples from 30 wastewater treatment plants [48].

The release of TBBPA and its derivatives leads to widespread contamination through multiple pathways, particularly when TBBPA interacts with co-contaminants. The co-presence of TBBPA with other compounds in water may impact the fate and behavior of zinc oxide NPs, but knowledge of the interaction mechanisms involved is currently lacking [25,28,49]. This research highlighted altered behavior of ZnO-NPs when interacting with TBBPA, an organic contaminant, under different environmentally relevant conditions in the water. Specifically, the study evaluated the changes in the physicochemical properties, particle size colloidal stability, and ZnO-NPs' zeta potential before and after they interacted with TBBPA in simulated aquatic environments containing electrolytes, bovine serum albumin (BSA), and humic acid (HA). The findings are crucial for evaluating the potential exposure to ZnO-NPs and associated contaminants under specific water conditions.

2. Materials and Methods

2.1. Chemicals

ZnO-NPs and electrolytes were acquired from Sigma-Aldrich Australia (Melbourne), and their properties were detailed in a previous study [28]. Briefly, most nanoparticles were less than 100 nm in diameter, with some particles reaching ≥100 nm due to agglomeration. X-ray diffraction (XRD) analysis confirmed that these samples exhibited a hexagonal wurtzite structure. TBBPA (3,3′,5,5–tetrabromobisphenol-A, 97%, CAS: 79-94-7, $C_{15}H_{12}Br_4O_2$, MW: 543.87 g/mol, mp: 178–181 °C), HA (humic acid technical, CAS: 53680-50G), and BSA (bovine serum albumin lyophilized powder, ≥96%, CAS: 9048-46-8, water soluble at 40 mg/mL for agarose gel electrophoresis) were also purchased from Sigma-Aldrich Australia and used in the study.

2.2. Interaction Between ZnO-NPs and TBBPA

To prepare the stock suspension of ZnO-NPs, 0.1 g of ZnO-NPs was added to 1 L of Milli-Q water and sonicated for 10 min. Different TBBPA concentrations were synthesized in Milli-Q water with the ZnO-NPs suspension (0.1 g/L), specifically, 0, 0.5, 1, 10, 50, 100, 200, and 500 μg/L, as well as 1, 5, and 10 mg/L. The increased levels of TBBPA were utilized to enhance the observable effects of interactions measurable by the zeta sizer and particle size analyzer, aiding in the comprehension of the interaction mechanisms. Detecting such alteration effects at environmentally relevant concentrations of TBBPA is challenging because of the constraints in characterization procedures. NPs underwent analysis through techniques including size assessment, zeta potential measurement, dissolution testing, adsorption analysis, TEM, and infrared, both before and following their association with TBBPA at different time points, specifically after one day, and one, two, and three weeks, to evaluate the alterations in behavior of associating NPs compared to the behavior of the pure ones.

Fixed concentrations (such as 5 mM) of two salts (such as NaCl and $CaCl_2$) were utilized to examine their impact on NPs' stability both independently and in conjunction with TBBPA. The influence of HA was also assessed with and without the presence of 10 and 500 μg/L of TBBPA. To prepare the stock solution, HA was added in 0.1 M NaOH solution. Zinc oxide NPs and HA mixers in a range of concentrations were prepared with or without TBBPA. Additionally, the effects of BSA, another organic compound found in nature, on the behavior of ZnO-NPs were studied regarding any modifications in size, shape, and zeta potential of NPs in relation to TBBPA, electrolytes, HA, and BSA.

The dispersed samples were subjected to centrifugation at 18,407 RCF (relative centrifugal force) for 30 min using an Eppendorf Centrifuge 5424 (Sigma-Aldrich, Taufkirchen, Germany). The supernatants were transferred into new 10 mL centrifuge tubes for subsequent analyses, including dissolved zinc and LC–MS analysis. The quantity of TBBPA that adhered to dissolved NPs was assessed after interactions of one day and two weeks. The amount (mg/g) of TBBPA that was adsorbed (Q_t) was calculated using the formula: $Q_t = \frac{(C_0 - C_t) * V}{W}$, where C_0 and C_t denote the concentrations of TBBPA in the aqueous solution (μg/L or mg/L) before and following the sorption experiments, respectively. Here, V (mL or L) indicates the volume of the solution, while W represents the mass (mg or g) of the ZnO-NPs. Sample pH was kept at 7 by employing a buffer made of potassium dihydrogen phosphate.

2.3. Characterization Techniques

To study morphological changes, including those in water suspensions and after exposure to TBBPA, HA, BSA, and salts, nanoparticles were deposited on TEM grids for observation. Surface areas of ZnO-NPs were analyzed using a Micromeritics TriStar II system, an X-ray diffraction (XRD) system (Empyrean Malvern Panalytical), and field emission scanning electron microscopy (FE-SEM), as described in previous research [28]. FTIR was utilized to detect interactions (functional groups). ZnO-NPs were isolated from the suspension through high-speed centrifugation before FTIR analysis. Changes in NPs' charge and size, individually and after exposure to TBBPA, HA, BSA, and electrolytes in Milli-Q water, were examined using a Malvern Panalytical Zetasizer. The dissolved zinc concentration was determined through inductively coupled plasma optical emission spectroscopy (ICP–OES; Agilent, Manchester, UK). Additionally, an Agilent LC–MS system was employed to quantify TBBPA adsorption and its associations.

3. Results and Discussion

3.1. Interaction Between ZnO-NPs and TBBPA

Changes in the surface structure, particle dimensions, and ZnO-NPs' surface charge were noted after they interacted with different concentrations of TBBPA. NPs' electric potential and size were measured after one day, one, two, and three weeks of exposure (see Figures 1 and S1). To verify interactions, TBBPA adsorption and FTIR analysis were conducted (refer to Figure 2a,b). The levels (concentrations, mg/L) of dissolved Zn were assessed both beforehand and following interactions with TBBPA (illustrated in Figure 2c). TEM analysis was carried out (after zero and one day of interaction) to examine the characteristics of NPs (depicted in Figures 3 and S2).

Size and surface charge: Changes in both surface charge and particle size were noted after interacting with different concentrations of TBBPA. When varying amounts of TBBPA were added to zinc oxide NPs (one day), the peak of the particle size distribution curve (PSDC) for the ZnO-NPs remained within the range of 143 to 222 nm, while the overall particle size varied from 106 to 955 nm (Figures 1a and S1a). Minor fluctuations in size at the peak of PSDC and across measured sizes after one day could result from the relatively short duration of contact between the nanoparticles and TBBPA. These results indicate that the interaction (ZnO-NPs + TBBPA) could depend on the duration of their coexistence. A notable elevation in size was perceived after a week of association (peak of PSDC rising to 166–222.5 nm). There was no observed upward trend in particle size from concentrations of 0.5 to 100 μg/L TBBPA. Nevertheless, there was a rise in the peak of size when the concentration of TBBPA rose from 200 μg/L to 10 mg/L. The variation in size distribution of nanoparticles was affected by TBBPA due to coating or adsorption of organic materials on the surface [24,25]. Throughout the interaction period, ZnO-NPs showed an increase in size, with notable growth observed (one day to three weeks). Enlargement could be due to the formation of large particles or subsequent sedimentation because of particles' associations with each other. Initially, NPs were uniform in size following suspension preparation, but after one week of interaction, the particles became nonuniform (polydisperse). TBBPA may block the active sites of the nanoparticles, keeping them dispersed [50,51], which could be reflected by the ZnO-NPs' sizes.

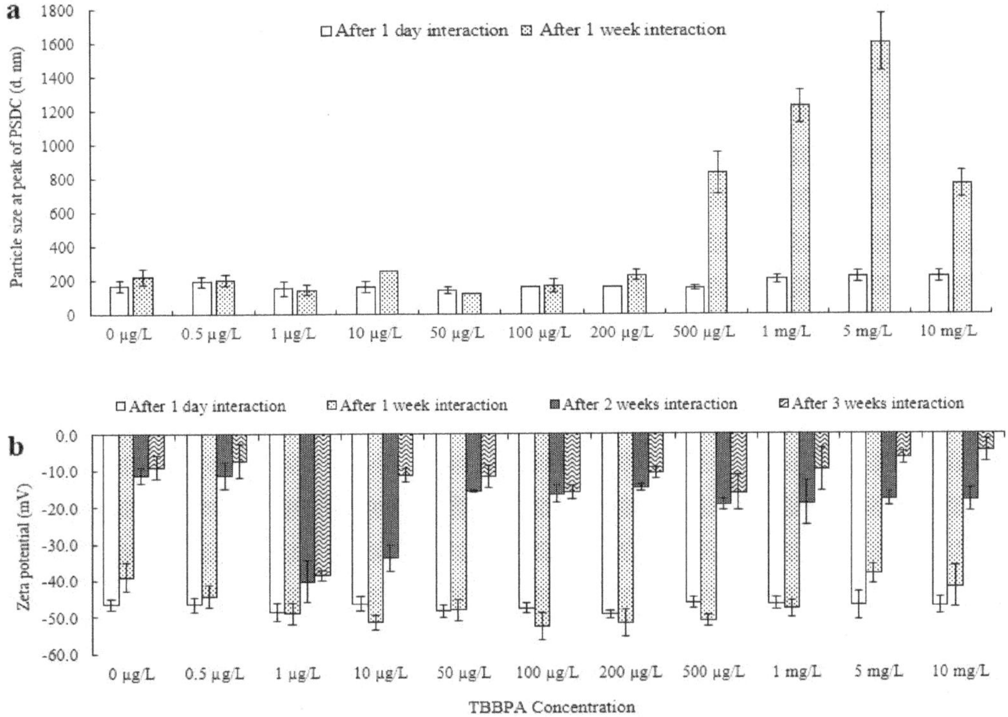

Figure 1. ZnO-NPs' sizes at the peak of PSDC (**a**) and electrical potential (**b**) after one day and one to three weeks of association with TBBPA. Original particle size at the peak of PSDC was 166 nm for one day and 222.5 nm for one week, and the zeta potential was −46.5 mV for one day, −39.0 mV for one week, −11.3 mV for two weeks, and −9.0 mV for three weeks. Where, "-" represents minus sign in Figure 1b.

The analysis of TEM showed NPs' dispersion and size reduction after 0 h and 1 day of interaction (refer to Figure 3). The particles observed between one and three weeks appeared in a polydisperse state, also indicated by their zeta potential (see Figure 1b). Adsorption analysis was conducted to measure the quantity of TBBPA that adhered to NPs, as detailed in the subsequent segment.

A decline in electrokinetic potential resulting from particle-to-particle interactions was noted from one day to three weeks of engagement. Nonetheless, the inclusion of different concentrations of TBBPA postponed the rapid reduction in electrokinetic potential over various periods. The changes in the electrical potential suggested that TBBPA covered the surface of the large structure of the ZnO-NPs and reversed their aggregation as the zeta potential increased. The values of zeta potential indicated that the ZnO-NPs showed increased stability in solution due to their distribution in the presence of TBBPA. This dispersion behavior could be linked to the extensive molecular structure of TBBPA, which likely coats the surface of the nanoparticles. This process is similar to the interaction between humic substances and NPs [25–28,52,53].

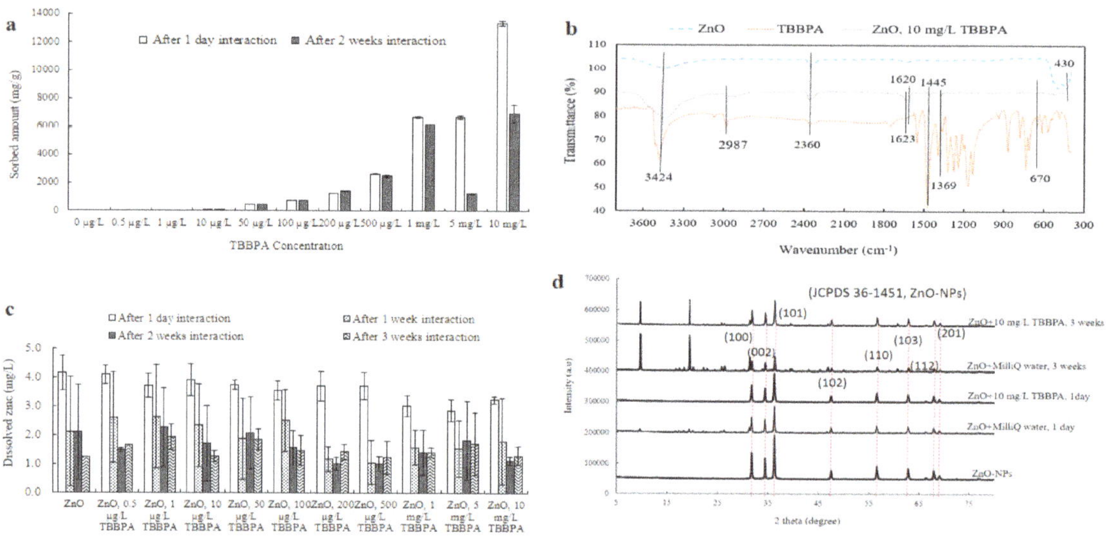

Figure 2. Adsorption (**a**), FTIR (**b**), dissolution (**c**), and XRD (**d**) analyses of ZnO-NPs after interaction with TBBPA after various time intervals.

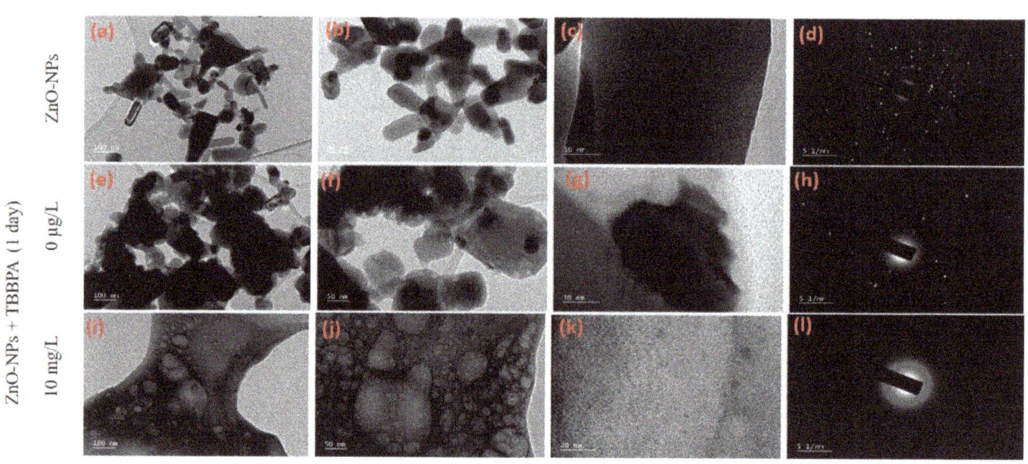

Figure 3. TEM analysis of the ZnO-NPs before and after one day of interaction with TBBPA. Bare ZnO-NPs (**a–d**), ZnO-NPs + 0 μg/L TBBPA after one day of interaction (**e–h**), and ZnO-NPs + 10 mg/L TBBPA after one day of interaction (**i–l**). From left to right, the scales of images are 100 nm, 50 nm, 20 nm and 5 1/nm.

However, an increase in size was observed with a relatively small decrease in zeta potential from 500 μg/L to 10 mg/L of TBBPA after one week. With aging, the polydispersity of NPs increased (unequal attachment of the large molecular structure of TBBPA, affecting the overall size). TBBPA is a molecule with two hydroxyl groups and four bromine atoms attached to a central phenyl ring. Owing to the ionization characteristics of TBBPA in water [54], it can undergo ionization due to the presence of its hydroxyl groups. The presence of more hydroxyl groups due to higher TBBPA concentrations restrained the overall surface charge to a greater degree than their size. The diffraction pattern (Figure 3l) of ZnO-NPs + 10 mg/L TBBPA after one day was more amorphous (less crystalline) than that of ZnO-NP + 0 μg/L TBBPA (Figure 3h). However, the peak of PSDC of zinc oxide NPs + 0 μg/L TBBPA particles was smaller (such as 166 nm) than that of the ZnO-NP + 10 mg/L TBBPA particles (such as 222 nm), which could be due to the sorption of TBBPA molecules onto the surface of the nanoparticles, resulting in less crystallinity (amorphous) after one day of interaction. Notably, some particles were sedimented/attached to the walls of the tubes from the time of preparation of the mixtures (samples) to three weeks of interaction; hence, whatever was present in the suspension form was analyzed.

Adsorption, FTIR, dissolution, and XRD analyses: The adsorption of TBBPA indicated the presence of interactions between NPs and TBBPA after one day and two weeks of associations (Figure 2a). Quantity of TBBPA adsorbed onto NPs' surface diminished after two weeks related to the amount observed after one day for higher concentrations (such as 500 μg/L, 1, 5, and 10 mg/L) of TBBPA. A range of interaction mechanisms may play a role in the adsorption of TBBPA onto the ZnO-NPs, including electrostatic interactions, hydrophobic interactions (such as π–π stacking, electron donor–acceptor interactions, and van der Waals forces), as well as hydrogen bonding [55].

When ZnO-NPs are exposed to water, hydrolysis commonly leads to generation of hydroxide layers ($Zn(OH)^{+}_{(aq)}$) on the nanoparticle surface. This occurs as water molecules are adsorbed onto the particles both chemically and physically [56,57]. This could result in development of many positive charges on NPs' surfaces, attracting deprotonated ($TBBPA^{-}$) forms of TBBPA, which carry a negative charge to be sorbed on NPs' surfaces. Initially (such as after one day of interaction), ZnO-NPs may have a high affinity for adsorbing TBBPA molecules because of the availability of active sites on their surface. However, the dispersion of NPs (polydispersed) due to adsorption of TBBPA (large molecular structure) led them to settle and decreased further adsorption with increasing time. Like HA, TBBPA may also form complexes with Zn ions released from the ZnO-NPs over time [58]. These complexes could alter the adsorption behavior of TBBPA and contribute to its desorption from the ZnO surface. This behavior resembles that of HA molecules, as the dispersion of ZnO-NPs might also be due to the complexation of zinc ions with anionic HA, leading to the creation of a larger complex structure [58] in which HA binds zinc ions.

FTIR analysis was performed on ZnO-NPs, TBBPA, and the ZnO + TBBPA mixture after one day of interaction to further investigate the interaction of TBBPA with the surface of the ZnO-NPs (Figure 2b). A peak at 430 cm^{-1} was observed, indicating Zn–O occurrence [28,59], which is typical of metal oxide spectra (400–600 cm^{-1}) [60,61]. For pure TBBPA powder, a vibrational peak between 500 and 700 cm^{-1} was identified, corresponding to the stretching vibration of (C–Br) bonds in the organic pollutant. The 670 cm^{-1} peak was attributed to C–X stretching in organic halogen compounds, where X represents Br [62]. These peaks were also present in both TBBPA and ZnO-NPs after exposure to TBBPA [28,63]. Furthermore, peaks at 1145 and 1620 cm^{-1} were linked to C–O stretching and the skeletal vibration of aromatic C=C bonds within the TBBPA structure [63]. C–H bending and C=O stretching were detected at 1369 and 1623 cm^{-1}, respectively [62,64]. Such peaks appeared in TBBPA and ZnO + TBBPA, confirming that bonding occurred between them. Additionally, peaks at 2987 cm^{-1} and 3424 cm^{-1} were associated with C–H/O–H stretching, C–H asymmetric stretching, and the water band [28,62]. FTIR data supported the existence of bonds in NPs after TBBPA interaction (one day), suggesting that TBBPA molecules accumulate on NPs' surfaces. This observation aligns with the elemental analysis, which detected Br and C in the nanoparticles. The concentrations of dissolved zinc from NPs alone and those combined with different concentrations of TBBPA were assessed in this research (Figure 2c). Following a day of engagement, the levels of dissolved zinc increased, when TBBPA was present compared to when it was absent, and this pattern continued to be evident even after three weeks of interaction. It can be inferred that TBBPA facilitated the dispersion of NPs after an extended interface period (2–3 weeks).

Figure 2d shows the XRD study. NPs exhibited distinct peaks at 2θ values of 31.84°, 34.6°, and 36.5°, confirming NPs' hexagonal wurtzite crystal structure with three specific orientations: (1 0 0), (0 0 2), and (1 0 1) [57]. These findings suggest that the crystal structure of NPs remained stable after exposure (one day). On the other hand, after three weeks of association (Figure 2d), peak strengths at (1 0 0), (0 0 2), and (1 0 1) decreased. Additionally, two crests appeared at 2θ angles of 9.68° and 19.40°, which might signify creation of different composites, zinc hydroxide dihydrate ($Zn_5(OH)_{10}{\cdot}2H_2O$) [65] and zinc phosphate [66], respectively. Peak intensities were lower for ZnO-NPs containing 10 mg/L of TBBPA compared to the sample that was in water, which could be because of TBBPA molecules covering ZnO-NPs' surfaces (Figure 2d).

TEM: The initial ZnO nanoparticles (ZnO-NPs), obtained from Sigma-Aldrich, were agglomerated (size: 100 nm or less) with various shapes (see Figure 3a,b). The lattice pattern observed (Figure 3c) along with the glittering spots/rings (Figure 3d) confirmed the crystalline nature of the ZnO-NPs. The EDAX analysis conducted confirmed the occurrence of zinc and oxygen [28]. The diffraction pattern (Figure 3d) revealed the crystalline/lattice arrangement of the ZnO-NPs [28,67–70] which was comparatively dull/had fewer bright spots (Figure 3h) after their interaction than the original ZnO-NPs in powder form. The presence of both zinc and oxygen was also observed (Figure S2).

Following the interaction with TBBPA, the ZnO-NPs displayed highly random and dispersed structures (Figure 3i,j), in contrast with the original particle arrangement (Figure 3a,b). A high-resolution TEM image (Figure 3k) illustrated the dispersion pattern of the ZnO-NPs after exposure to the large and complex molecular structures of TBBPA (Figure 3c, and with 0 μg/L of TBBPA in Figure 3g). In contrast to the pure ZnO-NPs, cloudy spots (Figure 3l) indicated altered NPs' morphology following their association with organic pollutant molecules (TBBPA). Elemental analysis indicated the existence of oxygen, zinc, bromine, and carbon atoms, as well as potassium and phosphorus, which came from the buffer (Figure S2).

3.2. *Influence of Salts on the Interaction Between ZnO-NPs and TBBPA*

Varying concentrations of cations and anions in ecosystem media can influence the physical and chemical characteristics of ZnO nanoparticles [19]. The stability of these systems is primarily determined by the charge present on the NPs' surfaces. Environmental elements, such as pH, ionic strength, and the existence of organic materials in the solution, also have an impact on the surface charge. To assess their impact on NPs' stability, both individually and in the existence of TBBPA, fixed concentrations (for instance, 5 mM) of NaCl and $CaCl_2$ were utilized. The following section explains the outcomes.

Hydrodynamic size: Changes in NPs' sizes were recorded both in Milli-Q H_2O and with TBBPA (at concentrations of 10 and 500 μg/L), as well as when combined with a consistent concentration (5 mM) of salts (NaCl and $CaCl_2$), over periods of interaction (see Figures 4a and S3). After one day, NPs' size at the peak of PSDC enlarged with the occurrence of NaCl and $CaCl_2$. This indicated that the salts had a predominant effect, likely due to the compression of the double layer resulting in a smaller hydrodynamic diameter and increased aggregation, even when TBBPA was also present. The introduction of 5 mM $CaCl_2$ caused a significant increase in particle size, exceeding the upper limit of the zeta sizer's measurement range (which goes up to 10 microns). Only the size of the measurable portion was reported (refer to Figures 4a and S3).

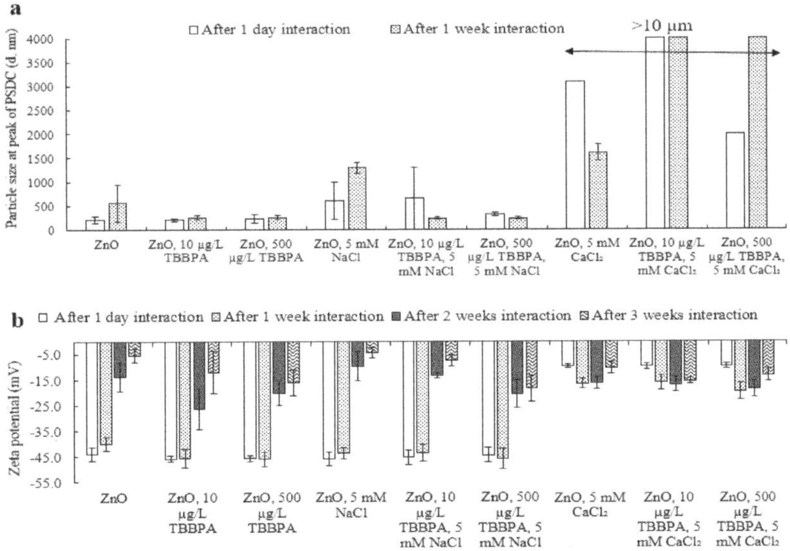

Figure 4. The size at the peak of PSDC for the ZnO-NPs (**a**) and zeta potential (**b**) after one day, one week, two weeks, and three weeks of interaction in the presence of electrolytes. The original particle sizes of the ZnO-NPs at the peak PSDC were 214 nm for one day and 560 nm for one week, and the zeta potentials were −43.9 mV for one day, −39.9 mV for one week, −13.7 mV for two weeks, and −5.3 mV for three weeks. Where, "-" represents minus sign in Figure 4b.

Compared with that after one day, NPs' size in buffered water was examined to increase after one week. However, no significant increase in size was examined in the presence of TBBPA. The presence of salts shifted NPs' size after one week. Compared with NPs alone, NPs with the existence of NaCl (5 mM) revealed a similar and steady increase in particle size, as the size increased from 560 nm (NPs in Milli-Q H_2O after one week) to 1290 nm (the size of the ZnO-NPs in the presence of 5 mM NaCl after one week). Various concentrations of $CaCl_2$ had the same effect on the size of the ZnO-NPs. The measurable size has been reported (Figures 4a and S3). These findings imply that NPs' sizes grew as cations gathered on negatively charged NPs' surfaces because of electrostatic forces. Hydrogen bonding and van der Waals forces further amplified this process, resulting in an overall increase in particle size [11,28,35].

The interaction of ZnO-NPs with low concentrations of TBBPA (e.g., 10 or 500 µg/L) was studied by introducing salts. The addition of these salts impacted the hydrodynamic properties of the ZnO-NPs, as more ions accumulated around the charged nanoparticles. Furthermore, when NaCl was present, the large TBBPA molecules coated and dispersed the nanoparticles after one week of exposure. NPs' size alone varied greatly with time because of the aggregation/polydispersity effect. It was challenging to observe the size behavior of ZnO-NPs at environmentally relevant concentrations of co-contaminants over time, as evidenced by the measurement of TBBPA in solution after interaction with the nanoparticles.

Surface charge: The electrokinetic potential of zinc oxide NPs diminished from −43.9 mV to −39.9, −13.7, and −5.3 mV after intervals of 1 day, 1, 2, and 3 weeks, respectively (Figure 4b). The noticeable decline in NPs' electric potential might be attributed to the agglomeration of multiple NPs due to hydrogen bonding, van der Waals forces, and hydrophobic interactions. A comparable reduction in the surface charge magnitude was noted in the presence of TBBPA (for concentrations of 10 and 500 μg/L). This zeta potential decrease was less pronounced than that observed with the ZnO-NPs, likely due to the dispersion effect resulting from TBBPA's large molecular structure. Nevertheless, the aging effect also played a substantial role. Likewise, the introduction of salts modified the NPs' electrokinetic potential (Figure 4b). It can be inferred that cation concentrations built up around the surfaces of nanoparticles that carried a negative charge, leading to a higher overall surface charge. A similar trend was noted regarding the dimensions of NPs, which grew larger in the existence of salts (Figures 4a and S3).

Dissolution: The concentration of dissolved zinc (mg/L) was assessed using ICP–OES (Figure 5). Over several weeks of association, NPs' size enhanced (aggregation followed by deposition), which led to a reduction in specific surface area and subsequently limited dissolution. Moreover, in the presence of $CaCl_2$, a decrease in dissolution was noted compared to all other samples, potentially linked to increased agglomeration caused by the bridging effect of Ca^{2+}. Additionally, nanoparticles tended to attach and settle within the low-density polyethylene tubes.

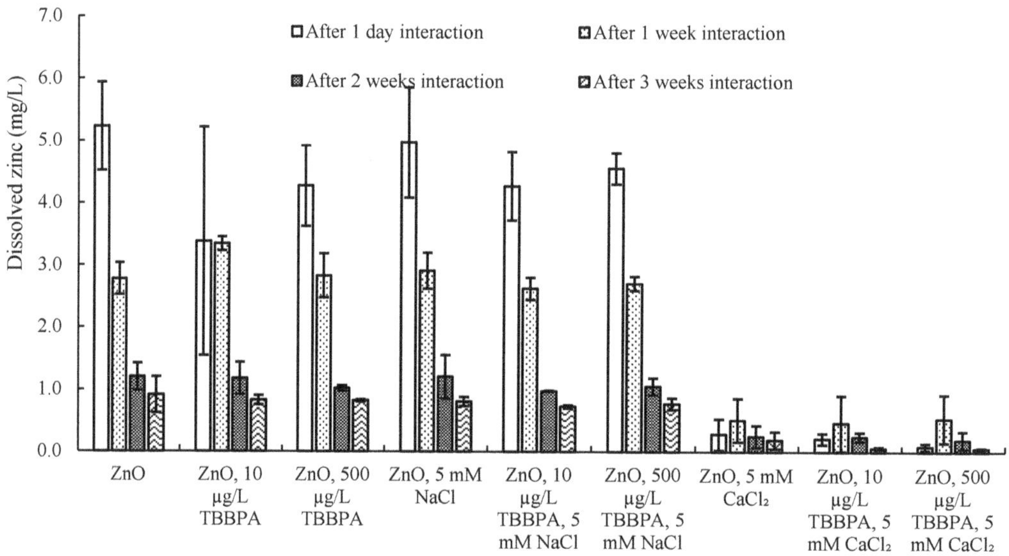

Figure 5. Dissolved zinc concentration (mg/L) after one day and after one, two, and three weeks of interaction.

TEM analysis: TEM analysis (Figures 6 and S4) and elemental mapping (Figure S4) were conducted to examine the behavior of ZnO-NPs in the presence of TBBPA and varying concentrations of $CaCl_2$ after 0 h and 1 day of interaction. A drop of the prepared solutions was directly placed on the TEM grid for observation. After one day of interaction, a dispersion effect, caused by the complex and aliphatic nature of TBBPA, was evident, with more nanoparticles being coated in the occurrence of 10 mg/L of TBBPA compared to 0 h (Figures 6a,b and S4). This finding aligns with the electrokinetic potential and particle size data from the zeta analyzer (Figure 4a,b). SAED images revealed a reduction in crystallinity after one day compared to the initial time point. Elemental mapping also detected Zn, P, O, K, Br, and C (Figure S4). Aggregation with thick or shaded layers of $CaCl_2$ was noted. After one day, both covered ($CaCl_2$ coatings) and scattered (due to TBBPA) ZnO-NP were observed at both 5 and 10 mM $CaCl_2$ concentrations. The diffuse diffraction patterns

(Figure S4) could be attributed to the thick layers of TBBPA and salts on the nanoparticle surfaces. The bigger TBBPA molecules facilitated the NPs' dispersion after one day.

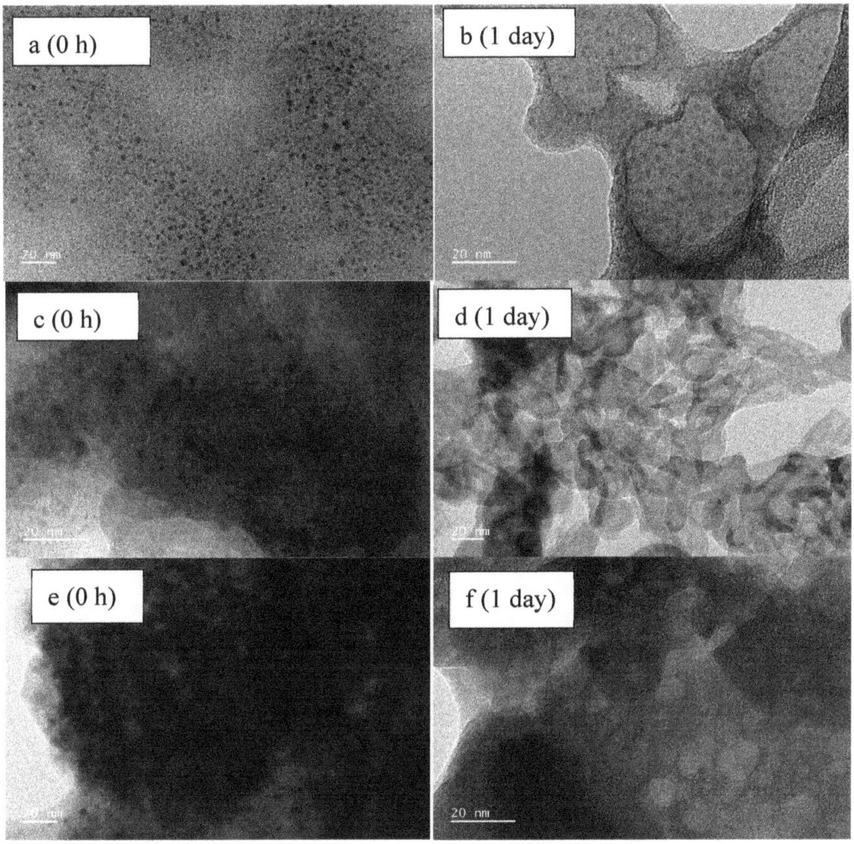

Figure 6. TEM images after 0 h and 1 day of interaction in solution (drops were taken on a TEM grid from the solution). ZnO, 10 mg/L TBBPA (**a**,**b**), ZnO, 10 mg/L TBBPA, 5 mM $CaCl_2$ (**c**,**d**), ZnO, 10 mg/L TBBPA, and 10 mM $CaCl_2$ (**e**,**f**).

3.3. *Influence of HA on TBBPA and ZnO-NPs' Interaction*

Particle size: The size of ZnO-NPs in buffered water, exposed to varying concentrations of TBBPA (10 or 500 μg/L), different levels of HA (1, 5, or 10 mg/L), and combinations of TBBPA and HA, was assessed after 1 day and 1–3 weeks (Figures 7a and S5). Size elevations were observed. This enlargement may be attributed to the formation of larger and/or sedimented particles, resultant from interactions between particles, as well as electrostatic and hydrophobic forces. Initially, the NPs were monodispersed. After one week, the nanoparticles displayed nonuniform characteristics (polydisperse). By the one to three weeks period, the particles were highly polydisperse, a pattern that was also revealed in their surface charge (Figure 7b). The distribution and dimensions of NPs in the occurrence of 1, 5, and 10 mg/L of humic acid diminished after one week of interaction, in contrast to the measurements recorded on day one (Figures 7a and S5). After two and three weeks, samples became highly polydisperse, and their size exceeded the measurement range of the dynamic light scattering analyzer, as some samples exhibited scattering behavior. A similar trend was observed for ZnO-NPs exposed to different concentrations of TBBPA (10 and 500 μg/L), when combined with HA (Figures 7a and S5). This dispersion is likely due to HA, with its large aliphatic carbon network, which may have capped the nanoparticle edges, promoting their dispersion. Coexistence of both TBBPA and HA influenced the NPs' size in a manner distinct from the effect seen with either TBBPA or HA individually. The distribution and NPs' size increased (resulting in agglomerated particles) when only ZnO-NPs or those in conjunction with TBBPA were present (an observable desorption pattern occurred over time (Figure 7a), which facilitated accumulation of NPs following several weeks of interaction). Conversely, when varying concentrations (1, 5, and 10 mg/L) of HA were present, NPs' size decreased (leading to polydispersal). This dispersal behavior of NPs might be linked to the occurrence of HA, which could mitigate their aggregation tendencies [32,44].

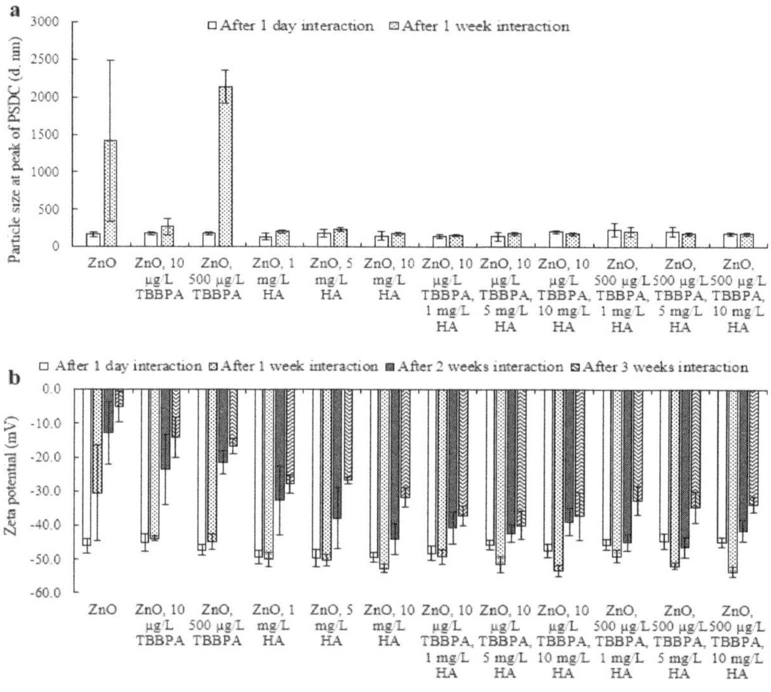

Figure 7. The size of ZnO-NPs at the peak of PSDC (**a**) and zeta potential (**b**) after one day and after one, two, and three weeks of interaction in the presence of various concentrations of TBBPA and HA at pH 7 and room temperature (i.e., 20 °C). The original particle size of ZnO-NPs at the peak of PSDC is 166 nm (one day) and 1413 nm (one week), and the zeta potential is −46.1 mV (one day), −30.5 mV (one week), −12.8 mV (two weeks), and −5.0 mV (three weeks). Where, "-" represents minus sign in Figure 7b.

Zeta potential: Electrokinetic potential results are shown (Figure 7b) and explained in this section. The strength of NPs' electrokinetic potential, both independently and with TBBPA, showed a decline (one day to three weeks). Nevertheless, the decline in surface charge was less pronounced for the ZnO-NPs in the presence of TBBPA compared to those without it (Figure 7b). Additionally, the presence of HA led to a further reduction in zeta potential, in comparison to the samples without HA. A higher concentration of HA played a significant role in mitigating the reduction in electric potential, unlike lower concentrations of HA, such as 1 mg/L HA (Figure 7b) [31].

The previously mentioned data on electrical potential indicated that the clustering behavior of pure ZnO nanoparticles in water may stem from van der Waals forces, electrostatic interactions, and hydrophobic effects. NPs' aging, both alone and alongside TBBPA, could modify the NPs' electrical potential, resulting in increased sedimentation in water due to a reduction in stability in aqueous conditions. When comparing the effects of pure ZnO nanoparticles with and without TBBPA, the existence of humic acid changed the surface charge in an opposing manner. HA compounds covered the NPs' surfaces and effective sites due to their higher aliphatic carbon content compared to TBBPA molecules, which led to a reduced likelihood of TBBPA adhering to the surfaces of the nanoparticles. This also promoted better dispersion of the nanoparticles.

Dissolution: ZnO-NPs' dissolution, whether by themselves or alongside different concentrations of TBBPA, HA, and their combination, was examined in Milli-Q water at a pH of 7 (Figure 8). The availability of zinc, either in its dissolved state or as an ionic form, poses potential toxicity to microorganisms, including microflora [58,71]. The existence of additional compounds in water can affect the dissolution of ZnO-NPs [72]. After one day, concentration of dissolved zinc was higher when various concentrations of HA were present compared to when HA was absent, and this pattern continued even after two weeks of interaction. It can be postulated that HA facilitated the dispersion of the nanoparticles after extended interaction periods (like two to three weeks), and the increased dispersion might result from van der Waals forces, electrostatic forces, and hydrophobic interactions. Additionally, this might be attributed to the complexation of zinc ions with the anionic HA, leading to the formation of a larger complex structure. These findings align with those posited in [58] that HA binds zinc ions.

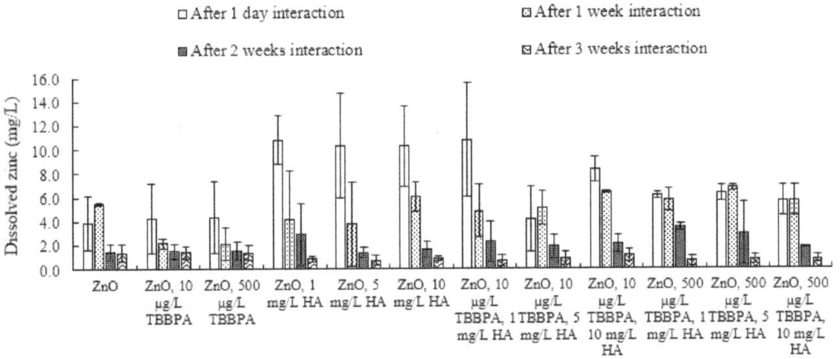

Figure 8. Dissolved zinc concentration (mg/L) after one day and after one, two, and three weeks of interaction in the presence of various concentrations of TBBPA and HA at pH 7 and room temperature (i.e., 20 °C).

3.4. Influence of BSA on the Behavior of ZnO-NPs

The effect of bovine serum albumin (BSA), a natural protein found in ecological water, on the stability of ZnO-NPs was investigated (Figures 9a,b and S6). BSA plays various physiological roles, including transporting, binding, and distributing fatty acids and steroids [73]. In this study, BSA was selected as a model protein due to its water-soluble properties.

Hydrodynamic size: Initially, after one day, NPs' sizes increased in the presence of BSA (Figure 9a). The BSA molecule, with a large molecular mass of 66,400 Da and consisting of approximately 583 amino acids linked in a single cross-linked chain with 17 cysteine residues [74], contributed to this increase. However, after one week of incubation, the presence of BSA reduced the size of the ZnO-NPs. The resulting dispersion effect was similar to that observed when other large molecular materials, like humic acid [28], interacted with the nanoparticles. As the concentration of BSA increased, NPs' sizes, measured at the peak of the PSDC, gradually decreased (Figures 9a and S6b). This behavior is consistent with the typical interactions between BSA and metal ions, which can lead to a decrease in the protein's configuration due to the disruption of disulfide bonds. This results in a partial loss of the α-helix structure, unfolding of the protein, or changes in the polarity of the surrounding environment, which may affect the exposure of tryptophan residues due to molecular interactions. These reactions include excited-state processes, molecular adjustments, energy transfer, complex formation, or collision quenching [74].

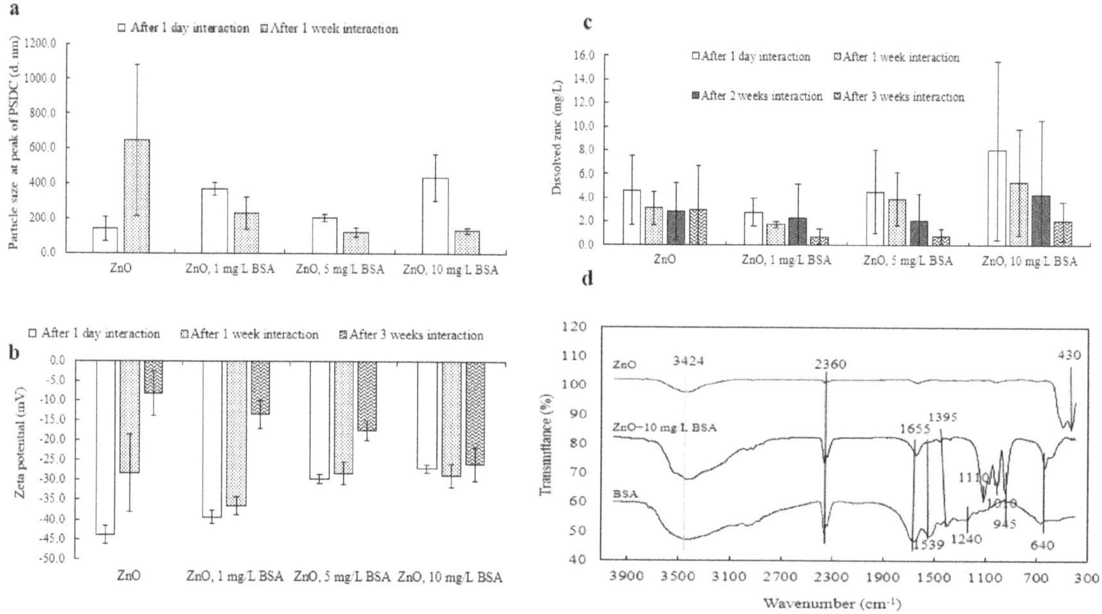

Figure 9. ZnO-NPs' sizes at the peak of PSDC (**a**), zeta potential (**b**) (after one day, one week, and three weeks), dissolution (**c**) (after one day, one week, two weeks, and three weeks), and FTIR (**d**) (after one day) of interactions with BSA. The original particle sizes of the ZnO-NPs at the peak PSDC were 140.7 nm after one day and 648.5 nm after one week, and the zeta potentials were −44.0 mV after one day, −28.3 mV after one week, and −8.0 mV after three weeks. Where, "-" represents minus sign in Figure 9b.

Zeta potential: Figure 9b shows the surface charge values on the surface of the ZnO-NPs before and after interactions with BSA at various time intervals. Compared with that after one day, the overall surface charge (magnitude) of the ZnO-NPs diminished after several weeks of interaction. Similarly, NPs' electrical potential with varying concentrations of BSA also decreased in magnitude after one day (Figure 9b), aligning with the findings from the particle size analysis (e.g., the size increased after one day; Figure 9a). This suggests that BSA, as a frothy substance, quickly coated the nanoparticles, causing the formation of large clusters, as confirmed by TEM analysis (only for 0 h and 1 day of interaction; Figures 10 and S7). However, after several weeks of interaction, the surface charge did not decrease compared with that of the ZnO-NPs, implying that the dispersion revealed the protein patterns of the BSA molecules. These findings were further supported by the size (Figure 9a) and TEM (Figures 10 and S7) analyses. The sizable and intricate molecular structure of the BSA protein molecules tended to envelop the ZnO-NPs, causing their dispersion, which in turn influenced their zeta potential.

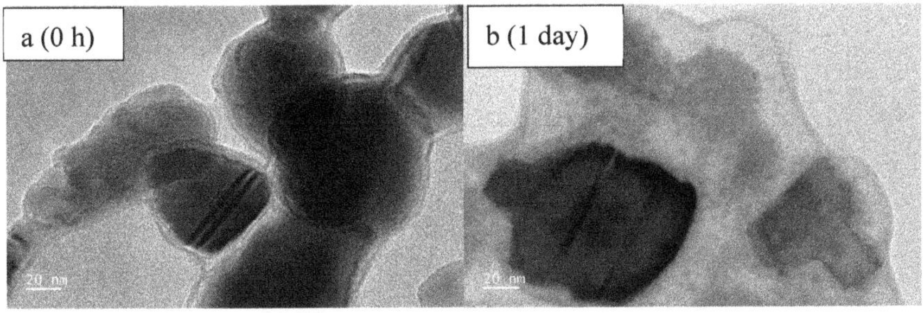

Figure 10. TEM images of ZnO + 10 mg/L BSA after 0 h (**a**) and 1 day (**b**) of interaction in solution (drop taken on a TEM grid from the solution).

Dissolution: The NPs' dissolution was evaluated both on its own and in conjunction with different concentrations of BSA in Milli-Q water at a pH of 7 (Figure 9c). After a day, the dissolved zinc concentration was higher when BSA was present at various levels (such as 5 and 10 mg/L) compared to when it was absent, and this pattern continued even after two weeks. It can be inferred that BSA aided in dispersing the nanoparticles following a prolonged interaction period (of 2–3 weeks), and the enhancement in this dispersion may be attributed to electrostatic forces, van der Waals interactions, and hydrophobic forces.

FTIR: FTIR spectra of ZnO-NPs with BSA were examined after one day (Figure 9d). The peak at 430 cm^{-1} in the spectrum of the pure ZnO-NPs confirmed the existence of Zn–O [28,59]. The peak at 430 cm^{-1} represented the existence of metal oxides (such as ZnO). This peak was visible for pure ZnO-NPs (Figure 9d). However, after interactions with BSA, the intensity of the peak was not detectable by infrared spectroscopy. It could be assumed that BSA molecules (10 mg/L) adsorbed on NPs' surface and generated coated layers that may have affected the detection of the ZnO-NP peak at 430 cm^{-1}. As a result, the BSA layer might absorb or scatter the incident light in a way that reduces/weakens the intensity of the ZnO peak at 430 cm^{-1}. The peak at 640 cm^{-1} in both samples was attributed to the secondary amide (N–H) [62] present in the BSA molecules. Notable bands in BSA included amide III at 1240 cm^{-1}, which was not observed at the same peak location in the ZnO + BSA sample, amide II at 1539 cm^{-1}, and amide I at 1655 cm^{-1} [75]. This could be due to the presence of interacting BSA molecules on NPs' surface. The peaks observed at 945 and 1010 cm^{-1} could be attributed to the stretching of metal (zinc) and nitrogen bonds present in ZnO + 10 mg/L BSA after their interaction [62]. The peaks at 1110 cm^{-1} and 1395 cm^{-1} likely corresponded to C–O stretching and C–H stretching found in the organic BSA protein [62]. The peaks at 2360 cm^{-1} and 3424 cm^{-1} were associated with C–H/O–H stretching, C–H asymmetric stretching, and the water band, respectively [28,62]. These observations suggest that BSA interacted with NPs' surfaces through π–π stacking as well as other molecular forces, including electrostatic and van der Waals interactions. The interactions further involved hydrophobic π–π stacking and hydrogen bonding between the active sites, such as oxygen-functionalized groups in water and oxygen/nitrogen groups within the protein molecules [76].

TEM: To assess the aggregation and dispersion behavior of ZnO-NPs in the presence of BSA at different time points, a TEM analysis was conducted by placing a drop of the solution directly onto the TEM grids (Figures 10 and S7). Initially, at 0 h, the nanoparticles remained undispersed, likely due to the binding forces exerted by the proteins, aided by hydrogen bonding and electrostatic and hydrophobic interactions. However, after one day of interaction with BSA molecules, the ZnO-NPs were observed to disperse (Figures 10 and S7), which could be attributed to the extensive coverage of nanoparticles by BSA molecules. Diffraction images (Figure S7) revealed bright spots at 0 h, indicating the crystalline nature of the ZnO-NPs, while after 1 day, the patterns appeared diffuse and cloudy, suggesting a reduction in NPs' purity and crystallinity due to the interaction with BSA. Elemental mapping (Figure S7) at both time points confirmed the presence of nitrogen, oxygen, zinc, and carbon in the ZnO + 10 mg/L BSA samples. The presence of potassium (K) and phosphorus (P) was traced back to the buffer solution, which was used to maintain the pH at 7.

4. Conclusions

This research illustrated the surface and structural characteristics of ZnO-NPs under different environmental conditions, both before and after their interaction with co-occurring electrolytes, an organic pollutant (TBBPA), HA, and BSA over different periods, including one day, one week, two weeks, and three weeks of interaction. After engaging with environmental agents, ZnO-NPs were not found in their original forms due to alterations in particle size and shape. The inclusion of electrolytes enhanced the aggregation of charged ZnO-NPs by reducing the level of surface charge. The interaction mechanisms could be attributed to electrostatic forces, van der Waals forces, and particle–particle interactions, such as cation bridging. The large molecular structures of HA, BSA, and TBBPA contributed to a decrease in the particle size of the ZnO-NPs due to a dispersion effect. Changes in the shape, size, and surface charge of the ZnO-NPs were noted following their interaction with the co-contaminants, affecting the dynamics and behavior of the ZnO-NPs in aquatic environments.

Supplementary Materials: The following supporting information can be downloaded at: https://www.mdpi.com/article/10.3390/toxics13030148/s1. Figure S1: Particle size distribution of ZnO-NPs after 1 day (a) and 1 week (b) of interaction with TBBPA; ZnO-NPs 0.1 g/L, pH 7. Figure S2: Compositional analysis of ZnO-NPs with TBBPA after 1 day of interaction. Figure S3: Particle size distributions of the ZnO-NPs after 1 day (a) and 1 week (b) of interaction in the presence of electrolytes at pH 7 and room temperature (i.e., 20 °C). Figure S4: TEM images (elemental mapping) of various contaminants after 0 h and 1 day of interaction in solution (drop taken on a TEM grid from the solution). Figure S5: Particle size distributions of the ZnO-NPs after 1 day (a) and 1 week (b) of interaction in the presence of various concentrations of TBBPA and HA at pH 7 and room temperature (i.e., 20 °C). Figure S6: Particle size distribution of the ZnO-NPs after 1 day (a) and 1 week (b) of interaction in the presence of various concentrations of BSA at pH 7 and room temperature (i.e., 20 °C). Figure S7: TEM images (elemental mapping) of ZnO + 10 mg/L BSA after 0 h and 1 day of incubation in solution (drops taken on a TEM grid from the solution).

Author Contributions: A.U.H.K., methodology development, experimental procedures and analysis, and initial drafting of the writing; Y.L., oversight, conceptual framework, visualization, and review and editing of the writing; R.N., oversight, securing funding, and review and editing of the writing; C.F., oversight, and review and editing of the writing; H.K.S., oversight, and review and editing of the writing. All authors have read and agreed to the published version of the manuscript.

Funding: This research received no external funding.

Institutional Review Board Statement: Not applicable.

Informed Consent Statement: Not applicable.

Data Availability Statement: The original contributions presented in this study are included in the article/Supplementary Materials. Further inquiries can be directed to the corresponding author(s).

Acknowledgments: The first author expresses gratitude to the University of Newcastle (UON), Australia, for offering a fully funded PhD (ECRHDR UNRS Central and UNIPRS) scholarship. The first author also thanks crcCARE for awarding a scholarship. The authors appreciate the financial support and research facilities provided by crcCARE and GCER, UON, which enabled the completion of this study. The authors recognize the contributions of Graeme Jameson and Kitty Tang from the Centre for Multiphase Processes, UON, along with Erica Wanless and Sonia Khandaker from the School of Environmental and Life Sciences, for offering the necessary facilities and training for nanoparticle size and surface analysis. The authors are thankful to Abhishek Sharma and Francisca Munyeza from GCER and UON for their assistance with ICP–OES and LC–MS, respectively.

Conflicts of Interest: The authors declare no conflicts of interest.

References

1. Bundschuh, M.; Filser, J.; Lüderwald, S.; Mckee, M.S.; Metreveli, G.; Schaumann, G.E.; Schulz, R.; Wagner, S. Nanoparticles in the environment: Where do we come from, where do we go to? *Environ. Sci. Eur.* **2018**, *30*, 6. [CrossRef]
2. Lead, J.R.; Batley, G.E.; Alvarez, P.J.J.; Croteau, M.N.; Handy, R.D.; McLaughlin, M.J.; Judy, J.D.; Schirmer, K. Nanomaterials in the environment: Behavior, fate, bioavailability, and effects—An updated review. *Environ. Toxicol. Chem.* **2018**, *37*, 2029–2063. [CrossRef] [PubMed]
3. Osmond, M.J.; McCall, M.J. Zinc oxide nanoparticles in modern sunscreens: An analysis of potential exposure and hazard. *Nanotoxicology* **2010**, *4*, 15–41. [CrossRef]
4. Stoller, M.; Ochando-Pulido, J.M. ZnO nano-particles production intensification by means of a spinning disk reactor. *Nanomaterials* **2020**, *10*, 1321. [CrossRef] [PubMed]
5. Ge, Y.; Schimel, J.P.; Holden, P.A. Evidence for negative effects of TiO_2 and ZnO nanoparticles on soil bacterial communities. *Environ. Sci. Technol.* **2011**, *45*, 1659–1664. [CrossRef]
6. Jeng, H.A.; Swanson, J. Toxicity of metal oxide nanoparticles in mammalian cells. *J. Environ. Sci. Health—Part A* **2006**, *41*, 2699–2711. [CrossRef] [PubMed]
7. Lin, D.; Story, S.D.; Walker, S.L.; Huang, Q.; Cai, P. Influence of extracellular polymeric substances on the aggregation kinetics of TiO_2 nanoparticles. *Water Res.* **2016**, *104*, 381–388. [CrossRef] [PubMed]
8. Peng, Y.H.; Tsai, Y.C.; Hsiung, C.E.; Lin, Y.H.; Shih, Y.H. Influence of water chemistry on the environmental behaviors of commercial ZnO nanoparticles in various water and wastewater samples. *J. Hazard. Mater.* **2017**, *322*, 348–356. [CrossRef]
9. Fairbairn, E.A.; Keller, A.A.; Mädler, L.; Zhou, D.; Pokhrel, S.; Cherr, G.N. Metal oxide nanomaterials in seawater: Linking physicochemical characteristics with biological response in sea urchin development. *J. Hazard. Mater.* **2011**, *192*, 1565–1571. [CrossRef]
10. Cupi, D.; Hartmann, N.B.; Baun, A. Influence of pH and media composition on suspension stability of silver, zinc oxide, and titanium dioxide nanoparticles and immobilization of *Daphnia magna* under guideline testing conditions. *Ecotoxicol. Environ. Saf.* **2016**, *127*, 144–152. [CrossRef] [PubMed]
11. Zhang, X.; Servos, M.R.; Liu, J. Ultrahigh nanoparticle stability against salt, pH and solvent with retained surface accessibility via depletion stabilization. *J. Am. Chem. Soc.* **2012**, *134*, 9910–9913. [CrossRef]
12. Li, L.Z.; Zhou, D.M.; Peijnenburg, W.J.G.M.; van Gestel, C.A.M.; Jin, S.Y.; Wang, Y.J.; Wang, P. Toxicity of zinc oxide nanoparticles in the earthworm, *Eisenia fetida* and subcellular fractionation of Zn. *Environ. Int.* **2011**, *37*, 1098–1104. [CrossRef] [PubMed]
13. Yung, M.M.N.; Wong, S.W.Y.; Kwok, K.W.H.; Liu, F.Z.; Leung, Y.H.; Chan, W.T.; Li, X.Y.; Djurišić, A.B.; Leung, K.M.Y. Salinity-dependent toxicities of zinc oxide nanoparticles to the marine diatom *Thalassiosira pseudonana*. *Aquat. Toxicol.* **2015**, *165*, 31–40. [CrossRef] [PubMed]
14. Lin, D.; Xing, B. Phytotoxicity of nanoparticles: Inhibition of seed germination and root growth. *Environ. Pollut.* **2007**, *150*, 243–250. [CrossRef]
15. Schiavo, S.; Oliviero, M.; Philippe, A.; Manzo, S. Nanoparticles based sunscreens provoke adverse effects on marine microalgae *Dunaliella tertiolecta*. *Environ. Sci. Nano* **2018**, *5*, 3011–3022. [CrossRef]
16. Sousa, V.S.; Teixeira, M.R. Metal-based engineered nanoparticles in the drinking water treatment systems: A critical review. *Sci. Total Environ.* **2020**, *707*, 136077. [CrossRef]
17. Gottschalk, F.; Sonderer, T.; Scholz, R.W.; Nowack, B. Modeled environmental concentrations of engineered nanomaterials (TiO_2, ZnO, Ag, CNT, fullerenes) for different regions. *Environ. Sci. Technol.* **2009**, *43*, 9216–9222. [CrossRef] [PubMed]
18. Goswami, L.; Kim, K.H.; Deep, A.; Das, P.; Bhattacharya, S.S.; Kumar, S.; Adelodun, A.A. Engineered nano particles: Nature, behavior, and effect on the environment. *J. Environ. Manag.* **2017**, *196*, 297–315. [CrossRef]
19. Bathi, J.R.; Moazeni, F.; Upadhyayula, V.K.K.; Chowdhury, I.; Palchoudhury, S.; Potts, G.E.; Gadhamshetty, V. Behavior of engineered nanoparticles in aquatic environmental samples: Current status and challenges. *Sci. Total Environ.* **2021**, *793*, 148560. [CrossRef]
20. Jeon, S.K.; Kim, E.J.; Lee, J.; Lee, S. Potential risks of TiO_2 and ZnO nanoparticles released from sunscreens into outdoor swimming pools. *J. Hazard. Mater.* **2016**, *317*, 312–318. [CrossRef] [PubMed]

21. Khan, R.; Inam, M.A.; Iqbal, M.M.; Shoaib, M.; Park, D.R.; Lee, K.H.; Shin, S.; Khan, S.; Yeom, I.T. Removal of ZnO nanoparticles from naturalwaters by coagulation-flocculation process: Influence of surfactant type on aggregation, dissolution and colloidal stability. *Sustainability* **2019**, *11*, 17. [CrossRef]
22. Keller, A.A.; McFerran, S.; Lazareva, A.; Suh, S. Global life cycle releases of engineered nanomaterials. *J. Nanopart. Res.* **2013**, *15*, 1692. [CrossRef]
23. Liu, W.S.; Peng, Y.H.; Shiung, C.E.; Shih, Y.H. The effect of cations on the aggregation of commercial ZnO nanoparticle suspension. *J. Nanopart. Res.* **2012**, *14*, 1259. [CrossRef]
24. Philippe, A.; Schaumann, G.E. Interactions of dissolved organic matter with natural and engineered inorganic colloids: A review. *Environ. Sci. Technol.* **2014**, *48*, 8946–8962. [CrossRef] [PubMed]
25. Khan, A.U.H.; Naidu, R.; Dharmarajan, R.; Fang, C.; Shon, H.; Dong, Z.; Liu, Y. The interaction mechanisms of co-existing polybrominated diphenyl ethers and engineered nanoparticles in environmental waters: A critical review. *J. Environ. Sci.* **2023**, *124*, 227–252. [CrossRef]
26. Adeleye, A.S.; Keller, A.A. Interactions between algal extracellular polymeric substances and commercial TiO_2 nanoparticles in aqueous media. *Environ. Sci. Technol.* **2016**, *50*, 12258–12265. [CrossRef]
27. Yu, S.; Liu, J.; Yin, Y.; Shen, M. Interactions between engineered nanoparticles and dissolved organic matter: A review on mechanisms and environmental effects. *J. Environ. Sci.* **2018**, *63*, 198–217. [CrossRef]
28. Khan, A.U.H.; Liu, Y.; Naidu, R.; Fang, C.; Dharmarajan, R.; Shon, H. Interactions between zinc oxide nanoparticles and hexabromocyclododecane in simulated waters. *Environ. Technol. Innov.* **2021**, *24*, 102078. [CrossRef]
29. Khan, R.; Inam, M.A.; Khan, S.; Park, D.R.; Yeom, I.T. Interaction between persistent organic pollutants and ZnO NPs in synthetic and natural waters. *Nanomaterials* **2019**, *9*, 472. [CrossRef]
30. Yang, K.; Lin, D.; Xing, B. Interactions of humic acid with nanosized inorganic oxides. *Langmuir* **2009**, *25*, 3571–3576. [CrossRef] [PubMed]
31. Bian, S.W.; Mudunkotuwa, I.A.; Rupasinghe, T.; Grassian, V.H. Aggregation and dissolution of 4 nm ZnO nanoparticles in aqueous environments: Influence of pH, ionic strength, size, and adsorption of humic acid. *Langmuir* **2011**, *27*, 6059–6068. [CrossRef]
32. Han, Y.; Kim, D.; Hwang, G.; Lee, B.; Eom, I.; Kim, J.P.; Tong, M.; Kim, H. Aggregation and dissolution of ZnO nanoparticles synthesized by different methods: Influence of ionic strength and humic acid. *Colloids Surf. A Physicochem. Eng. Asp.* **2014**, *451*, 7–15. [CrossRef]
33. Shrestha, S.; Wang, B.; Dutta, P. Nanoparticle processing: Understanding and controlling aggregation. *Adv. Colloid Interface Sci.* **2020**, *279*, 102162. [CrossRef] [PubMed]
34. Sasidharan, N.P.; Chandran, P.; Khan, S.S. Interaction of colloidal zinc oxide nanoparticles with bovine serum albumin and its adsorption isotherms and kinetics. *Colloids Surf. B Biointerfaces* **2013**, *102*, 195–201. [CrossRef] [PubMed]
35. Keller, A.A.; Wang, H.; Zhou, D.; Lenihan, H.S.; Cherr, G.; Cardinale, B.J.; Miller, R.; Zhaoxia, J.I. Stability and aggregation of metal oxide nanoparticles in natural aqueous matrices. *Environ. Sci. Technol.* **2010**, *44*, 1962–1967. [CrossRef] [PubMed]
36. Majedi, S.M.; Kelly, B.C.; Lee, H.K. Combined effects of water temperature and chemistry on the environmental fate and behavior of nanosized zinc oxide. *Sci. Total Environ.* **2014**, *496*, 585–593. [CrossRef]
37. Zhou, H.; Yin, N.; Faiola, F. Tetrabromobisphenol A (TBBPA): A controversial environmental pollutant. *J. Environ. Sci. (China)* **2020**, *97*, 54–66. [CrossRef]
38. Covaci, A.; Voorspoels, S.; Abdallah, M.A.E.; Geens, T.; Harrad, S.; Law, R.J. Analytical and environmental aspects of the flame retardant tetrabromobisphenol-A and its derivatives. *J. Chromatogr. A* **2009**, *1216*, 346–363. [CrossRef] [PubMed]
39. Ni, H.G.; Zeng, H. HBCD and TBBPA in particulate phase of indoor air in Shenzhen, China. *Sci. Total Environ.* **2013**, *458–460*, 15–19. [CrossRef] [PubMed]
40. Takigami, H.; Suzuki, G.; Hirai, Y.; Sakai, S.I. Brominated flame retardants and other polyhalogenated compounds in indoor air and dust from two houses in Japan. *Chemosphere* **2009**, *76*, 270–277. [CrossRef]
41. Guerra, P.; Eljarrat, E.; Barceló, D. Simultaneous determination of hexabromocyclododecane, tetrabromobisphenol A, and related compounds in sewage sludge and sediment samples from Ebro River basin (Spain). *Anal. Bioanal. Chem.* **2010**, *397*, 2817–2824. [CrossRef]
42. Huang, D.Y.; Zhao, H.Q.; Liu, C.P.; Sun, C.X. Characteristics, sources, and transport of tetrabromobisphenol A and bisphenol A in soils from a typical e-waste recycling area in South China. *Environ. Sci. Pollut. Res.* **2014**, *21*, 5818–5826. [CrossRef] [PubMed]
43. Kowalski, B.; Mazur, M. The simultaneous determination of six flame retardants in water samples using SPE pre-concentration and UHPLC-UV method. *Water Air Soil Pollut.* **2014**, *225*, 1866. [CrossRef]
44. Shi, Z.; Zhang, L.; Zhao, Y.; Sun, Z.; Zhou, X.; Li, J.; Wu, Y. Dietary exposure assessment of Chinese population to tetrabromobisphenol-A, hexabromocyclododecane and decabrominated diphenyl ether: Results of the 5th Chinese Total Diet Study. *Environ. Pollut.* **2017**, *229*, 539–547. [CrossRef]

45. Malkoske, T.; Tang, Y.; Xu, W.; Yu, S.; Wang, H. A review of the environmental distribution, fate, and control of tetrabromobisphenol A released from sources. *Sci. Total Environ.* **2016**, *569–570*, 1608–1617. [CrossRef]
46. Lee, I.S.; Kang, H.H.; Kim, U.J.; Oh, J.E. Brominated flame retardants in Korean river sediments, including changes in polybrominated diphenyl ether concentrations between 2006 and 2009. *Chemosphere* **2015**, *126*, 18–24. [CrossRef]
47. Yang, S.; Wang, S.; Wu, F.; Yan, Z.; Liu, H. Tetrabromobisphenol A: Tissue distribution in fish, and seasonal variation in water and sediment of Lake Chaohu, China. *Environ. Sci. Pollut. Res.* **2012**, *19*, 4090–4096. [CrossRef]
48. Song, S.; Song, M.; Zeng, L.; Wang, T.; Liu, R.; Ruan, T. Occurrence and profiles of bisphenol analogues in municipal sewage sludge in China. *Environ. Pollut.* **2014**, *186*, 14–19. [CrossRef]
49. Zhou, Y.; Fang, X.; Gong, Y.; Xiao, A.; Xie, Y.; Liu, L.; Cao, Y. The interactions between zno nanoparticles (NPs) and α-linolenic acid (LNA) complexed to BSA did not influence the toxicity of ZnO NPs on HepG2 cells. *Nanomaterials* **2017**, *7*, 91. [CrossRef] [PubMed]
50. Kroll, A.; Behra, R.; Kaegi, R.; Sigg, L. Extracellular polymeric substances (EPS) of freshwater biofilms stabilize and modify CeO_2 and Ag nanoparticles. *PLoS ONE* **2014**, *9*, e110709. [CrossRef]
51. Wang, X.; Adeleye, A.S.; Wang, H.; Zhang, M.; Liu, M.; Wang, Y.; Li, Y.; Keller, A.A. Interactions between polybrominated diphenyl ethers (PBDEs) and TiO_2 nanoparticle in artificial and natural waters. *Water Res.* **2018**, *146*, 98–108. [CrossRef] [PubMed]
52. Domingos, R.F.; Rafiei, Z.; Monteiro, C.E.; Khan, M.A.K.; Wilkinson, K.J. Agglomeration and dissolution of zinc oxide nanoparticles: Role of pH, ionic strength and fulvic acid. *Environ. Chem.* **2013**, *10*, 306–312. [CrossRef]
53. Domingos, R.F.; Tufenkji, N.; Wilkinson, K.J. Aggregation of titanium dioxide nanoparticles: Role of a fulvic acid. *Environ. Sci. Technol.* **2009**, *43*, 1282–1286. [CrossRef] [PubMed]
54. AlSalem, H.S.; Algethami, F.K.; Al-Goul, S.T.; Shahat, A. Adsorption and Removal of Tetrabromobisphenol A by Adsorption on Functionalized Mesoporous Silica Nanotubes: Isotherms, Kinetics, Thermodynamics, and Optimization via Box-Behnken Design. *ACS Omega* **2023**, *8*, 20125–20137. [CrossRef] [PubMed]
55. Ighalo, J.O.; Yap, P.S.; Iwuozor, K.O.; Aniagor, C.O.; Liu, T.; Dulta, K.; Iwuchukwu, F.U.; Rangabhashiyam, S. Adsorption of persistent organic pollutants (POPs) from the aqueous environment by nano-adsorbents: A review. *Environ. Res.* **2022**, *212*, 113123. [CrossRef] [PubMed]
56. Blok, L.; Bruyn, P.L.D. The ionic double layer at the ZnO solution interface. I. The experimental point of zero charge. *J. Colloid Interface Sci.* **1970**, *32*, 518–526. [CrossRef]
57. Khan, A.U.H.; Liu, Y.; Fang, C.; Naidu, R.; Shon, H.K.; Rogers, Z.; Dharmarajan, R. A comprehensive physicochemical characterization of zinc oxide nanoparticles extracted from sunscreens and wastewaters. *Environ. Adv.* **2023**, *12*, 100381. [CrossRef]
58. Ouyang, K.; Yu, X.Y.; Zhu, Y.; Gao, C.; Huang, Q.; Cai, P. Effects of humic acid on the interactions between zinc oxide nanoparticles and bacterial biofilms. *Environ. Pollut.* **2017**, *231*, 1104–1111. [CrossRef]
59. Srivastava, V.; Gusain, D.; Sharma, Y.C. Synthesis, characterization and application of zinc oxide nanoparticles (n-ZnO). *Ceram. Int.* **2013**, *39*, 9803–9808. [CrossRef]
60. Chandrasekar, M.; Panimalar, S.; Uthrakumar, R.; Kumar, M.; Saravanan, M.E.R.; Gobi, G.; Matheswaran, P.; Inmozhi, C.; Kaviyarasu, K. Preparation and characterization studies of pure and Li+ doped ZnO nanoparticles for optoelectronic applications. *Mater. Today Proc.* **2021**, *36 Pt 2*, 228–231. [CrossRef]
61. Gharagozlou, M.; Naghibi, S. Sensitization of ZnO nanoparticle by vitamin B12: Investigation of microstructure, FTIR and optical properties. *Mater. Res. Bull.* **2016**, *84*, 71–78. [CrossRef]
62. Stuart, B.H. Infrared spectroscopy: Fundamentals and applications. In *Infrared Spectroscopy: Fundamentals and Applications*; Hunt, B.H., Ed.; John Wiley Sons Ltd.: Hoboken, NJ, USA, 2004; pp. 1–224. [CrossRef]
63. Zhang, Y.; Tang, Y.; Li, S.; Yu, S. Sorption and removal of tetrabromobisphenol A from solution by graphene oxide. *Chem. Eng. J.* **2013**, *222*, 94–100. [CrossRef]
64. Lin, H.; Wang, Y.; Niu, J.; Yue, Z.; Huang, Q. Efficient Sorption and Removal of Perfluoroalkyl Acids (PFAAs) from Aqueous Solution by Metal Hydroxides Generated in Situ by Electrocoagulation. *Environ. Sci. Technol.* **2015**, *49*, 10562–10569. [CrossRef]
65. Gordeeva, A.; Hsu, Y.J.; Jenei, I.Z.; Carvalho, P.H.B.B.; Simak, S.I.; Andersson, O.; Haüssermann, U. Layered Zinc Hydroxide Dihydrate, $Zn_5(OH)_{10}{\cdot}2H_2O$, from Hydrothermal Conversion of ϵ-$Zn(OH)_2$ at Gigapascal Pressures and its Transformation to Nanocrystalline ZnO. *ACS Omega* **2020**, *5*, 17617–17627. [CrossRef]
66. Rao, M.S.; Satyavathi, K.; Bhaskararao, Y.N.; Cole, S. Structural and spectral investigations of undoped and Mn^{2+} ion doped $Zn_3(PO_4)_2ZnO$ nanocrystalline phosphor materials. *J. Alloys Compd.* **2016**, *682*, 7–13. [CrossRef]
67. Gupta, A.; Bhatti, H.S.; Kumar, D.; Verma, N.K.; Tandon, R.P. Nano and bulk crystals of ZnO: Synthesis and characterization. *J. Nanomater. Biostruct.* **2006**, *1*, 1–9. Available online: http://www.chalcogen.infim.ro/Agupta.pdf (accessed on 14 January 2025).
68. Khoshhesab, Z.M.; Sarfaraz, M.; Asadabad, M.A. Preparation of ZnO nanostructures by chemical precipitation method, Synth. React. Inorganic. *Met. Nano-Metal Chem.* **2011**, *41*, 814–819. [CrossRef]

69. Talam, S.; Karumuri, S.R.; Gunnam, N. Synthesis, characterization, and spectroscopic properties of ZnO nanoparticles. *ISRN Nanotechnol.* **2012**, *2012*, 1–6. [CrossRef]
70. Zak, A.K.; Razali, R.; Majid, W.H.A.; Darroudi, M. Synthesis and characterization of a narrow size distribution of zinc oxide nanoparticles. *Int. J. Nanomed.* **2011**, *6*, 1399–1403. [CrossRef]
71. Xia, T.; Kovochich, M.; Liong, M.; Mädler, L.; Gilbert, B.; Shi, H.; Yeh, J.I.; Zink, J.I.; Nel, A.E. Comparison of the mechanism of toxicity of zinc oxide and cerium oxide nanoparticles based on dissolution and oxidative stress properties. *ACS Nano* **2008**, *2*, 2121–2134. [CrossRef]
72. Li, M.; Lin, D.; Zhu, L. Effects of water chemistry on the dissolution of ZnO nanoparticles and their toxicity to *Escherichia coli*. *Environ. Pollut.* **2013**, *173*, 97–102. [CrossRef] [PubMed]
73. Ravindran, A.; Singh, A.; Raichur, A.M.; Chandrasekaran, N.; Mukherjee, A. Studies on interaction of colloidal Ag nanoparticles with Bovine Serum Albumin (BSA). *Colloids Surf. B Biointerfaces* **2010**, *76*, 32–37. [CrossRef] [PubMed]
74. Topală, T.; Bodoki, A.; Oprean, L.; Oprean, R. Bovine serum albumin interactions with metal complexes. *Clujul Med.* **2014**, *87*, 5. [CrossRef] [PubMed]
75. Wei, K.C.; Lin, F.W.; Huang, C.Y.; Ma, C.C.M.; Chen, J.Y.; Feng, L.Y.; Yang, H.W. 1,3-Bis(2-Chloroethyl)-1-Nitrosourea-Loaded Bovine Serum Albumin Nanoparticles with Dual Magnetic Resonance–Fluorescence Imaging for Tracking of Chemotherapeutic Agents. *Int. J. Nanomed.* **2016**, *11*, 4065–4075. [CrossRef]
76. Emadi, F.; Amini, A.; Gholami, A.; Ghasemi, Y. Functionalized Graphene Oxide with Chitosan for Protein Nanocarriers to Protect against Enzymatic Cleavage and Retain Collagenase Activity. *Sci. Rep.* **2017**, *7*, 42258. [CrossRef]

toxics

MDPI

Article

Long-Term Retrospective Predicted Concentration of $PM_{2.5}$ in Upper Northern Thailand Using Machine Learning Models

Sawaeng Kawichai [1,†], Patumrat Sripan [1,†], Amaraporn Rerkasem [1], Kittipan Rerkasem [1,2,*] and Worawut Srisukkham [3,*]

1 Research Institute for Health Sciences, Chiang Mai University, Chiang Mai 50200, Thailand; sawaeng.kaw@cmu.ac.th (S.K.); patumrat.sripan@cmu.ac.th (P.S.); amaraporn.rer@cmu.ac.th (A.R.)
2 Clinical Surgical Research Center, Department of Surgery, Faculty of Medicine, Chiang Mai University, Chiang Mai 50200, Thailand
3 Department of Computer Science, Faculty of Science, Chiang Mai University, 239 Huay-Kaew Road, Suthep, Muang, Chiang Mai 50200, Thailand
* Correspondence: kittipan.r@cmu.ac.th (K.R.); worawut.s@cmu.ac.th (W.S.)
† These authors contributed equally to this work.

Abstract: This study aims to build, for the first time, a model that uses a machine learning (ML) approach to predict long-term retrospective $PM_{2.5}$ concentrations in upper northern Thailand, a region impacted by biomass burning and transboundary pollution. The dataset includes PM_{10} levels, fire hotspots, and critical meteorological data from 1 January 2011 to 31 December 2020. ML techniques, namely multi-layer perceptron neural network (MLP), support vector machine (SVM), multiple linear regression (MLR), decision tree (DT), and random forests (RF), were used to construct the prediction models. The best ML prediction model was selected considering root mean square error (RMSE), mean prediction error (MPE), relative prediction error (RPE) (the lower, the better), and coefficient of determination (R^2) (the bigger, the better). Our study found that the ML model-based RF technique using PM_{10}, CO_2, O_3, fire hotspots, air pressure, rainfall, relative humidity, temperature, wind direction, and wind speed performs the best when predicting the concentration of $PM_{2.5}$ with an RMSE of 6.82 $\mu g/m^3$, MPE of 4.33 $\mu g/m^3$, RPE of 22.50%, and R^2 of 0.93. The RF prediction model of $PM_{2.5}$ used in this research could support further studies of the long-term effects of $PM_{2.5}$ concentration on human health and related issues.

Keywords: $PM_{2.5}$ prediction; retrospective prediction; long-term prediction; machine learning; fire hotspots

1. Introduction

Air pollution has been an issue in upper northern Thailand for several years, and the haze situation that has been caused by forest fires is a significant factor in this problem [1–4]. This burning causes a pollution problem throughout the dry season, which typically lasts from the end of February to the middle of April [1–4]. It is likely that exposure to specific environmental pollutants could have long-term effects and risk factors that contribute to an increased probability of sustaining lung cancer [5]. During this period, the amount of particulate matter smaller than 10 and 2.5 microns (PM_{10} and $PM_{2.5}$) in the atmosphere exceeds the standards of Thailand. Moreover, the $PM_{2.5}$ concentrations measured during the sampling period of 24 h exceeded the $PM_{2.5}$ Thailand Ambient Air Quality Standard (50 $\mu g/m^3$) by less than 30.60% (112 days in 2019) [6].

In this region, there is a significant gap in research on the long-term health effects of $PM_{2.5}$ exposure, primarily due to the lack of comprehensive, long-term retrospective data

https://doi.org/10.3390/toxics13030170

on $PM_{2.5}$ concentrations. The absence of data complicates the understanding of the impact of long-term air pollution exposure on health, particularly on lung cancer, cardiovascular diseases (CVDs), and chronic obstructive pulmonary disease (COPD) [5]. The lack of comprehensive long-term datasets limits the development of reliable evidence-based public health policies. In Chiang Mai, upper northern Thailand, $PM_{2.5}$ monitoring was started in 2011, while $PM_{2.5}$ data later became available in Lampang in 2018 and in other provinces in 2019, including Chiang Rai, Lamphun, Phayao, Phrae, Nan, and Mae Hong Son, owing to the limitations of resources. The importance of this work extends beyond regional limits, as the modeling methodology may be modified for application in other areas facing comparable problems with air quality. The combination of multiple sources of data, such as air pollutant concentrations, fire hotspot information, and meteorological variables, provides a thorough methodology for fulfilling the essential requirement for historical $PM_{2.5}$ data in environmental health research. In upper northern Thailand, PM_{10} has been widely monitored for more than 20 years [7]. Predicting the results of $PM_{2.5}$ values using PM_{10} [8–10] and fire hotspot data [11,12] with critical meteorological data [13,14] allows the study of the long-term effects of past exposure to $PM_{2.5}$ on various health problems.

In previous studies, predictive methods have used multivariate statistical analysis, but in the last two decades, artificial intelligence technology using machine learning (ML) has been applied to create a model for forecasting or predicting air quality with an ability to predict results that are better than operational air quality measurements [15,16]. As a result, a wide range of research studies have been conducted that have applied various machine learning techniques such as artificial neural networks (ANNs) and support vector machines. Random forests classification has also been used to create a model for air quality prediction [15,16]. Studies on the model for predicting $PM_{2.5}$ and PM_{10} concentrations revealed that various machine learning algorithms, capable of managing intricate and non-linear relationships among air quality variables, can effectively predict the value of new, unseen data with remarkable efficiency and precision [15–19]. ANNs are techniques for machine learning that mimic the neural activity of the human brain, appearing like nodes arranged in one or more layers. The nodes communicate with each other and store information in the form of the weight of each line connecting the nodes. This technique can retain knowledge that it has acquired and has been used in many tasks, including pattern recognition, bioinformatics, prediction, and other applications in many fields [19]. MLP neural networks are also widely used in predictive modeling. ML models provide highly accurate prediction results for $PM_{2.5}$ and PM_{10} dust content [15,17–19]. This is the first time a model has been built that uses an ML approach to predict long-term retrospective $PM_{2.5}$ concentrations in upper northern Thailand, a region impacted by biomass burning and transboundary pollution. The modeling framework developed here could not only be applied to northern Thailand, but also adapted to other regions with similar air quality issues. Furthermore, the retrospective study of $PM_{2.5}$ data usually encounters spatial and temporal limitations due to the absence of government-provided monitoring stations for $PM_{2.5}$ measurement in upper northern Thailand over the past decade. Therefore, several critical factors warrant the development of an ML model to predict retrospective $PM_{2.5}$. Additionally, this integrative strategy not only bridges gaps in direct monitoring, but also advances our understanding of how exposure to these environmental factors could have long-term effects and risk factors that cause health issues. The main objective of this study is to apply ML methods to create a model for predicting retrospective $PM_{2.5}$ values using air pollutant concentrations, fire hotspot data, and meteorological data.

2. Materials and Methods

2.1. Descriptions of the Data

Air pollutant concentrations including $PM_{2.5}$, PM_{10}, CO_2, SO_2, NO_2, and O_3; fire hotspot data; and critical meteorological data including air pressure, rainfall, relative humidity, temperature, wind direction, and wind speed from 1 January 2011 to 31 December 2020 were collected from eight of the upper northern provinces of Thailand—Chiang Mai, Lampang, Chiang Rai, Lamphun, Phayao, Phrae, Nan, and Mae Hong Son—using the official database of the Pollution Control Department (PCD). In the PCD's monitoring station, $PM_{2.5}$ and PM_{10} concentrations were measured via the tapered element oscillating microbalance method (TEOM) and then averaged at the data center to produce a time series of the daily mean of air pollutant concentrations and meteorological data. Air quality and meteorological data were collected as daily mean values from fixed monitoring stations operated by the PCD in each province. All datasets were aggregated on a daily basis and assigned a location number, e.g., 35t and 36t represent Chiang Mai province. This synchronization ensured that the model used co-located information from all data sources for each day. The PCD's monitoring stations, located in eight of the provinces of upper northern Thailand, are shown in Figure 1.

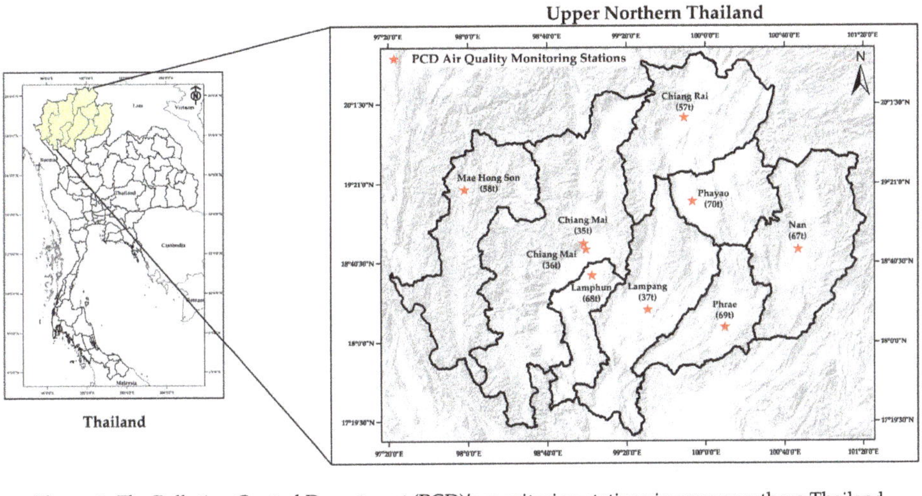

Figure 1. The Pollution Control Department (PCD)'s monitoring stations in upper northern Thailand.

The daily fire hotspot number was retrieved from NASA's Fire Information for Resource Management System (FIRMS). In this research, we obtained fire hotspot data from the MODIS Terra and Aqua Collection 6.1 via the NASA Level-1 and atmospheric archive and distribution system [20].

2.2. Predictive Model

The $PM_{2.5}$ prediction models were constructed by employing twelve input parameters: (1) PM_{10}, (2) CO_2, (3) SO_2, (4) NO_2, (5) O_3, (6) fire hotspots, (7) air pressure, (8) rainfall, (9) relative humidity, (10) temperature, (11) wind direction, and (12) wind speed. The $PM_{2.5}$ data were collected during different time periods for each province—in 2011 for Chiang Mai, 2018 for Lampang, and 2019 for Chiang Rai, Lamphun, Phayao, Phrae, Nan, and Mae Hong Son. The SO_2 data were not available for Mae Hong Son province, and the NO_2 data in this province were incomplete and inconsistent. So, we built the predictive model of $PM_{2.5}$ based on the data from Chiang Mai province, which has the oldest, longest, and most complete air quality and meteorological data from two monitoring stations (Figure 1).

Among 6974 records from Chiang Mai province used as training data, the amounts of missing data were 0.06% for NO_2, 0.20% for SO_2, 0.68% for rainfall, 1.15% for CO_2, 1.15% for O_3, and 47.57% for air pressure. We compared the performance of the model when using different numbers of features: (1) all 12 input features, (2) 11 input features (without SO_2), and (3) 10 input features (without SO_2 and NO_2).

The predictive models were built using supervised ML. The models were trained on a labeled dataset, meaning that each input data point had a corresponding output, which was the $PM_{2.5}$ concentration. The goal was for the model to learn the relationship between each input feature and the output so that it could predict the output for unseen data. One of the ML techniques used in this study was multi-layer perceptron (MLP), which is one of the most popular supervised neural network modelling techniques. It has been widely used in pattern recognition, bioinformatics, and computer vision and control systems [21]. MLP is a modern feed-forward neural network that consists of fully connected neurons or nodes. The nodes have a non-linear activation function, which is responsible for processing and giving answers to the next connecting nodes. It is usually trained using the backpropagation algorithm. Additionally, MLP consists of at least three layers of node networks: an input layer, a hidden layer, and an output layer [22]. In this study, we constructed MLP models using both a single hidden layer and two hidden layers. We employed the Levenberg–Marquardt backpropagation learning technique for both MLP models. We employed the Sigmoid activation function for the hidden layer and the Linear activation function for the output layer in a single hidden-layer configuration. The learning rate was 0.1, the momentum rate was 0.8, and the number of nodes varied from 1 to 50. Furthermore, for the MLP with two hidden layers, we employed the Sigmoid activation function for the hidden levels and the Linear activation function for the output layer. The learning rate was set at 0.1 and the momentum rate at 0.8, and the quantity of hidden-layer nodes varied from 1 to 30. We also used support vector machine (SVM), which is a kernel-based classification method. In general, it has to compute a linear function in a higher dimensional feature space, where the lower dimensional input data are mapped using a kernel function. It is used extensively in many fields such as prediction, pattern recognition, and classification [23]. This study involved constructing SVM models that utilize three distinct kernels: Linear, Polynomial, and Radial Basis functions. We employed grid search as the optimization method. The maximum objective evaluation and the maximum iteration were both set at 100. Multiple linear regression (MLR), which we also used, is a statistical model that estimates the relationships between one dependent variable and more independent variables by fitting multiple lines to the observed data. MLR extends a simple linear regression to include more than one explanatory variable to predict the outcome of a response variable [24,25]. Moreover, it is a widely used technique in many fields such as social sciences research, econometrics, and financial inference. In this study, we constructed the MLR model by using the "regress()" function for the multiple linear regression with a 95% confidence interval and setting epsilon (ε) to 0. Another technique we used was decision tree (DT), which is one of the general-purpose computationally intensive statistical algorithms for prediction and classification, artificial intelligence, machine learning, and knowledge discovery. DT has to do with using a procedure or rule repeatedly to generate subsetting of the target subject of data according to the values of associated input subjects to make partitions, and associated descendent leaves or nodes of the tree, that contain progressively similar intra-node target values and progressively dissimilar inter-node values at any given height of the tree [26]. This study involved the construction of a DT regression model utilizing a grid search optimizer, with the objective evaluation maximum set as 30. Random forests (RF) is another technique we used and is one of the famous ensemble machine learning techniques. Researchers have widely used it due to its good performance and simple usage [27,28]. This technique uses multiple decision trees. The trees' predictors are taught using random sample data, and the distribution is the same for the predictors of all trees in the forest. The primary voting method is used to choose the greatest number of identical answers (majority voting). In this study, we developed an RF model with the following hyperparameters: the number of trees was set as 10, the maximum depth of the trees was set as 5, and the number of learning cycles for the trees was set as 200. The experiments in this work were performed using MATLAB R2018a. A summary of the hyperparameter settings for the machine learning models for the dataset from Chiang Mai province (air quality data, meteorological data, and fire hotspots) is shown in Table S1.

2.3. Model Validation

The best ML prediction model was selected considering the root mean square error (RMSE), mean prediction error (MPE), relative prediction error (RPE) (the lower, the better), and coefficient of determination (R^2) (the bigger, the better) [29–31]. Additionally, in the evaluation of each run, 10-fold cross-validation [32] was implemented, dividing the dataset into 10 equally sized folds. In each iteration, 1 fold was designated as the validation set, with the remaining 9 folds used for training. This procedure was repeated until every fold had been used as the validation set once. This validation was applied to the dataset, which was divided into 70% for training and 30% for testing.

2.4. Prediction Evaluation Visual Check

The predicted $PM_{2.5}$ was compared with the observed data to evaluate the performance of the model. Visual inspection was used to confirm that the predicted and observed data were aligned as their two-way plots were on the identical line. The RMSE, R^2, MPE, and RPE were provided in addition to the graphical check. We performed the evaluation for both seen data (data for training model) and unseen data (data for testing model). This research used data from Chiang Mai province, which has the oldest, longest, and most complete data, for the visual check of the training data. Additionally, data from eight provinces were employed for the visual check of the unseen data.

3. Results

The air quality data, meteorological data, and data on fire hotspots in Chiang Mai province were separated into two parts, a dataset for training and a dataset for testing. The performances of the ML models—MLP, SVM, MLR, DT, and RF—are shown in Table 1.

Table 1. Performances of ML models for $PM_{2.5}$ prediction using different numbers of features.

Methods	Prediction Performances											
	10 Features (Without SO_2 and NO_2)				11 Features (Without SO_2)				12 Features			
	RMSE	R^2	MPE	RPE	RMSE	R^2	MPE	RPE	RMSE	R^2	MPE	RPE
MLP (1 Hidden Layer)	7.2287	0.9211	4.7944	23.88	7.2136	0.9214	4.8845	23.84	7.1802	0.9221	4.8121	23.73
MLP (2 Hidden Layers)	7.3328	0.9181	4.8184	24.24	7.3265	0.9189	4.8822	24.19	7.2367	0.9210	4.8854	23.91
SVM (Linear Kernel)	10.7402	0.8223	8.2057	35.51	11.1608	0.8111	8.4791	36.79	11.7684	0.7752	9.2530	38.99
SVM (Polynomial Kernel)	12.9420	0.7367	10.2391	42.70	12.5287	0.7602	9.8174	41.38	12.1642	0.7621	8.9826	40.08
SVM (RBF Kernel)	12.0770	0.7748	8.9261	39.91	12.6847	0.7539	9.6007	41.82	12.1247	0.7755	9.0067	40.01
MLR	7.7423	0.9103	5.2223	25.55	7.7415	0.9103	5.2270	25.55	7.7056	0.9111	5.2141	25.44
DT	9.0747	0.8762	5.8224	29.96	8.7843	0.8840	5.5989	29.01	8.9378	0.8800	5.6141	29.52
RF	6.8242	0.9306	4.3296	22.50	6.8234	0.9306	4.2499	22.49	6.7615	0.9318	4.1954	22.29

Note: RMSE: root mean square error ($\mu g/m^3$); R^2: coefficient of determination; MPE: mean prediction error ($\mu g/m^3$); RPE: relative prediction error (%).

When using all features, the RF model is the best model, considering its lowest RMSE at 6.7615 μg/m^3, MPE at 4.1954 μg/m^3, RPE at 22.29%, and highest R^2 at 0.9318. Therefore, the RF model was selected and used for the next steps of this study.

3.1. Performance of RF Model with Different Features

Based on the dataset from Chiang Mai province, after removing SO_2, the RMSE, MPE, and RPE slightly increased to 6.8234 μg/m^3, 4.2499 μg/m^3, and 22.49%, respectively, and R^2 was reduced to 0.9306. The performance of the model without SO_2 was not different when compared to the full features model ($p > 0.05$). When NO_2 was removed from the reduced model, the RMSE, MPE, and RPE increased to 6.8242 μg/m^3, 4.3296 μg/m^3, and 22.50%, respectively, while the R^2 did not change. No significant change in model performance was observed when the number of features was reduced to 10. The performance of the model without SO_2 and NO_2 was not different when compared to the full model ($p > 0.05$) (Table 2).

Table 2. Comparison between $PM_{2.5}$ prediction performances of RF model with different features.

Performances	12 Features	11 Features	*p*-Value	10 Features	*p*-Value
Average RMSE	6.7859	6.8110	0.8798	6.8216	0.9397
Average R^2	0.9313	0.9308	0.9397	0.9307	0.8798
Average MPE	4.3290	4.2533	0.2568	4.1944	0.3258
Average RPE	22.3740	22.4551	0.9397	22.4884	1.0000

3.2. Performance of RF During $PM_{2.5}$ Prediction in Eight Provinces in Upper Northern Thailand

The predictive model without SO_2 and NO_2 was used to predict $PM_{2.5}$ in eight provinces in Northern Thailand. Figure 2 demonstrates the performance of the RF model on the training data, employing data from Chiang Mai province.

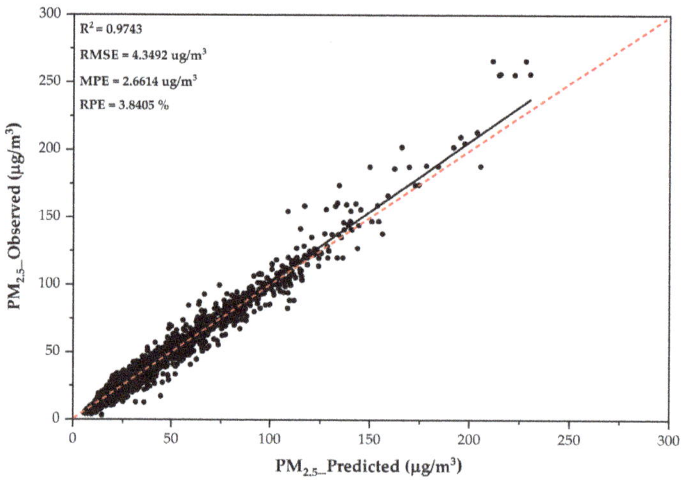

Figure 2. The performance of the RF model during $PM_{2.5}$ prediction using training data from Chiang Mai province.

The value of R^2 for the model used is 0.9743, while the R^2 ranged from 0.8797 to 0.9783 for the testing data (Figure 3). The R^2 was highest in Mae Hong Son province and lowest in Nan province. The performance of the model considering RMSE, MPE, and RPE indicated the same direction.

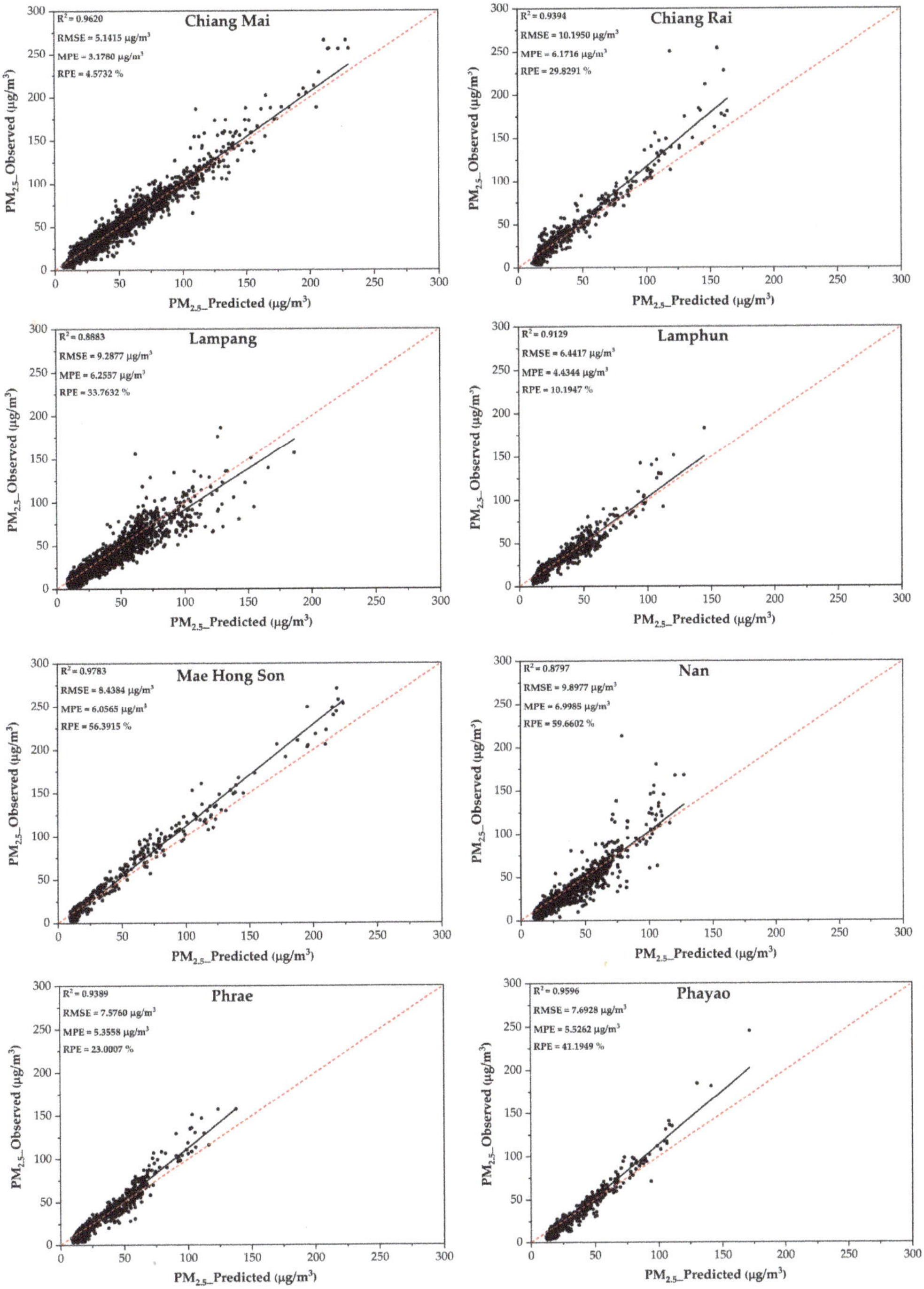

Figure 3. The performance of the RF model during $PM_{2.5}$ prediction using testing data from eight provinces in upper northern Thailand.

4. Discussion

From this study, it was found that the RF model was the most effective and had the highest accuracy in predicting $PM_{2.5}$ concentrations compared to the other models. This is consistent with a study by Chen [33] that predicted $PM_{2.5}$ using eight types of air quality data, as well as five types of meteorological data. Chen's study found that using the RF model was the most efficient way to predict $PM_{2.5}$, which had a relatively high R^2 value of 0.94. In 2023, Vignesh et al. performed a study that employed ML techniques. There were nine models, and a tool was established to evaluate their performance and accuracy when predicting $PM_{2.5}$ concentrations. The research, which was conducted in the United States, employed air pollution data collected over a period of five years, from 2017 to 2021 [34]. The investigation revealed that the RF model demonstrated high effectiveness when predicting concentrations of $PM_{2.5}$, with an R^2 of 0.77.

The higher concentrations of particulate pollution observed during the dry season are most likely a result of significant biomass burning, specifically from agricultural activities performed in preparation for the next agricultural season. Another issue is the transboundary transport of air pollution originating from neighboring countries such as Laos, Vietnam, and Myanmar, which is influenced by meteorological situations. These elements increase the problem of air quality in upper northern Thailand [35,36]. This study built a prediction model with data from Chiang Mai province. The unique model can be applied to eight provinces because of their comparable area characteristics. The primary factors contributing to air pollution in northern Thailand include biomass combustion, meteorological conditions, and geographical characteristics [37]. Moreover, a wide range of parameters have a significant impact on R^2, with geographical data being particularly important. Various variables influence the value of R^2, with Mae Hong Son province having the most significant influence. Mae Hong Son is a small province surrounded by forest. Mae Hong Son is located close to the Thailand–Myanmar border. Mae Hong Son province experiences multiple sources of air pollution, including transboundary effects, forest fires, and biomass burning, as indicated in prior research conducted by Kliengchuay et al. [38]. Forest fires play an important part in the emission of $PM_{2.5}$ in this area. The complicated relationships between meteorological conditions and $PM_{2.5}$ levels affect the numerous relations between $PM_{2.5}$ and meteorology [39], while the lower R^2 in an area like Nan, which is a large province, may be influenced by various meteorological parameters such as rainfall, wind direction, wind speed, temperature, relative humidity, and air pressure. Moreover, the lack of clarity regarding development across large areas could be contributing to the lower R^2 [40,41].

Additionally, this study shows that the RF model is the most effective in predicting $PM_{2.5}$, consistent with the findings of research conducted by Chen et al. [33] and Vignesh et al. [34]. However, our study differs from those studies in that we have included fire hotspots as a feature in our ML model for $PM_{2.5}$ prediction, in addition to factors such as the environment, the climate, and geographical characteristics. The topography, meteorological data, and agriculture of Southeast Asia significantly contributes to the prevalence of monoculture farming, resulting in an important number of fire hotspots. These regions in Southeast Asia show evidence of biomass burning, which emits $PM_{2.5}$ pollutants [1–4]. Thus, our study employs the number of fire hotspots as one of the important features for modeling the $PM_{2.5}$ predictor.

The retrospective $PM_{2.5}$ data from our research could be helpful for studying the long-term effects of $PM_{2.5}$ concentrations on human health issues such as lung cancer, cardiovascular diseases (CVDs), and chronic obstructive pulmonary disease (COPD). There are relatively few studies on the impact of exposure to $PM_{2.5}$ on lung cancer incidence rates among Asian populations. There are literature review studies that aim to explore the relationship between $PM_{2.5}$ and lung cancer incidence and mortality. One review study shows that some studies have found a significant relationship between $PM_{2.5}$ and the incidence of lung cancer, but other studies did not find this relationship [5]. The limited number of studies on the impact of exposure to $PM_{2.5}$ on lung cancer may be caused by the limited long-term $PM_{2.5}$ data, particularly in low- and middle-income countries. Our predicted $PM_{2.5}$ data, as shown in Supplementary Table S2, could be included in epidemiological health outcome prediction models to clarify $PM_{2.5}$-related health risks in upper northern Thailand. Additionally, investigating the relationships between socio-economic characteristics, healthcare accessibility, and health outcomes associated with $PM_{2.5}$ would enhance comprehension of the differences in disease burden. Long-term cohort studies that monitor individuals over time, considering both environmental exposures and personal health data, could be essential for enhancing exposure response models and guiding public health strategies to reduce the harmful effects of air pollution. Finally, the $PM_{2.5}$ concentration values from the past 10 years (2011 to 2020) that were predicted during our study could be used to investigate the long-term impact of $PM_{2.5}$ on acute and chronic respiratory diseases, as well as to study other health-related effects of $PM_{2.5}$ in eight provinces in the upper northern region of Thailand, an area where $PM_{2.5}$ concentration levels are reported to exceed Thailand's ambient air quality standard every year.

5. Conclusions

This study found that the RF model was the most effective in predicting $PM_{2.5}$ concentrations, outperforming other models in terms of accuracy. It also highlights the significant impact of biomass burning and fire hotspots on the prediction of $PM_{2.5}$ concentrations. Our study illustrates the significance of employing numerous data sources and effective methods for modeling in environmental studies. Moreover, future work could apply the RF model, which is a highly effective tool for predicting long-term $PM_{2.5}$ concentrations (RMSE of 6.82 $\mu g/m^3$, MPE of 4.33 $\mu g/m^3$, RPE of 22.50%, and R^2 of 0.93), to other regions or countries with similar environmental conditions, including biomass burning and transboundary pollution. The predicted $PM_{2.5}$ concentrations could lead to improved air quality management strategies and more informed public health policies. Furthermore, our research suggests that the prediction model of prolonged $PM_{2.5}$ concentrations could offer a foundation for further epidemiological studies on the long-term effects of $PM_{2.5}$ concentrations on human health and related problems.

Supplementary Materials: The following supporting information can be downloaded at: https://www.mdpi.com/article/10.3390/toxics13030170/s1, Table S1: The hyperparameters of the machine learning models for the dataset from Chiang Mai province (air quality data, meteorological data, and fire hotspots); Table S2: The predicted $PM_{2.5}$ concentrations in 8 provinces in northern Thailand, 2011–2020.

Author Contributions: Conceptualization, S.K., P.S., K.R. and W.S.; data curation, S.K., P.S. and W.S.; formal analysis, P.S. and W.S.; funding acquisition, K.R. and W.S.; investigation, A.R.; methodology, P.S. and W.S.; supervision, K.R. and W.S.; validation, P.S.; writing—original draft, S.K., P.S., A.R., K.R. and W.S.; writing—review and editing, S.K., P.S., A.R., K.R. and W.S. All authors will be updated at each stage of manuscript processing, including submission, revision, and revision reminder, via emails from our system or the assigned Assistant Editor. All authors have read and agreed to the published version of the manuscript.

Funding: This research was supported by the Coordinating Center for Thai Government Science and Technology Scholarship Students (CSTS) and the National Science and Technology Development Agency (NSTDA). It was partially supported by fundamental fund 2567 (FF67) from Chiang Mai University, and the $PM_{2.5}$ project was supported by Chiang Mai University.

Institutional Review Board Statement: Not applicable.

Informed Consent Statement: Not applicable.

Data Availability Statement: The datasets used and/or analyzed during the current study are available from the corresponding author upon reasonable request.

Acknowledgments: The authors express their gratitude to the Coordinating Center for Thai Government Science and Technology Scholarship Students (CSTS); the National Science and Technology Development Agency (NSTDA) THAILAND; the Department of Computer Science, Faculty of Science, Chiang Mai University, Chiang Mai, Thailand; and the Research Institute for Health Sciences, Chiang Mai University, Chiang Mai, Thailand.

Conflicts of Interest: The authors declare that they have no competing interests.

Abbreviations

The following abbreviations are used in this manuscript:

ANNs	Artificial neural networks
COPD	Chronic obstructive pulmonary disease
CVDs	Cardiovascular diseases
DT	Decision tree
FIRMS	Fire Information for Resource Management System
ML	Machine learning
MLP	Multi-layer perceptron neural network
MLR	Multiple linear regression
MODIS	Moderate Resolution Imaging Spectroradiometer
NASA	National Aeronautics and Space Administration
PCD	Pollution Control Department
$PM_{2.5}$	Particulate matter smaller than 2.5 microns
PM_{10}	Particulate matter smaller than 10 microns
RF	Random forests
SVM	Support vector machine
TEOM	Tapered element oscillating microbalance method

References

1. Chansuebsri, S.; Kraisitnitikul, P.; Wiriya, W.; Chantara, S. Fresh and aged $PM_{2.5}$ and their ion composition in rural and urban atmospheres of Northern Thailand in relation to source identification. *Chemosphere* **2021**, *286*, 131803. [CrossRef]
2. Kawichai, S.; Prapamontol, T.; Cao, F.; Song, W.; Zhang, Y. Source Identification of $PM_{2.5}$ during a smoke haze period in Chiang Mai, Thailand, using stable carbon and nitrogen isotopes. *Atmosphere* **2022**, *13*, 1149. [CrossRef]
3. Kawichai, S.; Prapamontol, T.; Cao, F.; Song, W.; Zhang, Y.L. Characteristics of carbonaceous species of $PM_{2.5}$ in Chiang Mai city, Thailand. *Aerosol Air Qual. Res.* **2024**, *24*, 230269. [CrossRef]
4. Song, W.; Hong, Y.; Zhang, Y.; Cao, F.; Rauber, M.; Santijitpakdee, T.; Kawichai, S.; Prapamontol, T.; Szidat, S.; Zhang, Y.L. Biomass burning greatly enhances the concentration of fine carbonaceous aerosols at an urban area in upper northern Thailand: Evidence from the radiocarbon-based source apportionment on size-resolved aerosols. *J. Geophys. Res. Atmos.* **2024**, *129*, e2023JD040692. [CrossRef]
5. Huang, F.; Pan, B.; Wu, J.; Chen, E.; Chen, L. Relationship between exposure to $PM_{2.5}$ and lung cancer incidence and mortality: A meta-analysis. *Oncotarget* **2017**, *8*, 43322–43331. [CrossRef]
6. Department of Pollution Control. *Manual Report: Air Quality Data Monitoring*; Pollution Control Department: Bangkok, Thailand, 2024; Available online: https://www.pcd.go.th/wp-content/uploads/2021/03/pcdnew-2021-04-07_06-54-58_342183.pdf (accessed on 10 January 2025).
7. Pollution Control Department. Air Quality and Noise. Available online: http://air4thai.pcd.go.th/webV3/#/Home (accessed on 10 January 2025).
8. Sirignano, C.; Riccio, A.; Chianese, E.; Ni, H.; Zenker, K.; D'Onofrio, A.; Meijer, H.A.J.; Dusek, U. High contribution of biomass combustion to $PM_{2.5}$ in the city centre of Naples (Italy). *Atmosphere* **2019**, *10*, 451. [CrossRef]
9. Xu, G.; Jiao, L.; Zhao, S.; Cheng, J. Spatial and temporal variability of $PM_{2.5}$ concentration in China. *Wuhan Univ. J. Nat. Sci.* **2016**, *21*, 358–368. [CrossRef]

10. Zhuang, Y.; Chen, D.; Li, R.; Chen, Z.; Cai, J.; He, B.; Gao, B.; Cheng, N.; Huang, Y. Understanding the influence of crop residue burning on $PM_{2.5}$ and PM_{10} concentrations in China from 2013 to 2017 using MODIS data. *Int. J. Environ. Res. Public Health* **2018**, *15*, 1504. [CrossRef]
11. Geng, G.; Murray, N.L.; Tong, D.; Fu, J.S.; Hu, X.; Lee, P.; Meng, X.; Chang, H.H.; Liu, Y. Satellite-based daily $PM_{2.5}$ estimates during fire seasons in Colorado. *J. Geophys. Res. Atmos.* **2018**, *123*, 8159–8171. [CrossRef]
12. Lee, H.H.; Iraqui, O.; Gu, Y.; Yim, S.H.L.; Chulakadabba, A.; Tonks, A.Y.M.; Yang, Z.; Wang, C. Impacts of air pollutants from fire and non-fire emissions on the regional air quality in southeast asia. *Atmos. Chem. Phys.* **2018**, *18*, 6141–6156. [CrossRef]
13. Chen, Z.; Chen, D.; Zhao, C.; Kwan, M.P.; Cai, J.; Zhuang, Y.; Zhao, B.; Wang, X.; Chen, B.; Yang, J.; et al. Influence of meteorological conditions on $PM_{2.5}$ concentrations across China: A review of methodology and mechanism. *Environ. Int.* **2020**, *139*, 105558. [CrossRef]
14. Li, Y.; Chen, Q.; Zhao, H.; Wang, L.; Tao, R. Variations in PM_{10}, $PM_{2.5}$ and $PM_{1.0}$ in an urban area of the Sichuan basin and their relation to meteorological factors. *Atmosphere* **2015**, *6*, 150–163. [CrossRef]
15. Suleiman, A.; Tight, M.R.; Quinn, A.D. Applying machine learning methods in managing urban concentrations of traffic-related particulate matter (PM_{10} and $PM_{2.5}$). *Atmos. Pollut. Res.* **2019**, *10*, 134–144. [CrossRef]
16. Zhang, G.; Rui, X.; Fan, Y. Critical review of methods to estimate $PM_{2.5}$ concentrations within specified research region. *ISPRS Int. J. Geo-Inf.* **2018**, *7*, 368. [CrossRef]
17. Biancofiore, F.; Busilacchio, M.; Verdecchia, M.; Tomassetti, B.; Aruffo, E.; Bianco, S.; Di Tommaso, S.; Colangeli, C.; Rosatelli, G.; Di Carlo, P. Recursive neural network model for analysis and forecast of PM_{10} and $PM_{2.5}$. *Atmos. Pollut. Res.* **2017**, *8*, 652–659. [CrossRef]
18. Chen, M.J.; Yang, P.H.; Hsieh, M.T.; Yeh, C.H.; Huang, C.H.; Yang, C.M.; Lin, G.M. Machine learning to relate $PM_{2.5}$ and PM_{10} concentrations to outpatient visits for upper respiratory tract infections in Taiwan: A nationwide analysis. *World J. Clin. Cases.* **2018**, *6*, 200–206. [CrossRef]
19. Gholizadeh, A.; Neshat, A.A.; Conti, G.O.; Ghaffari, H.R.; Aval, H.E.; Almodarresi, S.A.; Aval, M.Y.; Zuccarello, P.; Taghavi, M.; Mohammadi, A.; et al. $PM_{2.5}$ concentration modeling and mapping in the urban areas. *Model. Earth Syst. Environ.* **2019**, *5*, 897–906. [CrossRef]
20. Fire Information for Resource Management System. Available online: https://firms.modaps.eosdis.nasa.gov/active_fire (accessed on 24 October 2024).
21. Hagan, M.; Demuth, H.; Beale, M. *Neural Network Design*; PWS Publishing: Boston, MA, USA, 1997.
22. Cybenko, G. Approximation by superpositions of a sigmoidal function. *Math. Cont. Sig. Syst.* **1989**, *2*, 303–314. [CrossRef]
23. Basak, D.; Pal, S.; Patranabis, D. Support vector regression. *Neural Inf. Process. -Lett. Rev.* **2007**, *11*, 203–224.
24. Freedman, D. *Statistical Models: Theory and Practice*; Cambridge University Press: Cambridge, UK, 2005.
25. Tranmer, M.; Murphy, J.; Elliot, M.; Pampaka, M. *Multiple Linear Regression*, 2nd ed.; Cathie Marsh Institute Working Paper 2020–01; Cathie Marsh Institute for Social Research: Manchester, UK, 2020; Available online: https://hummedia.manchester.ac.uk/institutes/cmist/archive-publications/working-papers/2020/multiple-linear-regression.pdf (accessed on 10 January 2025).
26. de Ville, B. Decision trees. *WIREs Comp Stats.* **2013**, *5*, 448–455. [CrossRef]
27. Breiman, L. Random Forests. *Mach. Learn.* **2001**, *45*, 5–32. [CrossRef]
28. González, S.; García, S.; Del Ser, J.; Rokach, L.; Herrera, F. A practical tutorial on bagging and boosting based ensembles for machine learning: Algorithms, software tools, performance study, practical perspectives and opportunities. *Inf. Fusion* **2020**, *64*, 205–237. [CrossRef]
29. Sun, Y.; Zeng, Q.; Geng, B.; Lin, X.; Sude, B.; Chen, L. Deep learning architecture for estimating hourly ground-level $PM_{2.5}$ using satellite remote sensing. *IEEE Geosci. Remote Sens. Lett.* **2019**, *16*, 1343–1347. [CrossRef]
30. Wang, W.; Zhao, S.; Jiao, L.; Taylor, M.; Zhang, B.; Xu, G.; Hou, H. Estimation of $PM_{2.5}$ concentrations in China using a spatial back propagation neural network. *Sci. Rep.* **2019**, *9*, 13788. [CrossRef]
31. Yun, E.; Tornero-Velez, R.; Purucker, S.; Chang, D.; Edginton, A. Evaluation of quantitative structure property relationship algorithms for predicting plasma protein binding in humans. *Comput. Toxicol.* **2020**, *17*, 100142. [CrossRef] [PubMed]
32. Jain, A.; Duin, R.; Mao, J. Statistical pattern recognition: A review. IEEE Trans. *Pattern Anal. Mach. Intell.* **2000**, *22*, 4–37. [CrossRef]
33. Chen, M.; Chen, Y.C.; Chou, T.Y.; Ning, F.S. $PM_{2.5}$ Concentration prediction model: A CNN-RF ensemble framework. *Int. J. Environ. Res. Public Health* **2023**, *20*, 4077. [CrossRef]
34. Vignesh, P.P.; Jiang, J.H.; Kishore, P. Predicting PM2.5 concentrations across USA using machine learning. *Earth Space Sci.* **2023**, *10*, e2023EA002911. [CrossRef]
35. Amnuaylojaroen, T.; Kreasuwun, J. Investigation of fine and coarse particulate matter from burning areas in Chiang Mai, Thailand using the WRF/CALPUFF. *Chiang Mai J. Sci.* **2011**, *39*, 311–326.
36. Punsompong, P.; Pani, S.; Wang, S.H.; Thao, P. Assessment of biomass-burning types and transport over Thailand and the associated health risks. *Atmos. Environ.* **2020**, *247*, 118176. [CrossRef]

37. Suriyawong, P.; Chuetor, S.; Samae, H.; Piriyakarnsakul, S.; Amin, M.; Furuuchi, M.; Hata, M.; Inerb, M.; Phairuang, W. Airborne particulate matter from biomass burning in Thailand: Recent issues, challenges, and options. *Heliyon* **2023**, *9*, e14261. [CrossRef] [PubMed]
38. Kliengchuay, W.; Meeyai, A.; Worakhunpiset, S.; Tantrakarnapa, K. Relationships between meteorological parameters and particulate matter in Mae Hong Son Province, Thailand. *Int. J. Environ. Res. Public Health* **2018**, *15*, 2801. [CrossRef] [PubMed]
39. Chen, Z.; Xie, X.; Cai, J.; Danlu, C.; Gao, B.; He, B.; Cheng, N.; Xu, B. Understanding meteorological influences on $PM_{2.5}$ concentrations across China: A temporal and spatial perspective. *Atmos. Chem. Phys.* **2018**, *18*, 5343–5358. [CrossRef]
40. Chang, C.H.; Hsiao, Y.L.; Hwang, C. Evaluating spatial and temporal variations of aerosol optical depth and biomass burning over southeast asia based on satellite data products. *Aerosol Air Qual. Res.* **2015**, *15*, 2625–2640. [CrossRef]
41. Mohammadi, F.; Teiri, H.; Hajizadeh, Y.; Abdolahnejad, A.; Ebrahimi, A. Prediction of atmospheric $PM_{2.5}$ level by machine learning techniques in Isfahan, Iran. *Sci. Rep.* **2024**, *14*, 2109. [CrossRef] [PubMed]

toxics

MDPI

Review

Polycyclic Aromatic Hydrocarbons (PAHs) in Freshwater Systems: A Comprehensive Review of Sources, Distribution, and Ecotoxicological Impacts

Pedro J. Berríos-Rolón, María C. Cotto * and Francisco Márquez *

Nanomaterials Research Group, Department of Natural Sciences and Technology, Division of Natural Sciences, Technology and Environment, Universidad Ana G. Méndez-Gurabo Campus, Gurabo, PR 00778, USA; berriospl1@uagm.edu

* Correspondence: mcotto48@uagm.edu (M.C.C.); fmarquez@uagm.edu (F.M.); Tel.: +1-787-743-7979 (ext. 4491) (M.C.C.); +1-787-743-7979 (ext. 4250) (F.M.)

Abstract: This comprehensive review offers new perspectives on the distribution, sources, and ecotoxicological impacts of polycyclic aromatic hydrocarbons (PAHs) in freshwater systems. Unlike previous reviews, this work integrates recent findings on PAH dynamics within environmental matrices and emphasizes spatiotemporal variability across geographic regions. It critically examines both anthropogenic and natural sources, as well as the physical, chemical, and biological mechanisms driving PAH transport and fate. Special attention is given to the ecotoxicological effects of PAHs on freshwater organisms, including bioaccumulation, endocrine disruption, and genotoxicity. Notably, this review identifies key knowledge gaps and proposes an interdisciplinary framework to assess ecological risk and guide effective monitoring and management strategies for the protection of freshwater ecosystems.

Keywords: polycyclic aromatic hydrocarbons; water pollution; ecotoxicology; environmental matrices; freshwater ecosystems

1. Introduction

1.1. Polycyclic Aromatic Hydrocarbons

Polycyclic aromatic hydrocarbons (PAHs) are semi-volatile organic pollutants composed of two or more fused aromatic rings. They are classified into low molecular weight PAHs (LMW-PAHs) (2–3 rings) and high molecular weight PAHs (HMW-PAHs) (4 or more rings), with the latter posing greater carcinogenic, mutagenic, and genotoxic risks due to their higher hydrophobicity and lipophilicity [1,2]. PAHs are derived from three main sources: pyrogenic, petrogenic, and biogenic processes, which result from either natural phenomena or human activities. Pyrogenic PAHs are the by-products of incomplete combustion and/or pyrolysis of organic matter (OM) under low or no oxygen conditions, such as fossil fuel burning, wildfires, volcanic activity, coal burning, and anthropogenic activities [3]. On the other hand, petrogenic PAHs result from the diagenesis of organic materials over geological timescales, typically associated with crude oil, petroleum products, and their derivatives [4–6]. Although less common, biogenic PAHs are synthesized by plants [7,8], bacteria, fungi, and phytoplankton in specific ecological niches without involving diagenesis processes [9,10]. These distinct sources of PAHs can be identified using diagnostic ratios and molecular markers, which are valuable tools for tracking PAH pollution in the environment [11–13].

Upon formation, PAHs enter the atmosphere either in gaseous form or bound to particulate matter [14]. LMW-PAHs remain in the vapor phase, while HMW-PAHs adsorb to particulate matter, facilitating long-range transport (LRT) [15,16]. These atmospheric PAHs are subsequently deposited onto terrestrial and aquatic environments through wet and dry deposition processes [3]. Figure 1 illustrates the dynamics of PAHs in the environment, highlighting their mobility upon deposition on terrestrial surfaces [17], driven by runoff and weathering processes, and their subsequent transport into freshwater systems [18].

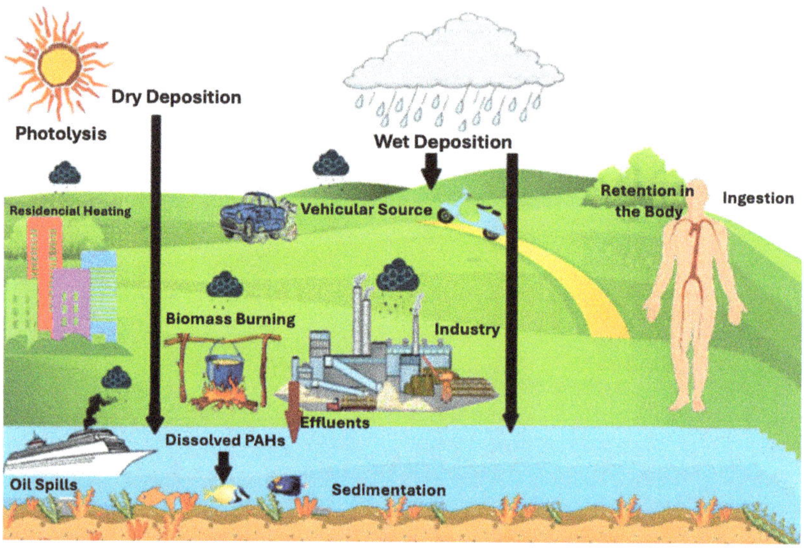

Figure 1. Distribution of PAHs in air, terrestrial, and aquatic environments. Reprinted with permission from ref. [17], Copyright 2019, Springer.

Globally, PAHs are a major concern due to their toxicity and widespread presence in air, water, and soil. In response, the United States Environmental Protection Agency (USEPA) designated 16 PAHs as priority pollutants due to their environmental prevalence [19] and associated human health risks [20,21]. Various international organizations, including the World Health Organization (WHO) [22], the Canadian Council of Ministers of the environment [23], and Greenpeace [24], have established monitoring frameworks and policies to address PAH contamination. Despite these efforts, mitigating PAH pollution remains a challenge, particularly in freshwater systems where their behavior is influenced by complex environmental interactions. Table S1 provides an overview of the physical and chemical properties of the 16 USEPA priority PAHs [25–28].

In aquatic environments, PAHs pose ecological and health risks to both aquatic organisms and humans [17]. Common pollution pathways include sewage discharges [29], industrial effluents [30], and runoff from urban [31,32] or agricultural areas [33]. In water, PAHs are distributed across dissolved phases, bound to OM, or adsorbed onto particulate matter or benthic sediments [34–37]. Their distribution within the water column is further influenced by the sediment–water partition coefficient [38], as their hydrophobic nature promotes adsorption onto suspended sediments, soil, and OM, contributing to their persistence and bioaccumulation in aquatic ecosystems [35,37].

PAHs exert toxic effects on a wide range of living organisms, including humans, animals, and microorganisms. Several PAHs are classified as genotoxic, mutagenic, carcinogenic, and teratogenic, thereby posing health risks to biological systems [39,40]. Their persistence is attributed to the dense π-electron system in their aromatic rings, which makes them resistant to nucleophilic attack [41]. HMW-PAHs are less volatile, more lipophilic, hydrophobic, environmentally persistent, and resistant to biodegradation [42,43]. These compounds readily bind to dissolved OM in sediments, accumulating in aquatic environments and increasing toxicity in benthic regions [36,44,45]. Additionally, rain runoff can transport soil-bound PAHs into water bodies, affecting flora, aquatic organisms, and food chains, ultimately posing direct risks to human health [46]. The lipophilic property of PAHs facilitates bioaccumulation across food webs, leading to human exposure [47–49]. Overall, the persistence and toxicity of PAHs in aquatic environments present significant risks to both human health and ecological systems.

1.2. Freshwater Systems

Freshwater systems cover only 2.5% of the Earth's surface and are among the most vital resources for human consumption and biodiversity [50–52]. These systems include lakes, ponds, reservoirs, rivers, streams, groundwater aquifers, estuaries, and wetlands, collectively accounting for 0.8% of the Earth's surface area [53–55]. Despite their importance, freshwater resources are under constant pressure, not only due to their geographical limitations but also from anthropogenic activities such as industrialization, agriculture, and urbanization, which contribute to the introduction of PAHs in these systems [54,56].

PAHs in freshwater systems present considerable risks to aquatic life, particularly to invertebrates, fish, and microorganisms. Their lipophilicity and hydrophobicity facilitate bioaccumulation, while their metabolisms can generate reactive intermediates, leading to oxidative stress, DNA damage, and endocrine disruption [43,57,58]. Even at low concentrations, PAHs negatively affect reproduction, growth, and survival in aquatic organisms, disrupting population dynamics and compromising ecosystem stability [45,59].

The occurrence, distribution, and fate of PAHs pose a global threat to freshwater ecosystems, regardless of geographical location [60–62]. The input of PAHs from point and non-point sources generates spatial and temporal heterogeneity in their distribution within these environments [63,64]. Through atmospheric deposition, surface runoff, and water discharges, PAHs enter water bodies where they either adsorb onto suspended sediments or persist dissolved in water [65]. Specifically, in lentic systems like lakes and wetlands, PAHs tend to accumulate in sediments due to slower water movement, whereas, in lotic systems such as rivers, they are transported over longer distances, promoting downstream pollution [66]. The resuspension of sediment-bound PAHs increases the exposure risk to aquatic organisms and facilitates LRT within watersheds, eventually reaching marine environments [67]. Although awareness of atmospheric PAHs has increased, significant gaps remain in understanding their transport, deposition, and dynamics within freshwater systems—particularly regarding their long-term ecological impacts [68].

1.3. Current Research Gaps and Emerging Perspectives in PAH Studies in Freshwater Systems

Despite significant progress in understanding the behavior and toxicity of PAHs in freshwater systems, key challenges remain unresolved. These include methodological inconsistencies [63,64], challenges in source identification and apportionment across spatial scales [69–71], and the resulting limitations in comparing findings or drawing generalizable conclusions. Additionally, the behavior of PAHs across environmental compartments—such as sediments, water, and biota—varies according to their physicochemical properties, further complicating efforts to predict their environmental fate and bioavailability [72–75]. While the toxicological mechanisms of PAHs in individual organisms are well-documented, their broader impacts at the community and ecosystem levels remain only partially understood. Understanding how PAHs affect trophic interactions and alter food web dynamics is essential to developing effective management and remediation strategies [76].

Numerous recent literature reviews have examined the environmental behavior and toxicological effects of PAHs across diverse matrices, including soil, sediments, air, and marine ecosystems [3,62,68,77–82]. However, most freshwater-specific reviews have focused on isolated components of the issue—such as sediment-associated contamination, species-specific toxicological responses, or individual ecosystem types—without integrating the complex interactions among PAH sources, transport pathways, environmental compartments, and ecological effects across diverse freshwater systems [45,46,59,79,81,83,84].

This review aims to provide a comprehensive analysis by critically synthesizing recent findings on PAH sources, environmental transport mechanisms, distribution patterns, and ecotoxicological impacts across various freshwater environments—including rivers, lakes, wetlands, and groundwater. Beyond compiling data, this work emphasizes spatial and temporal variability, ecosystem-level responses, and the influence of multiple stressors, such as climate change and co-contaminants. In doing so, it offers a comprehensive framework to interpret past findings and guide future monitoring, risk assessment, and management strategies. This review proposes an analysis by integrating ecological complexity and highlighting emerging research directions in the study of PAHs in freshwater systems.

2. Sources of PAHs in Freshwater Systems

PAHs enter freshwater systems through natural and anthropogenic processes. Source apportionment is commonly performed using techniques such as molecular diagnostic ratios, principal component analysis (PCA), and positive matrix factorization (PMF), which help differentiate between pyrogenic sources—such as fossil fuel combustion—and petrogenic sources, including crude oil or petroleum derivatives [85]. The emission and distribution of PAHs are influenced by a combination of geographic features, hydrological conditions, land use patterns, and human activities [69–71,86,87]. Furthermore, the physical and chemical properties of freshwater bodies, combined with climate variability, affect the behavior, transport, and accumulation of PAHs [88]. The unique characteristics of rivers, streams, lakes, wetlands, groundwater systems, and glaciers shape the contribution of different PAH sources and influence the resulting contamination dynamics in each environment [3,89,90].

2.1. Rivers

Rivers act as primary transport pathways for PAHs, carrying these pollutants from upstream sources to lakes, wetlands, or estuaries [91,92]. In the Buffalo River Estuary in South Africa, PAH concentrations in sediments reached up to 22,310 μg/kg and, in water, up to 206 μg/L, with diagnostic ratios indicating predominantly pyrogenic sources from automobiles, industrial effluents, and urban runoff [93]. Similarly, in the Bonny Estuary of the Niger Delta (Africa), sediment cores revealed total PAH concentrations ranging from 8699 to 22,528 μg/kg, with deeper layers enriched in pyrogenic PAHs and surface sediments showing elevated petrogenic levels linked to recent oil spills [94]. This pattern of mixed PAH sources is also observed in other rivers, such as the Niger Delta and the Amazon Basin, where pyrogenic and petrogenic PAHs have been identified—reaching up to 19,800 μg/kg in Niger Delta sediments and 163 ng/L in Amazon surface waters—originating from biomass burning, fossil fuel combustion, and oil-related activities [95,96]. In another study from Nigeria, surface water samples from the Ekulu River showed PAH concentrations as high as 3.17 mg/L, with a prevalence of HMW-PAHs derived from pyrogenic sources linked to combustion-related anthropogenic activities [97]. In the Middle East, the Euphrates River (Iraq) shows PAHs in water and sediments dominated by HMW species derived from petroleum product combustion, with carcinogenic PAHs comprising up to 55% of the total [98]. In contrast, surface sediments from the San Joaquin River in

California (USA) revealed that surface layers were largely influenced by pyrogenic PAHs (~70%), reflecting inputs from recent urban landscape alterations, while deeper layers exhibited higher proportions of biogenic PAHs [99].

Rivers in China, such as the Yangtze and Haihe, are significantly impacted by combustion-related PAHs, especially from coal, coke, vehicle fuel, and biomass burning. In the Yangtze River Estuary, PAH concentrations in sediments ranged from 34.9 to 580.3 ng/g, with source apportionment revealing major contributions from vehicle emissions (38.4%) and coal combustion (15.8%) [100]. Sediments in the middle-lower Yangtze also act as secondary sources of LMW-PAHs through resuspension, while retaining HMW-PAHs in the lower reaches due to reduced sediment discharge caused by dam impoundment [74]. In the Haihe River, sewage discharge was identified as a major source of PAHs, significantly influencing their partitioning behavior between sediments and pore water; parent PAHs exhibited strong sorption capacity, with a logarithmic organic carbon normalized partition coefficient (log K_{OC}) averaging 4.04 ± 0.80 [101]. Similarly, the River Benue in Nigeria exhibits mixed pyrogenic and petrogenic PAH sources, with sediment concentrations ranging from 55 to 382 μg/kg; source apportionment identified petrogenic burning (35%) and wood combustion (27%) as dominant contributors [102]. In Taiwan, sediment PAH concentrations across 30 major rivers reached up to 7.44 mg/kg, with pyrogenic sources prevailing and seasonal variations linked to wastewater discharge and combustion activities [103]. These findings emphasize the dual role of rivers as conduits and secondary sources of PAHs, with combustion-related emissions frequently dominating due to regional industrial activities and seasonal variability in land-use patterns.

2.2. Streams

Streams, especially those flowing through urban and forested landscapes, are strongly impacted by PAHs derived from combustion-related activities, primarily of pyrogenic origin. In China, sediment samples from urban streams in the Suzhou Industrial Park revealed total PAH concentrations ranging from 180 to 81,000 ng/g, with four-ring PAHs being the most abundant ($42 \pm 12\%$) and source apportionment pointing to coal and biomass combustion (61%) as the primary contributor, followed by vehicular emissions (18%) [104]. In the United States, a study across 10 urban watersheds found that streambed sediment PAH levels were significantly higher in regions where coal-tar-based pavement sealants were widely used—up to 54.3 mg/kg in the Northeast—underscoring pavement dust and sealed surfaces as major vectors of PAH transport to urban streams [105]. In Poland, sediment from retention tanks along the Oliwski and Strzyza streams showed PAH concentrations up to 20.4 mg/kg, with 4- and 5-ring PAHs predominating due to traffic emissions, heating systems, and vehicular abrasion, and presenting moderate- to high-risk levels to benthic organisms in certain locations [106]. These urban examples contrast with forested stream systems, where PAH contamination is more episodic. For instance, post-wildfire runoff in Portugal has been shown to increase PAH loading in streams due to soil erosion and ash transport following fire events [107]. Evidence from recent reviews indicates that wildfires are a diffuse but significant source of PAHs, mobilizing them into surface waters and sediments through post-fire runoff and other transport processes, with impacts depending on fire severity, vegetation type, and hydrological conditions [46,108]. While streams are influenced by both petrogenic and pyrogenic PAH sources, recent studies suggest that pyrogenic inputs—particularly those linked to combustion activities and urban expansion—tend to dominate, with some contributions originating from distant atmospheric sources [109,110].

2.3. Lakes

Lakes face challenges from a variety of PAH emission sources, particularly in urban areas, where petroleum, biomass, heavy oils, and natural gas consumption dominate [111]. Despite successful policies in several countries aimed at reducing point source pollution [112,113], non-point sources, such as atmospheric deposition and surface runoff, remain significant contributors to PAH contamination in lake systems [114]. Studies suggest that LMW-PAHs are found dissolved in the water column, and HMW-PAHs accumulate in lake sediments (Figure 2) [115,116]. In urban lakes, local non-point atmospheric deposition from pyrogenic activities often surpasses regional sources, as observed in Ontario Lakes in Canada [117]. Similarly, in Lake Baikal, PAHs primarily originate from biomass and fossil fuel combustion, with lighter PAHs concentrated in the water column and heavier ones sequestered in sediments [75]. Urbanization intensifies PAH concentrations, particularly in lakes exposed to high runoff from impervious surfaces, further amplifying the issue [111,118,119].

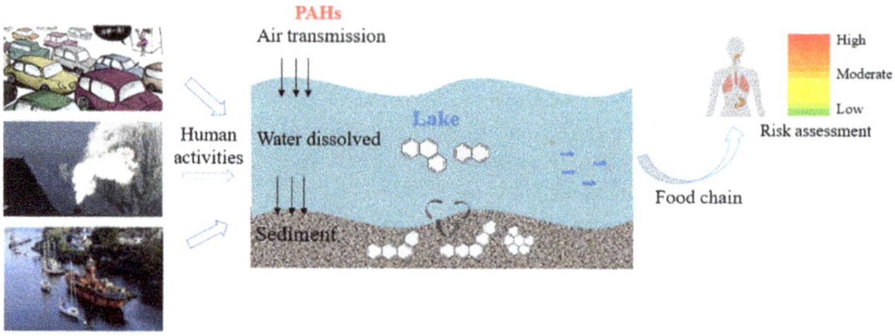

Figure 2. Sources, distribution, and transfer of PAHs through environmental compartments in freshwater lakes until human exposure is achieved. Reprinted with permission from ref. [115]; copyright 2019, Elsevier.

In contrast, remote lakes, which are less impacted by direct anthropogenic activities, often receive PAHs through atmospheric LRT and wildfires, which contribute to PAH deposition [46,120,121]. The rise in human activity, Gross Domestic Product, and population in lake catchment areas has also led to increased PAH levels, primarily from coal combustion and other pyrogenic sources [122–124]. Overall, PAH pollution in lakes results from both urbanization and natural processes, with urban lakes primarily affected by human activities, while the impact on remote lakes largely stems from natural phenomena, such as wildfires. These variations in source inputs and environmental dynamics between urban and remote lakes align with patterns observed across other freshwater systems worldwide. A comparative summary of PAH contamination, primary sources, and associated ecological effects reported in selected riverine and lacustrine systems is presented in Table 1, illustrating spatial variability and common risk factors across freshwater environments.

Table 1. Summary of PAH contamination in selected freshwater systems (rivers and lakes).

River System (Region)	Main PAHs Detected	Sources	Ecological Effects	References
Pearl River (China)	Fluoranthene, Pyrene, Benzo[a]pyrene	Industrial discharges, urban runoff	Bioaccumulation in fish; risk to benthic fauna	[125]
Yangtze River (China)	Phenanthrene, Fluoranthene, Chrysene	Domestic sewage, agricultural runoff	Chronic exposure in mollusks; endocrine disruption	[126]

Table 1. Cont.

River System (Region)	Main PAHs Detected	Sources	Ecological Effects	References
Tigris River (Iraq)	Benzo[a]anthracene, Chrysene	Wastewater discharges, oil refinery pollution	Decreased diversity of benthic invertebrates; moderate ecological risk	[127]
Ogun River (Nigeria)	Naphthalene, Acenaphthene	Industrial effluents	Risk to biota	[128]
Taihu Lake (China)	Phenanthrene, Pyrene, Benzo[a]anthracene	Urban-industrial interface	Oxidative stress in fish and invertebrates	[129]
Chaohu Lake (China)	Fluoranthene, Pyrene, Benzo[a]anthracene	Industrial discharges, urban runoff, atmospheric deposition	Moderate ecological risk; potential impact on aquatic organisms	[130]

2.4. Wetlands

Beyond lakes, freshwater wetlands also represent critical and vulnerable systems affected by diverse PAH sources. Freshwater wetlands are highly productive ecosystems that are particularly vulnerable to PAH pollution from various sources [19,131,132]. These ecosystems, recognized for their vital ecological functions, are increasingly degraded by oil exploitation and human disturbances [133,134]. The sources of PAHs in wetlands vary by region and proximity to human activities. For instance, pollution in Iran's Anzali Wetland is attributed to wastewater discharge, tourism, and oil-related activities [135–137], whereas China's Northeastern wetlands experience contamination primarily from coal burning and vehicle emissions [138]. Seasonal variations also play a significant role, such as the predominance of petrogenic sources during the wet season in Wang Lake Wetland, China [139]. Furthermore, constructed and urban wetlands, essential for water treatment, also face PAH contamination from wastewater discharge and fossil fuel combustion, with PAH distribution strongly influenced by molecular weight (Figure 3) [140]. In wetlands, PAHs in the water column are often derived from petroleum sources, while sediment-bound PAHs are typically pyrogenic, highlighting the complex distribution of PAHs in these environments (Figure 4) [141].

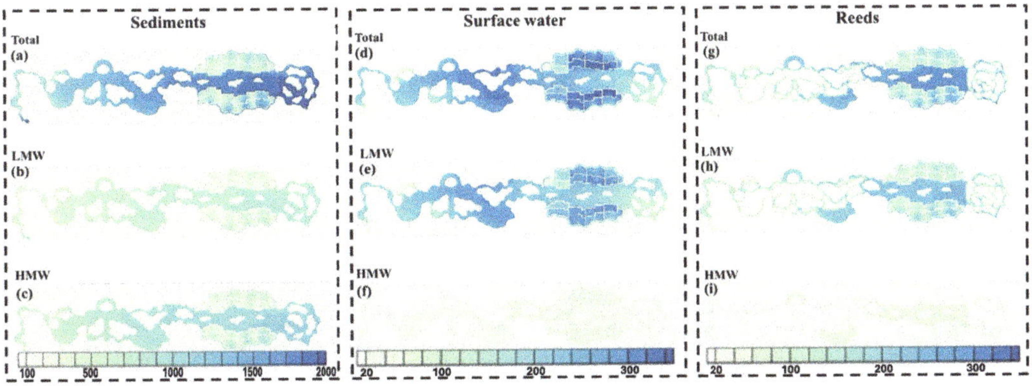

Figure 3. Distribution of total LMW-PAHs and HMW-PAHs in sediments, surface water, and reeds in a constructed wetland that receives water from a wastewater treatment plant. Simulated distribution results of sediments (**a–c**), surface water (**d–f**) and reeds (**g–i**) for Total PAHs. Reprinted with permission from ref. [140]; copyright 2021, Elsevier.

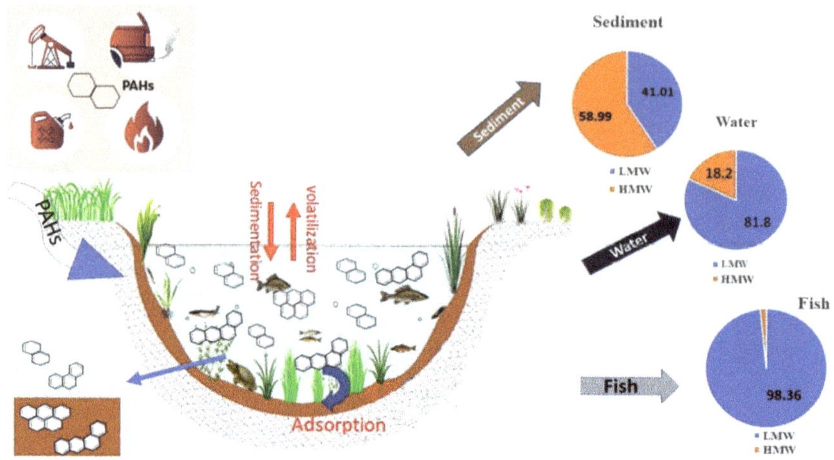

Figure 4. PAH sources and distribution in environmental matrices of a freshwater wetland. Reprinted with permission from ref. [141]; copyright 2023, Elsevier.

In Ramsar-designated wetlands and other protected areas, mixed sources of PAHs, especially from oil-related activities, present ongoing challenges [142]. Wetlands near oil extraction sites, such as Shadegan Wetland in Iran and Hoor-Al-Azim Wetland in Lower Mesopotamia, are often contaminated by both pyrogenic and petrogenic sources, with oil spills and combustion-related emissions being primary contributors [19,143,144]. Lastly, peatlands, marshes, and bogs, such as those in Northeast China and Iraq's Al-Hammar marshes, serve as environmental archives, reflecting historical pollution from both anthropogenic and natural sources, including wildfires and industrial activities [145–147]. Freshwater wetlands, critical for their ecological functions, face complex challenges for PAH pollution derived from mixed sources.

2.5. Groundwater

Groundwater, as a primary freshwater source for domestic, agricultural, and industrial purposes, is increasingly contaminated with PAHs introduced through diverse pathways [148,149]. PAHs leach from soils into aquifers or are transported via surface runoff, often impacting both surface and groundwater systems [68]. The strong hydrological connectivity between surface water and groundwater facilitates the migration of PAHs, particularly in closely linked systems [86,150]. Due to their higher mobility, LMW-PAHs can migrate more readily in soil, resulting in relatively uniform PAH concentrations across interconnected water systems [151].

In the Yellow River Delta, China, pyrogenic PAHs from petroleum combustion dominate groundwater pollution, entering aquifers primarily through surface water infiltration [150]. Similarly, in karst terrains, the rapid vertical migration of PAHs is enabled by geological features like conduits and sinkholes, transporting contaminants from the combustion of grass, wood, and coal [86,152,153]. Local factors within the river's catchment area significantly influence PAH sources in such environments [151,154].

Pyrogenic sources are a primary contributor to PAH contamination in groundwater, often infiltrating aquifers through soil leaching and various transport mechanisms [155]. Agricultural practices, such as long-term irrigation with wastewater or reclaimed water, are the main source of groundwater PAHs in farming areas. Wildfires also impact groundwater quality by introducing pyrogenic PAHs during post-wildfire events, particularly in areas connected to public water supplies, resulting in risks to human health [156]. Furthermore, industrial activities near abandoned complexes and oil-producing regions contribute to both pyrogenic and petrogenic PAHs, resulting in mixed-source profiles due to vertical migration over extended periods of time [157,158]. The complex and diverse sources of PAH pollution in groundwater underscore the need to address both localized and long-term contamination pathways.

Economic development and changing anthropogenic activities shape the evolution of PAH emission sources. In the groundwater of karstic regions in Southwest China, a shift from coal combustion to vehicular emissions over a decade has sustained a predominantly pyrogenic PAH profile [159]. Geological factors, such as complex lithology and hydrostratigraphic systems, further influence PAH origins in groundwater, with carbon and fuel combustion serving as major pollution sources in these environments [160]. Even in oil fields such as the upper Brahmaputra Valley of India, surface combustion activities dominate groundwater contamination, surpassing petrogenic contributions from oil spills [155]. Despite regional variations, pyrogenic PAHs consistently represent a significant and persistent source of groundwater pollution.

2.6. Glaciers

Recent research highlights the sources of PAHs in various glacial freshwater environments. Attribution ratios in Arctic freshwater systems suggest that PAHs predominantly originate from atmospheric deposition and combustion-related sources [68]. Fresh meltwater samples from the eastern Tibetan Glacier show PAHs associated with incomplete coal combustion and coking that are deposited via LRT and are also affected by local environmental conditions [161]. These melting glaciers act as secondary sources of PAHs, leading to elevated concentrations in sediment cores of proglacial lakes and glacial-fed streams, establishing PAHs as the most dominant of the persistent organic pollutants in the Arctic environment [162]. This retention is facilitated by the OM content in proglacial soils, which promotes the accumulation of PAHs in these environments [163].

In the case of the Kongsfjorden Glacier in Norway, PAH concentrations decline with increasing distance from the glacier, with local petrogenic combustion processes, atmospheric LRT, and historical coal mining identified as key sources [164]. Additionally, dissolved PAHs in the Kongsfjorden Glacier meltwater further indicate major contributions from grass, wood, and coal combustion, with their distribution influenced by ocean currents and glacier runoff [165].

In Antarctica, the most remote continent and holder of the largest freshwater reservoir, PAH sources have been studied to better understand their dynamics in glacial environments. Reports suggest that Antarctic waters receive a mix of locally consumed fuel combustion and globally transported PAHs via LRT [166]. On King George Island, PAHs accumulate in terrestrial soils, primarily linked to electricity generators and light-duty gasoline consumption associated with scientific research and tourism activities [167]. However, terrestrial PAH pollution levels on King George Island are significantly lower than in other parts of the world [168]. Reducing fossil fuel consumption in these regions is recommended to mitigate PAH emissions and minimize their impact on Arctic and Antarctic glacial freshwater environments.

Overall, the presence of PAHs in freshwater systems—including lakes, rivers, streams, groundwater, wetlands, and glaciers—is shaped by diverse emission sources (see Table S2). Lakes face significant challenges from both point and non-point sources, with atmospheric deposition and surface runoff being primary contributors. Rivers act as major transport pathways for PAHs, reflecting mixed emission sources, while wetlands, groundwater, and glacial freshwater systems exhibit distinct patterns of PAH contamination. These patterns are supported by numerous case studies conducted across Asia, Africa, Europe, and the Americas [19,74,86,91–93,97,98,102–104,106,107,109–111,119–121,135,140–143,146, 150,156–161,164–167,169–181]. These findings highlight the critical need for integrated management, robust monitoring, focused research, and effective policy strategies to address PAH pollution across freshwater systems.

3. Distribution of PAHs in Freshwater Systems

The distribution of PAHs in freshwater systems is influenced by various transport, mechanisms, environmental factors, and interactions with environmental matrices. Understanding these processes is essential for addressing the environmental distribution and fate of PAHs in these systems. This section explores the primary transport mechanisms and environmental factors influencing the distribution of PAHs, emphasizing their interactions with various environmental matrices in freshwater systems.

3.1. Transport Mechanisms

Polycyclic aromatic hydrocarbons (PAHs) reach freshwater systems through a range of environmental pathways, including atmospheric deposition, surface runoff, riverine transport, groundwater infiltration, sediment resuspension, and biological activity. These inputs are shaped by both local and distant anthropogenic sources—particularly combustion-related emissions—and modulated by climatic and land-use dynamics [61,182–184]. Once PAHs enter aquatic systems, their transport and eventual fate are governed by a combination of physical, chemical, and biological mechanisms that determine their partitioning, mobility, and persistence [185,186]. Understanding these mechanisms is essential to accurately assess the spatial and temporal distribution of PAHs, predict exposure scenarios, and inform risk assessments and management strategies. The following subsections explore each of these transport pathways in detail, emphasizing how system-specific factors such as flow regimes, sediment composition, and water chemistry influence PAH movement and accumulation in freshwater environments.

3.1.1. Physical Transport Mechanisms

PAHs are physically transported in freshwater systems through advection, diffusion, and sediment dynamics, all of which are shaped by hydrological and environmental conditions. Among these, advection—the transport of pollutants via flowing water—is a predominant mechanism in lotic systems [74,187–189]. This process enables the downstream redistribution of PAHs from both local and distant sources. In the Yangtze River (China), Zhao et al. [74] demonstrated that water runoff from megacities—driven by coal and coke combustion—served as a major advective input, with the river current transporting PAHs downstream, while reduced sediment discharge from upstream dams increased their retention in lower reaches. Modeling of the Yangtze River Delta under future land-use scenarios further confirmed that advection is the principal mechanism governing PAH redistribution across environmental compartments [190]. In Dianshan Lake, a shallow urban system, Du et al. [189] found that atmospheric advection was the main source of PAHs, with hydrodynamic movement driving their transfer from water to sediment. In another report, in Lake Qinghai (China), PAH concentrations were highest at sediment sites near river inlets, supporting the role of river inflows as advective carriers of local emissions [188]. On the other hand, in groundwater systems, advective flow accelerated by irrigation pumping, significantly increased the subsurface mobility of low to medium molecular weight PAHs [191]. Additionally, in the Columbia River Estuary (USA), hydrodynamic trapping processes enhanced the downstream transport and retention of combustion-derived PAHs, even at distances far from original inputs [192]. Collectively, these studies highlight advection as a system-specific but ubiquitous mechanism that mobilizes PAHs across spatial gradients in both lotic and lentic freshwater environments.

Diffusive transport refers to the passive movement of PAH molecules from regions of high to low concentration until equilibrium is achieved. This process is particularly relevant in low-flow freshwater environments—such as lakes and wetlands—where water circulation is minimal and vertical gradients persist. For instance, in Panguipulli Lake (Chile), net volatilization of LMW-PAHs and deposition of heavier congeners were observed across the air–water interface, demonstrating how hydrophobicity and hydrodynamic stagnation influence the direction and magnitude of diffusive fluxes [193]. In the Willamette River (USA), paired samplers revealed that sediments acted as a long-term source of 4- and 5-ring PAHs, diffusing into the water column and volatilizing into the atmosphere despite prior remediation efforts [194]. In Dianchi Lake (China), model simulations showed that diffusion, alongside degradation, was a dominant removal mechanism for PAHs from sediments, counteracting accumulation from atmospheric deposition and waterborne inputs [195]. Likewise, observations across the Yangtze River Delta (China) indicated that PAHs such as naphthalene (NaP) and benzo[a]pyrene (BaP) were subject to resuspension and diffusion at sediment–water interfaces, with transport direction driven by compound-specific properties and local hydrological conditions [196]. In Poyang Lake (China), seasonal dry–wet cycles were found to regulate sediment-to-water PAH exchanges, with LMW congeners volatilizing during periods when sediments were exposed to air, and HMW compounds showing varied flux directions when submerged under water [197]. Altogether, these studies highlight diffusion as a dynamic and compound-specific process that governs the long-term distribution, bioavailability, and environmental persistence of PAHs in stagnant or weakly mixed freshwater systems.

The hydrophobic nature of PAHs facilitates their attachment to suspended particulate matter (SPM) and sediments, enhancing their persistence and transport in freshwater environments [198]. SPM serves as a mobile phase that carries PAHs across aquatic systems before settling through deposition, while sediments function as temporary reservoirs [122]. However, these reservoirs are not static and resuspension events triggered by storms, flooding, or human activities can remobilize sediment-bound PAHs into the water column, promoting downstream transport and increasing their bioavailability [123,199]. The extent and efficiency of PAH partitioning are influenced by multiple factors, including particle size, organic carbon (OC) content, hydrodynamics, and sediment characteristics [200,201]. Fine particles, in particular, remain suspended longer after resuspension events, allowing PAHs to be transported farther from their original source areas [200]. In contrast, sediments rich in OC tend to exhibit stronger PAH sorption and retention capabilities, reinforcing their role as long-term sinks [201]. Over time, sediments act as dynamic reservoirs—both absorbing and re-releasing PAHs—thus functioning as both sinks and secondary sources in rivers [36], lakes [202], and wetlands [203]. Figure 5 illustrates this dual role in the Yangtze River (China), where PAH transport is shaped by runoff, sediment discharge, deposition, and resuspension under the influence of urbanization and industrial emissions [74]. Overall, sediment dynamics critically shape the environmental fate of PAHs, determining whether these compounds remain stored or become remobilized under changing environmental conditions.

The physical transport of PAHs in freshwater systems is strongly modulated by hydrological factors, which govern their mobilization, redistribution, and persistence across diverse aquatic environments. Rainfall, for instance, facilitates the wash of PAHs from urban and industrial surfaces into water bodies, particularly during intense storm events that generate high surface runoff and sediment discharge [204]. In systems with seasonal hydrological variation, dry-season conditions with lower dilution capacity led to elevated

PAH concentrations in water, while intense wet-season rainfall increased SPM, enhancing PAH sorption and downstream transport [205]. Similarly, flooding events, such as Hurricane Harvey in Houston, Texas (USA), caused large-scale PAH redistribution into Galveston Bay, with peak concentrations and biological responses observed shortly after flood onset [206]. Resuspension of contaminated sediments during floods has also been observed in the Niger Delta (Nigeria), intensifying ecological and health risks due to high levels of HMW-PAHs [79]. Seasonal hydrological changes further influence phase exchange processes; in tide-dominated estuaries, temperature and marine currents modulate PAH diffusion across sediment–water interfaces, with source contributions varying by season [116]. In major rivers such as the Yangtze River (China), temperature and discharge variations during the wet season favor the accumulation of HMW-PAHs, driven by increased flow and suspended solids [207]. On the other hand, in lentic systems, such as the Shuikou Dam (China), impoundment reduces flow and increases PAH retention, while wet-season surface runoff elevates PAH concentrations downstream of industrial zones [208]. Karst systems present unique hydrological pathways, where PAHs infiltrate rapidly through fissures and conduits, promoting swift subsurface transport from surface sources to spring-fed waters [152,153]. Finally, hydrogeological conditions—such as permeability, vadose zone thickness, and water table fluctuations—determine PAH distribution in aquifers, with variations observed across regions like Du'an, Lanzhou, and Golmud (China) [176]. In summary, hydrological processes such as rainfall, flooding, temperature shifts, and groundwater flow directly influence how PAHs are transported and where they accumulate in freshwater environments.

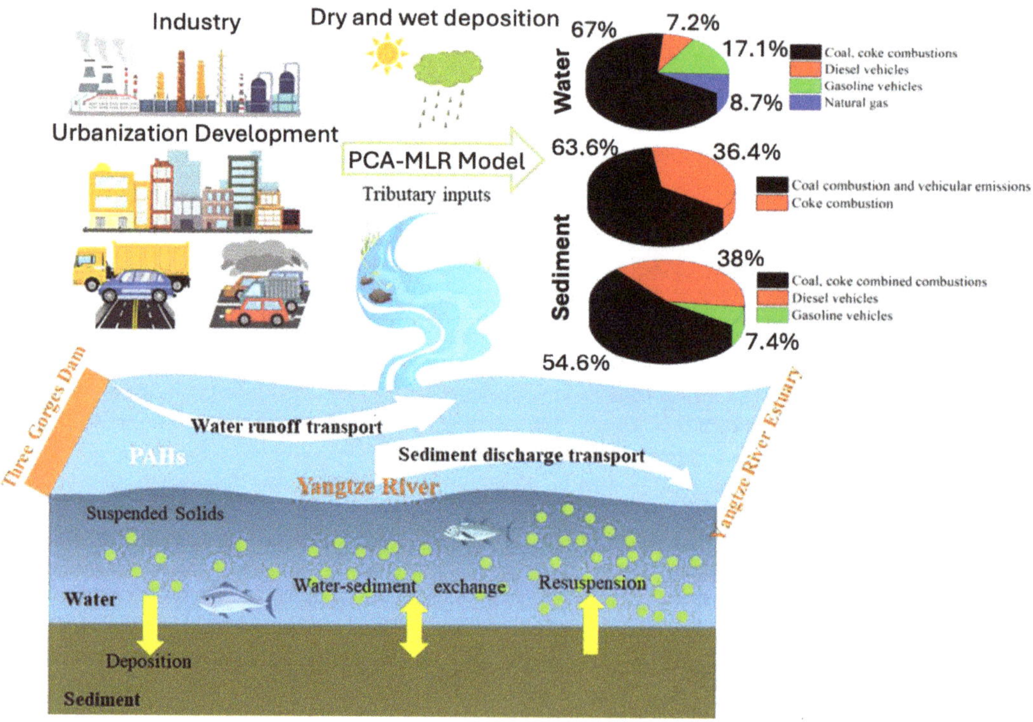

Figure 5. Water runoff from megacities, driven by coal and coke combustion sources, contributes to the transport and increased catchment retention of PAHs along the Yangtze River, China. Reprinted with permission from ref. [74]; copyright 2021, Elsevier.

As detailed throughout this subsection, the physical transport of PAHs in freshwater systems is governed by three interrelated mechanisms—advection, diffusion, and sediment dynamics—and is strongly modulated by hydrological factors such as rainfall, flow regime, and seasonal variations. Advective transport, driven by river flow, groundwater movement, and internal circulation in lentic environments, facilitates the broad dispersal of PAHs across spatial gradients. Diffusion becomes particularly relevant in stagnant or low-flow zones, where the physicochemical properties of individual PAHs and prevailing environmental conditions control their gradual redistribution across phases. Sediment dynamics, including deposition and resuspension, mediate the long-term fate of PAHs by acting as both sinks and secondary sources, with their behavior influenced by sediment composition, organic content, and external hydrological forces. These mechanisms do not operate in isolation; rather, they interact within the unique hydrological frameworks of rivers, lakes, and wetlands, shaping the spatial and temporal dynamics of PAH transport. Figure 6 illustrates this complexity in Dianchi Lake (China), where model simulations have enabled the quantification of PAH fluxes across compartments, highlighting how different processes such as advection, diffusion, and sedimentation collectively determine the fate of pollutants in freshwater systems.

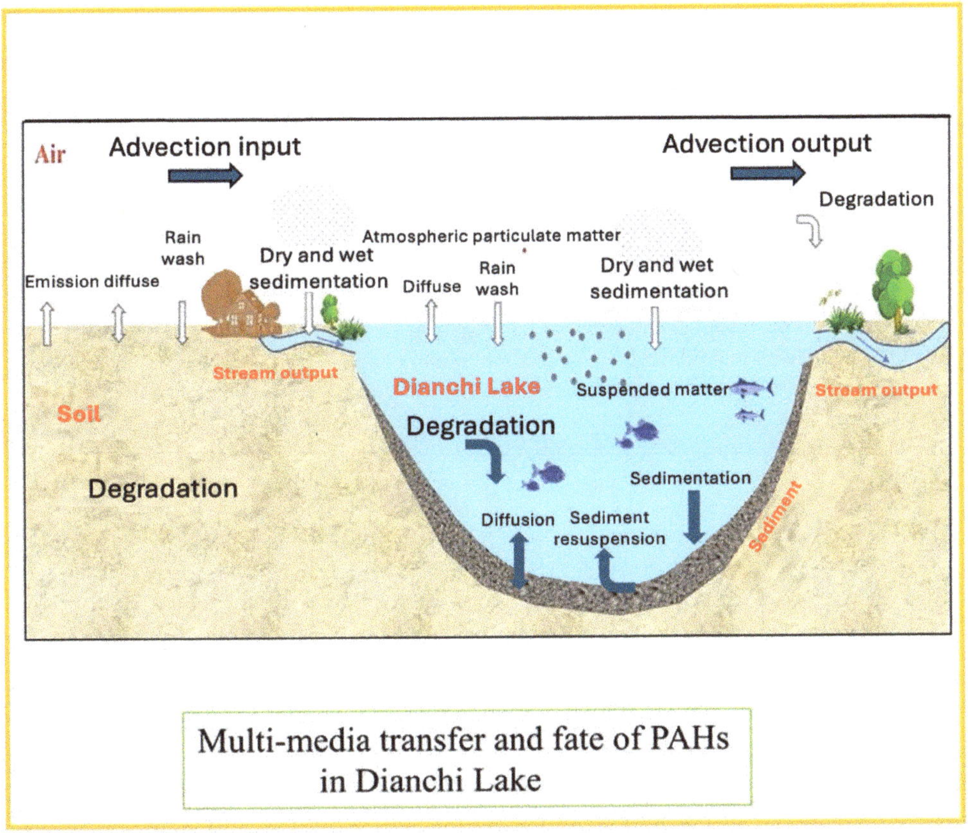

Figure 6. Multimedia transfer of PAHs in Dianchi Lake, China, illustrating advective, diffusive, and sediment transport mechanisms. The diagram shows the pathways of PAH movement through air, water, and sediment compartments, emphasizing advective inflows and outflows, diffusion dynamics, degradation processes, and interactions at the sediment–water interface. Reprinted with permission from ref. [195]; copyright 2024, Elsevier.

3.1.2. Chemical Transport Mechanisms

The transport and distribution of PAHs in freshwater systems are influenced by chemical processes such as sorption and desorption, both of which are affected by environmental and chemical conditions. Sorption occurs when PAHs adhere to sediments or suspended matter, a process governed by molecular hydrophobicity, molecular weight, and steric factors [209,210]. Skic et al. [209] demonstrated that sorption in bottom sediments is facilitated by steric hindrance and the presence of negatively dissociating functional groups, especially for HMW-PAHs, with finer pores (<5 μm) serving as preferential sequestration sites. OM composition and structure also play a dominant role, as shown by Huang et al. [211], who reported that the heterogeneity of sediment OM—including humic substances, black carbon, and kerogen—results in nonlinear sorption behaviors such as hysteresis and slow desorption kinetics. Dissolved and particulate OC enhance PAH affinity across colloidal, dissolved, and particulate phases in underground river systems [44,212]. In riparian zones and riverine sediments, PAH accumulation is intensified by high OC content, where sorption is influenced by π-π interactions and hydrophobic affinity [213,214]. Duttagupta et al. [191] showed that the mineralogical composition of sediments—especially the dominance of quartz and kaolinite—also determines PAH sorption preferences across different land-use zones.

Desorption—the release of PAHs back into the water column—is influenced by sediment aging, environmental gradients, and physicochemical interactions. Lu et al. [215] demonstrated that sediment aging reduces desorption efficiency due to increased molecular bonding and compaction within sediment matrices. Their kinetic and thermodynamic analyses revealed that desorption follows a three-stage process (fast, slow, and smooth) and is inhibited under strongly acidic/basic pH or high salinity conditions while being promoted by low oxygen and small particle size. Thermal desorption experiments further confirmed that water content and OM reduce PAH release, and that desorption is thermally activated [216]. Additionally, Oyo-Ita et al. [205] reported seasonal variability in salinity and suspended solids as key controls on PAH phase exchange, where higher salinity can inhibit desorption and promote retention in suspended matter. An expansion on these findings showed that both biodegradable and non-biodegradable microplastics sorb and release petroleum hydrocarbons, following reversible pseudo-second-order kinetics [217].

Freshwater chemical conditions—such as pH, redox potential, and ionic strength—also shape PAH behavior by influencing their partitioning and transformation. For instance, Salowsky et al. [73] showed that transient redox conditions in groundwater systems affect PAH retention and biodegradation through variable oxygen availability and iron-reducing microbial processes. Sediments may function as sinks or sources of PAHs depending on molecular structure and surrounding chemical gradients, with fugacity fraction analyses revealing the re-emission of lighter PAHs and the accumulation of heavier congeners in riverine environments [74]. Almouallem et al. [218] demonstrated that increasing ionic strength enhances the sorption of certain PAHs by promoting stronger interactions with sediment particles, whereas low liquid–solid ratios increase adsorption capacity.

In summary, the chemical transport of PAHs in freshwater systems is driven by the interplay of sorption and desorption processes, modulated by sediment composition, OM properties, PAH molecular structure, and dynamic environmental factors such as pH, ionic strength, temperature, and redox potential. Understanding these chemical interactions is essential for predicting the long-term fate and mobility of PAHs in freshwater environments.

3.1.3. Biological Transport Mechanisms

Biological transport mechanisms influence the redistribution of PAHs in freshwater ecosystems through interactions with aquatic organisms and their associated processes.

Seasonal algae blooms can enhance the biological pump effect, increasing the transfer of PAHs from the water column to SPM and sediments, especially during high-productivity periods in eutrophic lakes [202,219]. This algae-driven sequestration is often associated with elevated concentrations and toxicity of PAHs in shallow, nutrient-rich environments. Similarly, phytoplankton biomass in hypereutrophic lakes can increase the bioavailability of PAHs in sediments, facilitating their uptake by benthic organisms such as clams and mussels [220]. In cyanobacteria- and macrophyte-dominated systems, the differential distribution of PAHs in riverine versus lacustrine sediments suggests biological community structure influences PAH fate [221]. PAHs can also be actively accumulated and biomagnified through food web interactions, with zooplankton and fish incorporating PAHs in their tissues depending on feeding strategies and habitat [222,223]. Freshwater invertebrates such as gastropods also accumulate PAHs in their tissues, as seen in *S. quadrata*, with levels influenced by local pollution and anthropogenic activity [224]. Moreover, microbial processes—including the downward migration of filamentous bacteria—may enhance vertical PAH migration in sediments [225]. Together, these biological mechanisms interact with hydrological and physicochemical processes to shape the spatiotemporal dynamics and ecological risks associated with PAHs in freshwater systems.

In summary, the transport of PAHs in freshwater systems is governed by a combination of physical, chemical, and biological mechanisms, each modulated by environmental and hydrological forces that shape their mobility, transformation, and persistence. These mechanisms interact within system-specific contexts—such as river flow, sediment composition, water chemistry, and biological activity—producing distinct transport patterns across aquatic compartments. Recognizing the interdependence among these processes is essential to understanding how PAHs are distributed across freshwater landscapes. The following section expands on these dynamics by examining the physicochemical, hydrological, anthropogenic, and natural factors that influence the spatial and temporal distribution of PAHs in freshwater systems.

3.2. Factors Affecting Distribution of PAHs

The distribution of PAHs in freshwater systems is influenced by a diverse range of environmental factors, including the physicochemical and hydrological characteristics unique to each system, environmental conditions, such as pH, redox potential, and temperature, anthropogenic activities, and associated pollutant discharges, as well as natural factors like soil composition, land cover, and vegetation. These factors interact in complex ways to influence the distribution, partitioning, and fate of PAHs in freshwater environments. This section explores the interaction among these influences, emphasizing their roles in shaping the spatial distribution of PAHs in freshwater systems.

The distribution of PAHs in freshwater systems is governed by their intrinsic physicochemical properties, such as hydrophobicity, solubility, and volatility, which vary with molecular weight and affect their behavior [72–75]. Seasonal temperature changes and the presence of OC in sediments lead to fluctuations in PAH concentrations, with higher levels observed during dry periods due to reduced dilution and an increased prevalence of HMW-PAHs under warmer conditions [226–229]. In lake systems, both external factors (e.g., seasonal salinity, temperature) and internal factors (e.g., sediment characteristics, particulate OC) control PAH partitioning, affecting their distribution among dissolved and particulate phases [29,101,205,230]. Karst springs, wetlands, and transitional estuarine areas display unique hydrological traits that influence PAH distribution, driven by the interplay of hydrodynamics, dilution, and physicochemical properties [139,153,231].

Environmental conditions in water, such as pH, redox potential, ionic strength, and temperature variations, impact PAH behavior in freshwater systems. Extreme pH levels and anaerobic conditions can increase PAH solubility and reduce degradation rates, while higher ionic strength enhances sorption to sediments [73,101,218]. On the other hand, temperature fluctuations affect PAH volatility, solubility, and microbial degradation [116], creating complex distribution patterns that are further intensified by anthropogenic activities.

Anthropogenic activities, including industrial discharges, wastewater and sewage discharges, fuel and coal combustion, vehicle emissions, and oil spills, contribute to PAH pollution, creating localized hotspots that alter their spatial distribution within freshwater systems [96,138,184,186]. These hotspots arise from concentrated pollutant inputs, which vary depending on the type and intensity of human activities in the surrounding region [123]. While nonpoint sources generate diffuse pollution over large areas, their cumulative effects in regions with intense anthropogenic activities often result in localized hotspots [232].

In addition to these localized hotspots, the co-occurrence of other pollutants, such as heavy metals and microplastics, further intensifies the complexity of PAH distribution dynamics. Heavy metals interact with PAHs, forming complexes that alter their solubility and distribution, often resulting in overlapping distribution patterns in hotspots where both pollutants share common anthropogenic sources and environmental processes [233]. Similarly, microplastics act as carriers for PAHs through surface adsorption mechanisms, influencing their distribution, persistence, and bioavailability [234–236]. Also, construction activities and regional economic developments near freshwater bodies complicate and alter these distribution dynamics of PAHs by introducing additional contaminants [123,237,238]. Collectively, these anthropogenic influences highlight the complex interaction of pollutants that shape the distribution of PAHs in freshwater systems.

Natural factors, such as soil composition, land cover, and vegetation, shape the distribution and fate of PAHs in freshwater environments. Organic-rich soils enhance PAH adsorption, reducing their mobility and serving as long-term reservoirs for these pollutants [239]. Similarly, riparian vegetation acts as a natural buffer, trapping runoff sediments and promoting microbial degradation of PAHs before reaching water bodies, thereby mitigating their impact [44,191,209,240]. In wetland environments, the interaction between OM and hydrodynamic processes affects the sequestration of PAHs, particularly of HMW-PAHs, within sediment layers [139,241]. Additionally, geological features such as karst formations influence PAH transport by facilitating rapid infiltration and storage in subsurface freshwater environments [86,152,153]. These natural processes interact with environmental conditions, such as seasonal temperature changes and hydrodynamic forces, further influencing PAH partitioning and mobility [74,207,228]. While natural factors often act as mitigating mechanisms, they also highlight the need for comprehensive monitoring strategies to evaluate the interplay between natural and anthropogenic influences on PAH dynamics. Understanding how these factors interact and influence PAH dynamics in freshwater systems requires a detailed examination of their interactions with environmental matrices, such as sediments, water, aquifers, and porewater, which ultimately shape their distribution, bioavailability, and ecological impact.

3.3. Interactions with Environmental Matrices

The distribution and fate of PAHs in freshwater systems rely heavily on their interactions with environmental matrices, including sediments, water, aquifers, and porewater. Sediments, serve as a major reservoir for PAHs, especially HMW-PAHs, as observed in stormwater retention ponds and natural lakes [242]. The partitioning of PAHs between sediments and water determines their bioavailability and toxicity to benthic organisms [45,243]. For example, PAH concentrations in the clam *Corbicula fluminea* strongly correlate with sediment PAH levels, indicating sediments as the primary source of bioavailable PAHs [244].

Also, the ability of sediments to act as both a sink and a secondary source of PAHs depends on environmental conditions, such as OM content, total OC levels, and hydrodynamic forces, with steric effects driving the sequestration of HMW-PAHs in fine sediment pores [209]. This mechanism, driven by hydrodynamic forces, resuspends sediment-bound PAHs into the water column, thereby influencing their distribution and ecological impact [123,198]. Consequently, sediment dynamics, including deposition and resuspension, shape the fate and distribution of PAHs in aquatic ecosystems.

In the water phase, PAHs exhibit varying behavior depending on their solubility and hydrophobicity, which govern their partitioning between dissolved and SPM phases [165]. LMW-PAHs, such as NaP, are more soluble and tend to remain in the dissolved phase, while HMW-PAHs, such as BaP, are more likely to adhere to SPM [245]. In the case of resuspension events tend to redistribute sediment-bound PAHs throughout water bodies [74,246]. The properties of SPM, shaped by location, season, and human activities, play a crucial role in PAH distribution by increasing sorption to OC [247]. The solubility and partitioning of PAHs within the water column can serve as predictive factors for their distribution and movement, particularly under the influence of local environmental conditions.

In subsurface environments, aquifers serve as reservoirs for PAHs, with strong adsorption processes in the vadose zone facilitating their retention. The distribution and retention of PAHs in aquifers are primarily governed by their interactions with fine particulate matter and OC, which promote the sequestration of HMW-PAHs [129,248]. This partitioning behavior varies at the sediment–water interface, driven by differences in molecular sizes, with larger PAHs exhibiting higher adsorption potential [249]. Furthermore, minerals such as quartz and kaolinite significantly enhance PAH retention in groundwater systems, highlighting the pivotal role of aquifer geological composition in shaping PAH distribution [191]. Collectively, these processes establish aquifers as essential long-term storage matrices that influence the mobility, fate, and ecological risk of PAHs in freshwater systems.

Porewater, located in the interstitial spaces between sediment particles, is a medium for the distribution of PAHs, enhancing their toxicity to aquatic organisms. LMW-PAHs are often released from sediment-associated OC into porewater, contributing to their bioavailability and toxicity [250]. Toxicological studies conducted in the amphipod *Hyalella azteca* and zebrafish embryos (*Danio rerio*) have demonstrated that PAH-contaminated porewater induces cardiovascular toxicity and decreases survival rates [251]. The freely dissolved fractions of PAHs in porewater, particularly during wet seasons, are readily bioaccumulated by benthic species, resulting in adverse health effects [251].

The interaction of PAHs with environmental matrices, including sediments, water, aquifers, and porewater, shape their distribution and ecological impact in freshwater systems. Sediments function as both a sink and source of PAHs, with processes such as resuspension and partitioning between sediment and water phases influencing their bioavailability and toxicity to aquatic organisms. In the water column, PAH behavior is governed by their physicochemical properties, which determine their partitioning between dissolved and SPM phases. On the other hand, the retention and distribution of PAHs in aquifers and porewater are influenced by interactions with environmental factors such as particulate matter, OC, and minerals. Larger PAH compounds exhibit stronger adsorption in groundwater systems, while freely dissolved PAHs in porewater contribute to toxicity and bioaccumulation in aquatic organisms. Finally, the complex interactions between PAHs and these environmental matrices dictate their long-term distribution, fate, mobility, and ecological risk within freshwater ecosystems.

4. Ecotoxicological Impacts of PAHs in Freshwater Systems

4.1. Molecular and Cellular Level Effects

PAHs interact with cellular components and induce oxidative stress (OS) through the generation of reactive oxygen species (ROS), either directly or via metabolic intermediates [252–254]. These mechanisms encompass three major pathways—radical-cation, diol-epoxide, and quinone—that contribute to DNA damage [255,256]. As illustrated in Figure 7, BaP undergoes metabolic activation via cytochrome P450 (CYP) enzymes and epoxide hydrolase, forming DNA adducts that promote genotoxicity [257]. This activation leads to oxidative metabolism, where CYP enzymes generate diol-epoxides, o-quinones, and radical cations, each with specific toxic effects [258–260]. Diol-epoxides form stable DNA adducts [260], radical cations produce genetic mutations [258], and o-quinones promote OS by generating ROS [261]. These ROS—including superoxide anions (O_2^-), hydrogen peroxide (H_2O_2), and hydroxyl radicals (·OH)—overwhelm antioxidant defenses such as glutathione [58], superoxide dismutase [262,263], and catalase [264,265], ultimately causing damage to lipids, proteins [266], and DNA, which triggers apoptosis or necrosis [265–268].

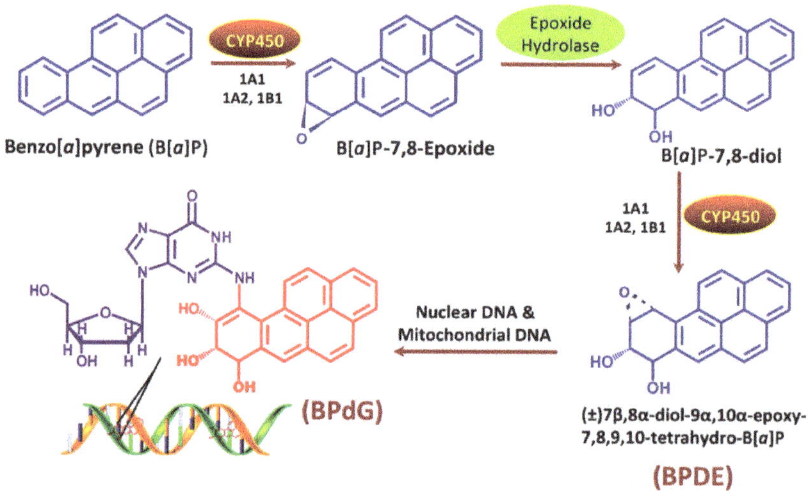

Figure 7. Primary mechanism of DNA binding by BaP, a pro-carcinogenic PAH. Reprinted with permission from ref. [257], under the terms of the Creative Commons Attribution License (CC BY). Copyright 2018, published by Portland Press.

DNA adduct formation by PAHs is closely associated with mutagenesis and carcinogenesis [269]. BaP is especially potent due to its stable DNA adducts, serving as a model compound for genotoxicity research [270,271]. The mutagenic potential of PAHs depends on their chemical structure and metabolic activation efficiency [253,258,272–274]. Furthermore, BaP induces epigenetic changes by altering DNA methylation—either hypo- or hypermethylation—depending on concentration and tissue type, disrupting gene regulation and potentially affecting offspring [275]. BaP also affects the biotin homeostasis pathway and circadian rhythm regulation, as observed in marine medaka, with heritable osteotoxicity in the F3 generation [275]. These effects are mediated by CpG–BPDE adducts and altered DNA methyltransferase and histone deacetylase activity [275]. CYP1A expression, used as a biomarker for PAH exposure, shows strong tissue-specific responses. For example, in *Gambusia affinis*, BaP exposure induces CYP1A mRNA expression in the testis, liver, and other organs, with distinct temporal patterns [276].

PAHs also act as endocrine disruptors by mimicking estrogen or interfering with hormone pathways [48,277]. Their metabolites bind estrogen receptors, altering signaling and disrupting physiological processes such as reproduction and development [278–283]. The endocrine-disruptive potential of PAHs varies among different compounds and is mediated by their chemical structures and metabolic products [284]. Additionally, specific PAHs exhibit distinct interactions with hormone receptors, further complicating the effects [253,278]. Endocrine disruption in aquatic species can result in diverse and profound health effects, potentially leading to long-term ecological consequences.

In summary, PAHs impact cellular and molecular functions through mechanisms involving OS, mutagenesis, epigenetic alteration, and endocrine disruption, affecting not only individual organism health but potentially the stability of entire aquatic ecosystems [285].

4.2. Impacts on Freshwater Biota

PAH contamination has been widely associated with reproductive and developmental disruptions in aquatic wildlife, often resulting in long-term ecological consequences that may extend across generations [286]. Chronic exposure, even at environmentally relevant concentrations, has been linked to persistent physiological and behavioral impairments in aquatic organisms [48], ranging from sub-lethal effects such as neurobehavioral disruptions [59] to severe tissue damage and mortality. In Caspian White fish (*Rutilus frissi*), exposure to environmentally relevant concentrations of BaP led to significant DNA damage, increased frequency of micronuclei in erythrocytes, and histopathological lesions in liver and gills [287]. Similarly, *Corbicula fluminea*, a native freshwater clam, exhibited PAH accumulation levels that closely correlated with sediment contamination, indicating sediment as a major exposure route through benthic feeding [244]. Experimental oil spills using diluted bitumen (dilbit) in limnocorrals demonstrated that insect emergence declined significantly in a dose-dependent manner, with notable shifts in benthic community structure despite no measurable loss in total abundance [288]. Sublethal effects were also observed in *Hyalella azteca* and *Chironomus riparius* exposed to dilbit-spiked sediments, with amphipods showing higher sensitivity and significant size reduction [289]. BaP has been reported to exhibit LC_{50} values as low as 5 μg/L for *Daphnia pulex* after 96 h of exposure and 5.6 μg/L for *Pimephales promelas* under UV-enhanced conditions, indicating its high acute toxicity to freshwater organisms [290]. These studies emphasize how PAH-rich unconventional oils compromise the physiological integrity and growth of benthic invertebrates.

Macroinvertebrate assemblages in the Persian Gulf's Nayband Bay, impacted by oil-related industrial activity, displayed significantly reduced abundance and species richness in stations with high levels of PAHs and total petroleum hydrocarbons (TPHs), revealing strong negative correlations with sedimentary pollutant levels [291]. Similarly, microbial communities in the Pearl River Estuary were disrupted by PAH contamination, which influenced microbial structure, diversity, and deterministic assembly processes across both water and sediment habitats [76]. Comparative studies of microbial communities in lake and river sediments have shown that riverine systems with higher degradation rates of PAHs support natural attenuation, while lacustrine sediments often require biostimulation due to lower functional activity [292]. These disruptions to microbial networks can cascade through the ecosystem, ultimately affecting nutrient cycling, bioremediation capacity, and the health and stability of macroinvertebrate communities.

Taken together, these studies highlight how PAHs exert diverse and sometimes cascading impacts across freshwater biota—affecting microbial assemblages, benthic organisms, and fish—ultimately impairing the structure and function of aquatic communities. These effects are often mediated by key toxicological mechanisms such as bioaccumulation, genotoxicity, mutagenesis, and carcinogenesis, which are further explored in the following subsections.

4.2.1. Bioaccumulation

The hydrophobic and lipophilic nature of PAHs promotes their accumulation in the fatty tissues of aquatic organisms, facilitating bioaccumulation throughout freshwater food webs. In Lake Chaohu (China), PAHs were detected in a variety of freshwater organisms including fish, with significant concentrations observed particularly in lipid-rich tissues, such as the brain and gills; this bioaccumulation posed notable carcinogenic risks to human consumers [272]. In the Ogbese River, Nigeria, shellfish like periwinkles, snails, and mussels exhibited higher biota–sediment accumulation factors compared to fish, with LMW-PAHs dominating the congener profiles [222]. Top predators are not exempt from exposure. In the Um Alnaaj Marsh, Iraq, muscle tissue of waterfowl such as *Anas platyrhynchos*, *A. crecca*, and *A. acuta*, revealed high concentrations of pyrogenic PAHs, suggesting that these birds bioaccumulate contaminants through trophic transfer and present a potential public health concern for consumers of bird meat [293]. In the Pearl River Delta, South China, trophic levels and aquatic productivity were found to significantly influence the distribution and bioconcentration of PAHs in algae and zooplankton, where high chlorophyll-a levels corresponded with greater PAH accumulation [294]. Moreover, a broader analysis in Laizhou Bay demonstrated that algae accumulated the highest concentrations of PAHs, followed by benthic fauna and fish, revealing a biodilution pattern along the food web—though HMW-PAHs still exhibited strong bioaccumulation factors [295].

PAH structure plays a crucial role in bioaccumulation potential. For example, fluoranthene and pyrene, both 3- and 4-ring PAHs, were dominant in the muscle tissues of demersal fish species from the Fengshan River system [234]. PAHs with intermediate log Kow values (5.0–5.6) have been identified as most prone to bioaccumulation [222,296]. A study of 30 rivers in Taiwan also found that up to 99.7% of PAHs bioaccumulated in fish were HMW-PAHs, especially BaP, which poses a substantial carcinogenic risk [103]. These findings collectively demonstrate how trophic level, lipid content, and PAH physicochemical properties influence the extent of bioaccumulation, ultimately compromising food web integrity and the sustainability of freshwater fisheries [297].

4.2.2. Genotoxic, Mutagenic, and Carcinogenic Effects

Several PAHs, especially BaP, are classified as human carcinogens, associated with cancers of the lung, cervix, and prostate [275,298]. Upon metabolic activation by CYP enzymes, particularly CYP1A, BaP forms reactive metabolites, such as BPDE, that bind to DNA, leading to mutations and initiation of carcinogenic processes. Consistent with cellular-level responses, CYP1A was significantly upregulated in the testis, liver, and brain of *Gambusia affinis* following BaP exposure, reflecting systemic genotoxic sensitivity to PAHs in aquatic vertebrates [276]. In juvenile haddock (*Melanogrammus aeglefinus*), both dietary and injected exposure to PAHs led to DNA adduct formation in the liver and intestines, skeletal deformities, and persistent genotoxic effects even after a two-month recovery period (See Figure 8) [57]. Similarly, hepatocellular fibrillar inclusions in European flounder (*Platichthys flesus*) from polluted UK estuaries have been linked to PAH metabolism, particularly in males, suggesting a sex-specific sensitivity to PAH-induced liver pathology [299].

Fish embryos represent another highly sensitive model. Research on teleost fish, including tunas and Japanese medaka (*Oryzias latipes*), revealed that early-stage exposure to PAHs caused developmental abnormalities, especially in cardiac morphology and function, with evidence of phototoxicity and mortality at trace concentrations [300–302]. In addition, exposure to NaP reduced acetylcholinesterase activity in the brain, liver, and gills of *Anabas testudineus*, demonstrating clear neurotoxic effects [303]. Other PAHs like phenanthrene and pyrene altered intracellular calcium regulation and suppressed sarco/endoplasmic reticulum Ca^{2+}-ATPase (SERCA) gene expression, contributing to cardiotoxicity and muscle impairment [304].

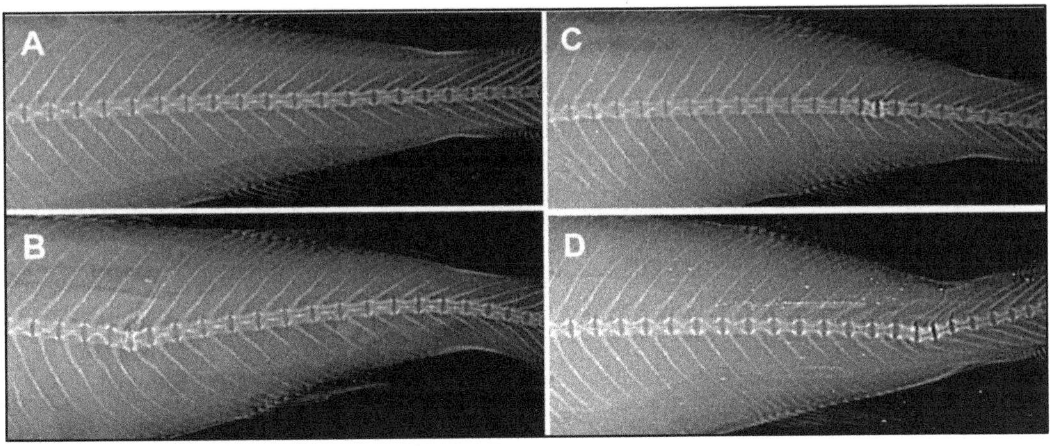

Figure 8. X-ray images of haddock (Melanogrammus aeglefinus) under different treatments: (**A**) control, (**B**) PAH exposure, and (**C**,**D**) oil exposure. The images show various vertebral malformations, including deformities in vertebrae. Reprinted from Meier et al. [57] under the terms of the Creative Commons Attribution License (CC BY). Copyright 2020, published by PLoS ONE.

Exposure to PAH mixtures exacerbates toxicity. For example, hepatocyte cultures of *Sparus aurata* exposed to mixtures of phenanthrene, BaP, and benzo[b]fluoranthene exhibited enhanced CYP1A expression and DNA strand breaks compared to single compound exposure, indicating synergistic genotoxicity [305]. In field studies, exposure to Σ16PAHs in vehicle-wash wastewater increased DNA damage in both fish and freshwater mussels, with mussels showing higher genotoxic sensitivity [306]. Similarly, water from the Dnieper River showed mutagenic potential in bioassays despite treatment at municipal plants, highlighting persistent contamination by legacy PAHs [307]. Long-term PAH pollution in Lithuanian rivers and lakes revealed genotoxicity gradients influenced by ecosystem type and hydrodynamics, with lacustrine systems exhibiting higher genotoxic risk but potential biotic adaptation over time [66]. Collectively, these reports underscore the widespread genotoxic, mutagenic, and carcinogenic effects of PAHs across freshwater species, influenced by environmental conditions, mixture composition, and exposure history.

5. Conclusions

This comprehensive review has examined the sources, distribution, and ecological impacts of PAHs in freshwater systems. Rivers, streams, lakes, wetlands, groundwater, and glaciers are influenced by a combination of natural and anthropogenic PAH sources. Rivers and streams function as conduits, transporting PAHs from upstream sources, while lakes, particularly in urban areas, accumulate LMW-PAHs in water columns and HMW-PAHs in sediments due to atmospheric deposition and surface runoff. Wetlands are especially vulnerable to wastewater discharge and oil-related activities, leading to significant accumulation of sediment-bound PAHs. On the other hand, groundwater systems face increasing contamination from PAHs leaching through soil and surface water infiltration, with agricultural and industrial regions being particularly at risk. Meanwhile, glaciers are impacted by long-range transported PAHs that deposit in these environments and are subsequently released during glacial melting. Effective monitoring and management strategies are essential across all freshwater systems to mitigate both historical and ongoing sources of PAH pollution.

The distribution of PAHs in freshwater systems is shaped by complex interactions among physical, chemical, biological, and environmental mechanisms. Physical mechanisms, such as advection, diffusion, and sediment transport, drive the movement of PAHs, while chemical mechanisms, including sorption and desorption, determine their mobility and persistence. These chemical mechanisms are further influenced by environmental conditions such as water pH and ionic strength. Biological mechanisms contribute to PAH transport through processes such as bioaccumulation and degradation by aquatic organisms. Additionally, environmental factors, such as temperature, sediment characteristics, and human activities, influence the partitioning and retention of PAHs within various freshwater matrices. These dynamic interactions highlight the complex nature of PAH distribution in freshwater systems, where comprehensive monitoring and management strategies are recommended.

The ecological impacts of PAHs in freshwater systems manifest at the molecular, cellular, and organismal levels, collectively posing significant risks to aquatic biodiversity. At the cellular level, PAHs induce OS, DNA damage, and endocrine disruption, with metabolites such as diol-epoxides forming DNA adducts that increase mutagenic and carcinogenic potential. These disruptions compromise genomic stability and affect reproduction and development in aquatic organisms. Chronic exposure to PAHs exacerbates these issues across freshwater ecosystems, particularly among sensitive species such as invertebrates, fish, and microbial communities. These impacts extend through the food web, with the bioaccumulation of PAHs in aquatic species consumed by humans indicating an urgent need for stringent and ongoing remediation efforts. Given the genotoxic, mutagenic, and carcinogenic properties of PAHs, an ecological approach is essential to understanding and mitigating their effects for the protection of biodiversity and ecosystem function.

In conclusion, this comprehensive review highlights the diverse sources, distribution mechanisms, and ecological impacts of PAHs in freshwater systems, emphasizing the extensive and complex interactions that drive their dynamics and persistence in these environments. PAH dynamics are influenced by local environmental factors and transport mechanisms that promote their persistence in sediments, water, and biota, leading to bioaccumulation and posing significant ecological and health risks to aquatic organisms and humans. The main toxicological effects include molecular and cellular damage, endocrine disruption, genotoxicity, mutagenicity, and carcinogenicity. While substantial progress has been made in understanding these processes, critical gaps remain in quantifying the long-term ecological effects of PAHs across different freshwater environments and species, as well as assessing the influence of climate change on their dynamics. Addressing these gaps through more comprehensive, ecosystem-level studies is essential for advancing PAH pollution research, management, and remediation strategies. These efforts should include monitoring across diverse environmental matrices, developing advanced and rapid analytical methodologies, and establishing robust regulatory frameworks that incorporate input from diverse stakeholders. A collaborative approach that integrates scientific research with policy action is vital to protecting and sustaining freshwater ecosystems for future generations.

Supplementary Materials: The following supporting information can be downloaded at https://www.mdpi.com/article/10.3390/toxics13040321/s1, Table S1: Physical properties and chemical structure of the 16 USEPA priority PAHs; Table S2: Origin and sources of PAHs in various types of freshwater systems.

Author Contributions: Conceptualization, P.J.B.-R., M.C.C. and F.M.; methodology, P.J.B.-R., M.C.C. and F.M.; writing—original draft preparation, P.J.B.-R., M.C.C. and F.M.; writing—review and editing, P.J.B.-R., M.C.C. and F.M.; supervision, M.C.C. All authors have read and agreed to the published version of the manuscript.

Funding: This research received no external funding.

Data Availability Statement: This study is a review article, and all data analyzed or cited are derived from publicly available studies and sources, as referenced within the manuscript. No new datasets were generated for this study. Any additional information can be obtained from the corresponding author upon reasonable request.

Conflicts of Interest: The authors declare no conflicts of interest.

Abbreviations

The following abbreviations are used in this manuscript:

PAHs	Polycyclic aromatic hydrocarbons
LMW-PAHs	Low molecular weight polycyclic aromatic hydrocarbons
HMW-PAH	High molecular weight polycyclic aromatic hydrocarbons
LRTs	Long-range transport
USEPA	United States Environmental Protection Agency
WHO	World Health Organization
SPM	Suspended particulate
OM	Organic matter
OC	Organic carbon
BaP	Benzo[a]pyrene
ROS	Reactive oxygen species
OS	Oxidative stress
CYP	Cytochrome P450
Log Kow	Octanol-water partitioning

References

1. Okere, U.V. Biodegradation of PAHs in 'Pristine' Soils from Different Climatic Regions. *J. Bioremed. Biodegrad.* **2011**, *S1*, 006. [CrossRef]
2. Edokpayi, J.; Odiyo, J.; Popoola, O.; Msagati, T. Determination and Distribution of Polycyclic Aromatic Hydrocarbons in Rivers, Sediments and Wastewater Effluents in Vhembe District, South Africa. *Int. J. Environ. Res. Public Health* **2016**, *13*, 387. [CrossRef] [PubMed]
3. Abdel-Shafy, H.I.; Mansour, M.S.M. A Review on Polycyclic Aromatic Hydrocarbons: Source, Environmental Impact, Effect on Human Health and Remediation. *Egypt. J. Pet.* **2016**, *25*, 107–123. [CrossRef]
4. Boehm, P.D.; Page, D.S.; Brown, J.S.; Neff, J.M.; Burns, W.A. Polycyclic Aromatic Hydrocarbon Levels in Mussels from Prince William Sound, ALASKA, USA, Document the Return to Baseline Conditions. *Environ. Toxicol. Chem.* **2004**, *23*, 2916–2929. [CrossRef]
5. Baali, A.; Yahyaoui, A.; Baali, A.; Yahyaoui, A. Polycyclic Aromatic Hydrocarbons (PAHs) and Their Influence to Some Aquatic Species. In *Biochemical Toxicology—Heavy Metals and Nanomaterials*; IntechOpen: London, UK, 2019; ISBN 978-1-78984-697-3.
6. Provencher, J.F.; Thomas, P.J.; Pauli, B.; Braune, B.M.; Franckowiak, R.P.; Gendron, M.; Savard, G.; Sarma, S.N.; Crump, D.; Zahaby, Y.; et al. Polycyclic Aromatic Compounds (PACs) and Trace Elements in Four Marine Bird Species from Northern Canada in a Region of Natural Marine Oil and Gas Seeps. *Sci. Total Environ.* **2020**, *744*, 140959. [CrossRef]
7. Azuma, H.; Toyota, M.; Asakawa, Y.; Kawano, S. Naphthalene—A Constituent of Magnolia Flowers. *Phytochemistry* **1996**, *42*, 999–1004. [CrossRef]
8. Jürgens, A.; Webber, A.C.; Gottsberger, G. Floral Scent Compounds of Amazonian Annonaceae Species Pollinated by Small Beetles and Thrips. *Phytochemistry* **2000**, *55*, 551–558. [CrossRef]
9. Hansen, D.J.; DiToro, D.M.; McGrath, J.A.; Swartz, R.C.; Mount, D.R.; Spehar, R.L.; Burgess, R.M.; Ozretich, R.J.; Bell, H.E.; Reiley, M.C.; et al. *Procedures for the Derivation of Equilibrium Partitioning Sediment Benchmarks (ESBs) for the Protection of Benthic Organisms: PAH Mixtures 2003*; EPA: Washington, DC, USA, 2003.

10. Dhar, K.; Subashchandrabose, S.R.; Venkateswarlu, K.; Krishnan, K.; Megharaj, M. Anaerobic Microbial Degradation of Polycyclic Aromatic Hydrocarbons: A Comprehensive Review. In *Reviews of Environmental Contamination and Toxicology*; de Voogt, P., Ed.; Reviews of Environmental Contamination and Toxicology; Springer International Publishing: Cham, Switzerland, 2020; Volume 251, pp. 25–108, ISBN 978-3-030-27149-7.
11. Wang, Z.; Yang, C.; Parrott, J.L.; Frank, R.A.; Yang, Z.; Brown, C.E.; Hollebone, B.P.; Landriault, M.; Fieldhouse, B.; Liu, Y.; et al. Forensic Source Differentiation of Petrogenic, Pyrogenic, and Biogenic Hydrocarbons in Canadian Oil Sands Environmental Samples. *J. Hazard. Mater.* **2014**, *271*, 166–177. [CrossRef]
12. Mali, M.; Ragone, R.; Dell'Anna, M.M.; Romanazzi, G.; Damiani, L.; Mastrorilli, P. Improved Identification of Pollution Source Attribution by Using PAH Ratios Combined with Multivariate Statistics. *Sci. Rep.* **2022**, *12*, 19298. [CrossRef]
13. Pang, S.Y.; Suratman, S.; Latif, M.T.; Khan, M.F.; Simoneit, B.R.T.; Mohd Tahir, N. Polycyclic Aromatic Hydrocarbons in Coastal Sediments of Southern Terengganu, South China Sea, Malaysia: Source Assessment Using Diagnostic Ratios and Multivariate Statistic. *Environ. Sci. Pollut. Res.* **2022**, *29*, 15849–15862. [CrossRef]
14. Bralewska, K.; Rakowska, J. Concentrations of Particulate Matter and PM-Bound Polycyclic Aromatic Hydrocarbons Released during Combustion of Various Types of Materials and Possible Toxicological Potential of the Emissions: The Results of Preliminary Studies. *Int. J. Environ. Res. Public Health* **2020**, *17*, 3202. [CrossRef] [PubMed]
15. Shimada, K.; Nohchi, M.; Yang, X.; Sugiyama, T.; Miura, K.; Takami, A.; Sato, K.; Chen, X.; Kato, S.; Kajii, Y.; et al. Degradation of PAHs during Long Range Transport Based on Simultaneous Measurements at Tuoji Island, China, and at Fukue Island and Cape Hedo, Japan. *Environ. Pollut.* **2020**, *260*, 113906. [CrossRef] [PubMed]
16. Zhang, L.; Yang, L.; Zhou, Q.; Zhang, X.; Xing, W.; Wei, Y.; Hu, M.; Zhao, L.; Toriba, A.; Hayakawa, K.; et al. Size Distribution of Particulate Polycyclic Aromatic Hydrocarbons in Fresh Combustion Smoke and Ambient Air: A Review. *J. Environ. Sci.* **2020**, *88*, 370–384. [CrossRef] [PubMed]
17. Hussain, K.; Hoque, R.R.; Balachandran, S.; Medhi, S.; Idris, M.G.; Rahman, M.; Hussain, F.L. Monitoring and Risk Analysis of PAHs in the Environment. In *Handbook of Environmental Materials Management*; Hussain, C.M., Ed.; Springer International Publishing: Cham, Switzerland, 2019; pp. 973–1007. ISBN 978-3-319-73645-7.
18. Bhatti, S.S.; Bhatia, A.; Bhagat, G.; Singh, S.; Dhaliwal, S.S.; Sharma, V.; Verma, V.; Yin, R.; Singh, J. PAHs in Terrestrial Environment and Their Phytoremediation. In *Bioremediation for Sustainable Environmental Cleanup*; CRC Press: Boca Raton, FL, USA, 2024; ISBN 978-1-003-27794-1.
19. Ashayeri, N.Y.; Keshavarzi, B.; Moore, F.; Kersten, M.; Yazdi, M.; Lahijanzadeh, A.R. Presence of Polycyclic Aromatic Hydrocarbons in Sediments and Surface Water from Shadegan Wetland—Iran: A Focus on Source Apportionment, Human and Ecological Risk Assessment and Sediment-Water Exchange. *Ecotoxicol. Environ. Saf.* **2018**, *148*, 1054–1066. [CrossRef]
20. Andersson, J.T.; Achten, C. Time to Say Goodbye to the 16 EPA PAHs? Toward an Up-to-Date Use of PACs for Environmental Purposes. *Polycycl. Aromat. Compd.* **2015**, *35*, 330–354. [CrossRef]
21. Keith, L.H. The Source of U.S. EPA's Sixteen PAH Priority Pollutants. *Polycycl. Aromat. Compd.* **2015**, *35*, 147–160. [CrossRef]
22. World Health Organization. *Human Health Effects of Polycyclic Aromatic Hydrocarbons as Ambient Air Pollutants: Report of the Working Group on Polycyclic Aromatic Hydrocarbons of the Joint Task Force on the Health Aspects of Air Pollution*; World Health Organization Regional Office for Europe: Copenhagen, Denmark, 2021; ISBN 978-92-890-5653-3.
23. Canadian Council of Ministers of the Environment. *Canadian Soil Quality Guidelines for the Protection of Environmental and Human Health: Polycyclic Aromatic Hydrocarbons 2010*; Canadian Council of Ministers of the Environment: Winnipeg, MB, Canada, 2010.
24. Kelly, C.; Santillo, D.; Johnston, P.; Fayad, G.; Baker, K.L.; Law, R.J. Polycyclic Aromatic Hydrocarbons in Oysters from Coastal Waters of the Lebanon 10 Months after the Jiyeh Oil Spill in 2006. *Mar. Pollut. Bull.* **2008**, *56*, 1215–1218. [CrossRef]
25. Sayers, E.W.; Bolton, E.E.; Brister, J.R.; Canese, K.; Chan, J.; Comeau, D.C.; Farrell, C.M.; Feldgarden, M.; Fine, A.M.; Sherry, S.T.; et al. National Center for Biotechnology Information PubChem Dataset 2023. *Nucleic Acids Res.* **2023**, *51*, D29–D38. [CrossRef]
26. Shen, H. *Polycyclic Aromatic Hydrocarbons*; Springer Theses; Springer: Berlin/Heidelberg, Germany, 2016; ISBN 978-3-662-49678-7.
27. Agency for Toxic Substances and Disease Registry. *Toxicological Profile for Polycyclic Aromatic Hydrocarbons*; U.S. Department of Health and Human Services: Washington, DC, USA, 1996.
28. Mackay, D.; Shiu, W.-Y.; Shiu, W.-Y.; Lee, S.C. *Handbook of Physical-Chemical Properties and Environmental Fate for Organic Chemicals*, 2nd ed.; CRC Press: Boca Raton, FL, USA, 2006; ISBN 978-0-429-15007-4.
29. Ji, G.; Zou, L.; Guan, W.; Yang, T.; Qiu, H.; Zhu, L. Partition, Transportation and Ecological Risks of Polycyclic Aromatic Hydrocarbons (PAHs) under Heavy Anthropogenic Estuary: A Case Study in the Xiaoqing River Estuary, North China. *Reg. Stud. Mar. Sci.* **2021**, *43*, 101664. [CrossRef]
30. Xiong, X.; Zhang, H.; Wang, R.; Tang, Z.; Yang, M. Distribution and Risk Assessment of 80 Polycyclic Aromatic Compounds in Surface Water and Industrial Effluent in the Upper Yangtze River, China. *ACS EST Water* **2024**, *4*, 2300–2308. [CrossRef]
31. Azimi, A.; Riahi Bakhtiari, A.; Tauler, R. Polycyclic Aromatic Hydrocarbon Source Fingerprints in the Environmental Samples of Anzali—South of Caspian Sea. *Environ. Sci. Pollut. Res.* **2020**, *27*, 32719–32731. [CrossRef] [PubMed]

32. Ouro-Sama, K.; Tanouayi, G.; Solitoke, H.D.; Barsan, N.; Mosnegutu, E.; Badassan, T.E.-E.; Agbere, S.; Adje, K.; Nedeff, V.; Gnandi, K. Polycyclic Aromatic Hydrocarbons (PAHs) Contamination in Chrysichthys Nigrodigitatus Lacépède, 1803 from Lake Togo-Lagoon of Aného, Togo: Possible Human Health Risk Suitable to Their Consumption. *Int. J. Environ. Res. Public Health* **2023**, *20*, 1666. [CrossRef] [PubMed]
33. Sahai, H.; García Valverde, M.; Murcia Morales, M.; Hernando, M.D.; Aguilera Del Real, A.M.; Fernández- Alba, A.R. Exploring Sorption of Pesticides and PAHs in Microplastics Derived from Plastic Mulch Films Used in Modern Agriculture. *Chemosphere* **2023**, *333*, 138959. [CrossRef] [PubMed]
34. Wetzel, D.L.; Van Vleet, E.S. Accumulation and Distribution of Petroleum Hydrocarbons Found in Mussels (*Mytilus galloprovincialis*) in the Canals of Venice, Italy. *Mar. Pollut. Bull.* **2004**, *48*, 927–936. [CrossRef]
35. Ambade, B.; Sethi, S.S. Health Risk Assessment and Characterization of Polycyclic Aromatic Hydrocarbon from the Hydrosphere. *J. Hazard. Toxic Radioact. Waste* **2021**, *25*, 05020008. [CrossRef]
36. Montuori, P.; De Rosa, E.; Di Duca, F.; De Simone, B.; Scippa, S.; Russo, I.; Sarnacchiaro, P.; Triassi, M. Polycyclic Aromatic Hydrocarbons (PAHs) in the Dissolved Phase, Particulate Matter, and Sediment of the Sele River, Southern Italy: A Focus on Distribution, Risk Assessment, and Sources. *Toxics* **2022**, *10*, 401. [CrossRef]
37. Santos, E.; Souza, M.R.R.; Junior, A.R.V.; da Silva Soares, L.; Frena, M.; Alexandre, M.R. Polycyclic Aromatic Hydrocarbons in Suspended Particulate Matter of a Region Influenced by Agricultural Activities in Northeast Brazil. *Reg. Stud. Mar. Sci.* **2023**, *57*, 102683. [CrossRef]
38. Yang, J.; Qadeer, A.; Liu, M.; Zhu, J.-M.; Huang, Y.-P.; Du, W.-N.; Wei, X.-Y. Occurrence, Source, and Partition of PAHs, PCBs, and OCPs in the Multiphase System of an Urban Lake, Shanghai. *Appl. Geochem.* **2019**, *106*, 17–25. [CrossRef]
39. Cathey, A.L.; Watkins, D.J.; Rosario, Z.Y.; Vélez Vega, C.M.; Loch-Caruso, R.; Alshawabkeh, A.N.; Cordero, J.F.; Meeker, J.D. Polycyclic Aromatic Hydrocarbon Exposure Results in Altered CRH, Reproductive, and Thyroid Hormone Concentrations during Human Pregnancy. *Sci. Total Environ.* **2020**, *749*, 141581. [CrossRef]
40. Sun, K.; Song, Y.; He, F.; Jing, M.; Tang, J.; Liu, R. A Review of Human and Animals Exposure to Polycyclic Aromatic Hydrocarbons: Health Risk and Adverse Effects, Photo-Induced Toxicity and Regulating Effect of Microplastics. *Sci. Total Environ.* **2021**, *773*, 145403. [CrossRef]
41. Gupte, A.; Tripathi, A.; Patel, H.; Rudakiya, D.; Gupte, S. Bioremediation of Polycyclic Aromatic Hydrocarbon (PAHs): A Perspective. *Open Biotechnol. J.* **2016**, *10*, 363–378. [CrossRef]
42. Méndez García, M.; García de Llasera, M.P. A Review on the Enzymes and Metabolites Identified by Mass Spectrometry from Bacteria and Microalgae Involved in the Degradation of High Molecular Weight PAHs. *Sci. Total Environ.* **2021**, *797*, 149035. [CrossRef] [PubMed]
43. Barbosa, F.; Rocha, B.A.; Souza, M.C.O.; Bocato, M.Z.; Azevedo, L.F.; Adeyemi, J.A.; Santana, A.; Campiglia, A.D. Polycyclic Aromatic Hydrocarbons (PAHs): Updated Aspects of Their Determination, Kinetics in the Human Body, and Toxicity. *J. Toxicol. Environ. Health B Crit. Rev.* **2023**, *26*, 28–65. [CrossRef] [PubMed]
44. Baran, A.; Klimkowicz-Pawlas, A.; Ukalska-Jaruga, A.; Mierzwa-Hersztek, M.; Gondek, K.; Szara-Bąk, M.; Tarnawski, M.; Spałek, I. Distribution of Polycyclic Aromatic Hydrocarbons (PAHs) in the Bottom Sediments of a Dam Reservoir, Their Interaction with Organic Matter and Risk to Benthic Fauna. *J. Soils Sediments* **2021**, *21*, 2418–2431. [CrossRef]
45. Jesus, F.; Pereira, J.L.; Campos, I.; Santos, M.; Ré, A.; Keizer, J.; Nogueira, A.; Gonçalves, F.J.M.; Abrantes, N.; Serpa, D. A Review on Polycyclic Aromatic Hydrocarbons Distribution in Freshwater Ecosystems and Their Toxicity to Benthic Fauna. *Sci. Total Environ.* **2022**, *820*, 153282. [CrossRef]
46. Campos, I.; Abrantes, N. Forest Fires as Drivers of Contamination of Polycyclic Aromatic Hydrocarbons to the Terrestrial and Aquatic Ecosystems. *Curr. Opin. Environ. Sci. Health* **2021**, *24*, 100293. [CrossRef]
47. Baumard, P.; Budzinski, H.; Michon, Q.; Garrigues, P.; Burgeot, T.; Bellocq, J. Origin and Bioavailability of PAHs in the Mediterranean Sea from Mussel and Sediment Records. *Estuar. Coast. Shelf Sci.* **1998**, *47*, 77–90. [CrossRef]
48. Honda, M.; Suzuki, N. Toxicities of Polycyclic Aromatic Hydrocarbons for Aquatic Animals. *Int. J. Environ. Res. Public Health* **2020**, *17*, 1363. [CrossRef]
49. Man, Y.B.; Mo, W.Y.; Zhang, F.; Wong, M.H. Health Risk Assessments Based on Polycyclic Aromatic Hydrocarbons in Freshwater Fish Cultured Using Food Waste-Based Diets. *Environ. Pollut.* **2020**, *256*, 113380. [CrossRef]
50. Shiklomanov, I. World Fresh Water Resources. In *Water in Crisis: A Guide to the World's Fresh Water Resources*; Gleick, P.H., Ed.; Oxford University Press: New York, NY, USA, 1993; ISBN 0-19-507628-1.
51. Shiklomanov, I.A. Appraisal and Assessment of World Water Resources. *Water Int.* **2000**, *25*, 11–32. [CrossRef]
52. Reid, A.J.; Carlson, A.K.; Creed, I.F.; Eliason, E.J.; Gell, P.A.; Johnson, P.T.J.; Kidd, K.A.; MacCormack, T.J.; Olden, J.D.; Ormerod, S.J.; et al. Emerging Threats and Persistent Conservation Challenges for Freshwater Biodiversity. *Biol. Rev.* **2019**, *94*, 849–873. [CrossRef] [PubMed]
53. Gleick, P.H. Water Resources. In *Encyclopedia of Climate and Weather*; Oxford University Press: New York, NY, USA, 1996; pp. 817–823.

54. Dudgeon, D.; Arthington, A.H.; Gessner, M.O.; Kawabata, Z.-I.; Knowler, D.J.; Lévêque, C.; Naiman, R.J.; Prieur-Richard, A.-H.; Soto, D.; Stiassny, M.L.J.; et al. Freshwater Biodiversity: Importance, Threats, Status and Conservation Challenges. *Biol. Rev.* **2006**, *81*, 163–182. [CrossRef] [PubMed]
55. Irfan, S.; Alatawi, A.M.M. Aquatic Ecosystem and Biodiversity: A Review. *Open J. Ecol.* **2019**, *9*, 1–13. [CrossRef]
56. Albert, J.S.; Destouni, G.; Duke-Sylvester, S.M.; Magurran, A.E.; Oberdorff, T.; Reis, R.E.; Winemiller, K.O.; Ripple, W.J. Scientists' Warning to Humanity on the Freshwater Biodiversity Crisis. *Ambio* **2021**, *50*, 85–94. [CrossRef]
57. Meier, S.; Karlsen, Ø.; Goff, J.L.; Sørensen, L.; Sørhus, E.; Pampanin, D.M.; Donald, C.E.; Fjelldal, P.G.; Dunaevskaya, E.; Romano, M.; et al. DNA Damage and Health Effects in Juvenile Haddock (*Melanogrammus aeglefinus*) Exposed to PAHs Associated with Oil-Polluted Sediment or Produced Water. *PLoS ONE* **2020**, *15*, e0240307. [CrossRef]
58. Yazdani, M. Comparative Toxicity of Selected PAHs in Rainbow Trout Hepatocytes: Genotoxicity, Oxidative Stress and Cytotoxicity. *Drug Chem. Toxicol.* **2020**, *43*, 71–78. [CrossRef]
59. Simão, F.C.P.; Gravato, C.; Machado, A.L.; Soares, A.M.V.M.; Pestana, J.L.T. Toxicity of Different Polycyclic Aromatic Hydrocarbons (PAHs) to the Freshwater Planarian Girardia Tigrina. *Environ. Pollut.* **2020**, *266*, 115185. [CrossRef]
60. Lu, X.-Y.; Zhang, T.; Fang, H.H.-P. Bacteria-Mediated PAH Degradation in Soil and Sediment. *Appl. Microbiol. Biotechnol.* **2011**, *89*, 1357–1371. [CrossRef]
61. Bai, Y.; Yu, H.; Shi, K.; Shang, N.; He, Y.; Meng, L.; Huang, T.; Yang, H.; Huang, C. Polycyclic Aromatic Hydrocarbons in Remote Lakes from the *Tibetan plateau*: Concentrations, Source, Ecological Risk, and Influencing Factors. *J. Environ. Manag.* **2022**, *319*, 115689. [CrossRef]
62. Vijayanand, M.; Ramakrishnan, A.; Subramanian, R.; Issac, P.K.; Nasr, M.; Khoo, K.S.; Rajagopal, R.; Greff, B.; Wan Azelee, N.I.; Jeon, B.-H.; et al. Polyaromatic Hydrocarbons (PAHs) in the Water Environment: A Review on Toxicity, Microbial Biodegradation, Systematic Biological Advancements, and Environmental Fate. *Environ. Res.* **2023**, *227*, 115716. [CrossRef]
63. House, W.A. Sampling Techniques for Organic Substances in Surface Waters. *Int. J. Environ. Anal. Chem.* **1994**, *57*, 207–214. [CrossRef]
64. Manoli, E.; Samara, C. Polycyclic Aromatic Hydrocarbons in Natural Waters: Sources, Occurrence and Analysis. *TrAC Trends Anal. Chem.* **1999**, *18*, 417–428. [CrossRef]
65. Frapiccini, E.; Panfili, M.; Guicciardi, S.; Santojanni, A.; Marini, M.; Truzzi, C.; Annibaldi, A. Effects of Biological Factors and Seasonality on the Level of Polycyclic Aromatic Hydrocarbons in Red Mullet (*Mullus barbatus*). *Environ. Pollut.* **2020**, *258*, 113742. [CrossRef] [PubMed]
66. Raudonytė-Svirbutavičienė, E.; Jokšas, K.; Stakėnienė, R.; Rybakovas, A.; Nalivaikienė, R.; Višinskienė, G.; Arbačiauskas, K. Pollution Patterns and Their Effects on Biota within Lotic and Lentic Freshwater Ecosystems: How Well Contamination and Response Indicators Correspond? *Environ. Pollut.* **2023**, *335*, 122294. [CrossRef]
67. Men, B.; He, M.; Tan, L.; Lin, C.; Quan, X. Distributions of Polycyclic Aromatic Hydrocarbons in the Daliao River Estuary of Liaodong Bay, Bohai Sea (China). *Mar. Pollut. Bull.* **2009**, *58*, 818–826. [CrossRef]
68. Balmer, J.E.; Hung, H.; Yu, Y.; Letcher, R.J.; Muir, D.C.G. Sources and Environmental Fate of Pyrogenic Polycyclic Aromatic Hydrocarbons (PAHs) in the Arctic. *Emerg. Contam.* **2019**, *5*, 128–142. [CrossRef]
69. Heintzman, L.J.; Anderson, T.A.; Carr, D.L.; McIntyre, N.E. Local and Landscape Influences on PAH Contamination in Urban Stormwater. *Landsc. Urban Plan.* **2015**, *142*, 29–37. [CrossRef]
70. Awonaike, B.; Lei, Y.D.; Parajulee, A.; Mitchell, C.P.J.; Wania, F. Polycyclic Aromatic Hydrocarbons and Quinones in Urban and Rural Stormwater Runoff: Effects of Land Use and Storm Characteristics. *ACS EST Water* **2021**, *1*, 1209–1219. [CrossRef]
71. Gong, X.; Zhao, Z.; Zhang, L.; Yao, S.; Xue, B. North-South Geographic Heterogeneity and Control Strategies for Polycyclic Aromatic Hydrocarbons (PAHs) in Chinese Lake Sediments Illustrated by Forward and Backward Source Apportionments. *J. Hazard. Mater.* **2022**, *431*, 128545. [CrossRef]
72. Rabodonirina, S.; Net, S.; Ouddane, B.; Merhaby, D.; Dumoulin, D.; Popescu, T.; Ravelonandro, P. Distribution of Persistent Organic Pollutants (PAHs, Me-PAHs, PCBs) in Dissolved, Particulate and Sedimentary Phases in Freshwater Systems. *Environ. Pollut.* **2015**, *206*, 38–48. [CrossRef]
73. Salowsky, H.; Schäfer, W.; Schneider, A.-L.; Müller, A.; Dreher, C.; Tiehm, A. Beneficial Effects of Dynamic Groundwater Flow and Redox Conditions on Natural Attenuation of Mono-, Poly-, and NSO-Heterocyclic Hydrocarbons. *J. Contam. Hydrol.* **2021**, *243*, 103883. [CrossRef] [PubMed]
74. Zhao, Z.; Gong, X.; Zhang, L.; Jin, M.; Cai, Y.; Wang, X. Riverine Transport and Water-Sediment Exchange of Polycyclic Aromatic Hydrocarbons (PAHs) along the Middle-Lower Yangtze River, China. *J. Hazard. Mater.* **2021**, *403*, 123973. [CrossRef] [PubMed]
75. Semenov, M.Y.; Marinaite, I.I.; Silaev, A.V.; Begunova, L.A. Composition, Concentration and Origin of Polycyclic Aromatic Hydrocarbons in Waters and Bottom Sediments of Lake Baikal and Its Tributaries. *Water* **2023**, *15*, 2324. [CrossRef]
76. Lin, W.; Fan, F.; Xu, G.; Gong, K.; Cheng, X.; Yuan, X.; Zhang, C.; Gao, Y.; Wang, S.; Ng, H.Y.; et al. Microbial Community Assembly Responses to Polycyclic Aromatic Hydrocarbon Contamination across Water and Sediment Habitats in the Pearl River Estuary. *J. Hazard. Mater.* **2023**, *457*, 131762. [CrossRef]

77. Kumar, V.; Kothiyal, N.C.; Saruchi; Vikas, P.; Sharma, R. Sources, Distribution, and Health Effect of Carcinogenic Polycyclic Aromatic Hydrocarbons (PAHs)—Current Knowledge and Future Directions. *J. Chin. Adv. Mater. Soc.* **2016**, *4*, 302–321. [CrossRef]
78. Dat, N.-D.; Chang, M.B. Review on Characteristics of PAHs in Atmosphere, Anthropogenic Sources and Control Technologies. *Sci. Total Environ.* **2017**, *609*, 682–693. [CrossRef]
79. Mojiri, A.; Zhou, J.L.; Ohashi, A.; Ozaki, N.; Kindaichi, T. Comprehensive Review of Polycyclic Aromatic Hydrocarbons in Water Sources, Their Effects and Treatments. *Sci. Total Environ.* **2019**, *696*, 133971. [CrossRef]
80. Patel, A.B.; Shaikh, S.; Jain, K.R.; Desai, C.; Madamwar, D. Polycyclic Aromatic Hydrocarbons: Sources, Toxicity, and Remediation Approaches. *Front. Microbiol.* **2020**, *11*, 562813. [CrossRef]
81. Ceschin, S.; Bellini, A.; Scalici, M. Aquatic Plants and Ecotoxicological Assessment in Freshwater Ecosystems: A Review. *Environ. Sci. Pollut. Res.* **2021**, *28*, 4975–4988. [CrossRef]
82. Robin, S.L.; Marchand, C. Polycyclic Aromatic Hydrocarbons (PAHs) in Mangrove Ecosystems: A Review. *Environ. Pollut.* **2022**, *311*, 119959. [CrossRef]
83. Behera, B.K.; Das, A.; Sarkar, D.J.; Weerathunge, P.; Parida, P.K.; Das, B.K.; Thavamani, P.; Ramanathan, R.; Bansal, V. Polycyclic Aromatic Hydrocarbons (PAHs) in Inland Aquatic Ecosystems: Perils and Remedies through Biosensors and Bioremediation. *Environ. Pollut.* **2018**, *241*, 212–233. [CrossRef] [PubMed]
84. Ben Othman, H.; Pick, F.R.; Sakka Hlaili, A.; Leboulanger, C. Effects of Polycyclic Aromatic Hydrocarbons on Marine and Freshwater Microalgae—A Review. *J. Hazard. Mater.* **2023**, *441*, 129869. [CrossRef] [PubMed]
85. Famiyeh, L.; Chen, K.; Xu, J.; Sun, Y.; Guo, Q.; Wang, C.; Lv, J.; Tang, Y.-T.; Yu, H.; Snape, C.; et al. A Review on Analysis Methods, Source Identification, and Cancer Risk Evaluation of Atmospheric Polycyclic Aromatic Hydrocarbons. *Sci. Total Environ.* **2021**, *789*, 147741. [CrossRef] [PubMed]
86. Sun, Y.; Zhang, S.; Lan, J.; Xie, Z.; Pu, J.; Yuan, D.; Yang, H.; Xing, B. Vertical Migration from Surface Soils to Groundwater and Source Appointment of Polycyclic Aromatic Hydrocarbons in Epikarst Spring Systems, Southwest China. *Chemosphere* **2019**, *230*, 616–627. [CrossRef]
87. Birks, S.J.; Cho, S.; Taylor, E.; Yi, Y.; Gibson, J.J. Characterizing the PAHs in Surface Waters and Snow in the Athabasca Region: Implications for Identifying Hydrological Pathways of Atmospheric Deposition. *Sci. Total Environ.* **2017**, *603–604*, 570–583. [CrossRef]
88. Marquès, M.; Mari, M.; Audí-Miró, C.; Sierra, J.; Soler, A.; Nadal, M.; Domingo, J.L. Climate Change Impact on the PAH Photodegradation in Soils: Characterization and Metabolites Identification. *Environ. Int.* **2016**, *89–90*, 155–165. [CrossRef]
89. Alegbeleye, O.O.; Opeolu, B.O.; Jackson, V.A. Polycyclic Aromatic Hydrocarbons: A Critical Review of Environmental Occurrence and Bioremediation. *Environ. Manag.* **2017**, *60*, 758–783. [CrossRef]
90. Akinpelumi, V.K.; Kumi, K.G.; Onyena, A.P.; Sam, K.; Ezejiofor, A.N.; Frazzoli, C.; Ekhator, O.C.; Udom, G.J.; Orisakwe, O.E. A Comparative Study of the Impacts of Polycyclic Aromatic Hydrocarbons in Water and Soils in Nigeria and Ghana: Towards a Framework for Public Health Protection. *J. Hazard. Mater. Adv.* **2023**, *11*, 100336. [CrossRef]
91. Wu, Y.; Wang, X.; Ya, M.; Li, Y.; Hong, H. Seasonal Variation and Spatial Transport of Polycyclic Aromatic Hydrocarbons in Water of the Subtropical Jiulong River Watershed and Estuary, Southeast China. *Chemosphere* **2019**, *234*, 215–223. [CrossRef]
92. Schwanen, C.A.; Kronsbein, P.M.; Balik, B.; Schwarzbauer, J. Dynamic Transport and Distribution of Organic Pollutants in Water and Sediments of the Rur River. *Water Air Soil. Pollut.* **2023**, *235*, 9. [CrossRef]
93. Adeniji, A.O.; Okoh, O.O.; Okoh, A.I. Levels of Polycyclic Aromatic Hydrocarbons in the Water and Sediment of Buffalo River Estuary, South Africa and Their Health Risk Assessment. *Arch. Environ. Contam. Toxicol.* **2019**, *76*, 657–669. [CrossRef] [PubMed]
94. Anyanwu, I.N.; Sikoki, F.D.; Semple, K.T. Risk Assessment of PAHs and N-PAH Analogues in Sediment Cores from the Niger Delta. *Mar. Pollut. Bull.* **2020**, *161*, 111684. [CrossRef] [PubMed]
95. Iwegbue, C.M.A.; Irerhievwie, G.O.; Tesi, G.O.; Olisah, C.; Nwajei, G.E.; Martincigh, B.S. Polycyclic Aromatic Hydrocarbons (PAHs) in Surficial Sediments from Selected Rivers in the Western Niger Delta of Nigeria: Spatial Distribution, Sources, and Ecological and Human Health Risks. *Mar. Pollut. Bull.* **2021**, *167*, 112351. [CrossRef] [PubMed]
96. Rizzi, C.; Villa, S.; Waichman, A.V.; de Souza Nunes, G.S.; de Oliveira, R.; Vighi, M.; Rico, A. Occurrence, Sources, and Ecological Risks of Polycyclic Aromatic Hydrocarbons (PAHs) in the Amazon River. *Chemosphere* **2023**, *336*, 139285. [CrossRef]
97. Umeh, C.T.; Nduka, J.K.; Omokpariola, D.O.; Morah, J.E.; Mmaduakor, E.C.; Okoye, N.H.; Lilian, E.-E.I.; Kalu, I.F. Ecological Pollution and Health Risk Monitoring Assessment of Polycyclic Aromatic Hydrocarbons and Heavy Metals in Surface Water, Southeastern Nigeria. *Environ. Anal. Health Toxicol.* **2023**, *38*, e2023007. [CrossRef]
98. Grmasha, R.A.; Abdulameer, M.H.; Stenger-Kovács, C.; Al-sareji, O.J.; Al-Gazali, Z.; Al-Juboori, R.A.; Meiczinger, M.; Hashim, K.S. Polycyclic Aromatic Hydrocarbons in the Surface Water and Sediment along Euphrates River System: Occurrence, Sources, Ecological and Health Risk Assessment. *Mar. Pollut. Bull.* **2023**, *187*, 114568. [CrossRef]
99. Wakeham, S.G.; Canuel, E.A. Biogenic Polycyclic Aromatic Hydrocarbons in Sediments of the San Joaquin River in California (USA), and Current Paradigms on Their Formation. *Environ. Sci. Pollut. Res. Int.* **2016**, *23*, 10426–10442. [CrossRef]

100. Liu, X.; Chen, Z.; Wu, J.; Cui, Z.; Su, P. Sedimentary Polycyclic Aromatic Hydrocarbons (PAHs) along the Mouth Bar of the Yangtze River Estuary: Source, Distribution, and Potential Toxicity. *Mar. Pollut. Bull.* **2020**, *159*, 111494. [CrossRef]
101. Liu, Y.; He, Y.; Han, B.; Liu, H.; Tao, S.; Liu, W. Sewage Discharge and Organic Matter Affect the Partitioning Behaviors of Different Polycyclic Aromatic Hydrocarbons in a River Surface Sediment-Pore Water System. *J. Hazard. Mater.* **2023**, *446*, 130757. [CrossRef]
102. Arowojolu, I.M.; Tongu, S.M.; Itodo, A.U.; Sodre, F.F.; Kyenge, B.A.; Nwankwo, R.C. Investigation of Sources, Ecological and Health Risks of Sedimentary Polycyclic Aromatic Hydrocarbons in River Benue, Nigeria. *Environ. Technol. Innov.* **2021**, *22*, 101457. [CrossRef]
103. Lee, C.-C.; Chen, C.S.; Wang, Z.-X.; Tien, C.-J. Polycyclic Aromatic Hydrocarbons in 30 River Ecosystems, Taiwan: Sources, and Ecological and Human Health Risks. *Sci. Total Environ.* **2021**, *795*, 148867. [CrossRef] [PubMed]
104. Yuan, Z.; He, B.; Wu, X.; Simonich, S.L.M.; Liu, H.; Fu, J.; Chen, A.; Liu, H.; Wang, Q. Polycyclic Aromatic Hydrocarbons (PAHs) in Urban Stream Sediments of Suzhou Industrial Park, an Emerging Eco-Industrial Park in China: Occurrence, Sources and Potential Risk. *Ecotoxicol. Environ. Saf.* **2021**, *214*, 112095. [CrossRef] [PubMed]
105. Van Metre, P.C.; Mahler, B.J.; Qi, S.L.; Gellis, A.C.; Fuller, C.C.; Schmidt, T.S. Sediment Sources and Sealed-Pavement Area Drive Polycyclic Aromatic Hydrocarbon and Metal Occurrence in Urban Streams. *Environ. Sci. Technol.* **2022**, *56*, 1615–1626. [CrossRef] [PubMed]
106. Nawrot, N.; Pouch, A.; Matej-Łukowicz, K.; Pazdro, K.; Mohsin, M.; Rezania, S.; Wojciechowska, E. A Multi-Criteria Approach to Investigate Spatial Distribution, Sources, and the Potential Toxicological Effect of Polycyclic Aromatic Hydrocarbons (PAHs) in Sediments of Urban Retention Tanks. *Environ. Sci. Pollut. Res.* **2023**, *30*, 27895–27911. [CrossRef]
107. Carvalho, F.; Pradhan, A.; Abrantes, N.; Campos, I.; Keizer, J.J.; Cássio, F.; Pascoal, C. Wildfire Impacts on Freshwater Detrital Food Webs Depend on Runoff Load, Exposure Time and Burnt Forest Type. *Sci. Total Environ.* **2019**, *692*, 691–700. [CrossRef]
108. Kieta, K.A.; Owens, P.N.; Petticrew, E.L.; French, T.D.; Koiter, A.J.; Rutherford, P.M. Polycyclic Aromatic Hydrocarbons in Terrestrial and Aquatic Environments Following Wildfire: A Review. *Environ. Rev.* **2023**, *31*, 141–167. [CrossRef]
109. Khiari, N.; Charef, A.; Atoui, A.; Azouzi, R.; Khalil, N.; Khadhar, S. Southern Mediterranean Coast Pollution: Long-Term Assessment and Evolution of PAH Pollutants in Monastir Bay (Tunisia). *Mar. Pollut. Bull.* **2021**, *167*, 112268. [CrossRef]
110. dos Santos, P.R.S.; Moreira, L.F.F.; Moraes, E.P.; de Farias, M.F.; Domingos, Y.S. Traffic-Related Polycyclic Aromatic Hydrocarbons (PAHs) Occurrence in a Tropical Environment. *Environ. Geochem. Health* **2021**, *43*, 4577–4587. [CrossRef]
111. Yao, K.; Xie, Z.; Zhi, L.; Wang, Z.; Qu, C. Polycyclic Aromatic Hydrocarbons in the Water Bodies of Dong Lake and Tangxun Lake, China: Spatial Distribution, Potential Sources and Risk Assessment. *Water* **2023**, *15*, 2416. [CrossRef]
112. Keiser, D.A.; Shapiro, J.S. Consequences of the Clean Water Act and the Demand for Water Quality. *Q. J. Econ.* **2019**, *134*, 349–396. [CrossRef]
113. Zhou, Z.; Liu, J.; Zhou, N.; Zhang, T.; Zeng, H. Does the "10-Point Water Plan" Reduce the Intensity of Industrial Water Pollution? Quasi-Experimental Evidence from China. *J. Environ. Manag.* **2021**, *295*, 113048. [CrossRef] [PubMed]
114. Dong, Y.; Guo, Y.; Wang, Y.; Zeng, W. Spatial and Temporal Evolution of the "Source–Sink" Risk Pattern of NPS Pollution in the Upper Reaches of Erhai Lake Basin under Land Use Changes in 2005–2020. *Water Air Soil Pollut.* **2022**, *233*, 202. [CrossRef]
115. Meng, Y.; Liu, X.; Lu, S.; Zhang, T.; Jin, B.; Wang, Q.; Tang, Z.; Liu, Y.; Guo, X.; Zhou, J.; et al. A Review on Occurrence and Risk of Polycyclic Aromatic Hydrocarbons (PAHs) in Lakes of China. *Sci. Total Environ.* **2019**, *651*, 2497–2506. [CrossRef] [PubMed]
116. Niu, L.; Luo, X.; Cai, H.; Liu, F.; Zhang, T.; Yang, Q. Seasonal Dynamics of Polycyclic Aromatic Hydrocarbons between Water and Sediment in a Tide-Dominated Estuary and Ecological Risks for Estuary Management. *Mar. Pollut. Bull.* **2021**, *162*, 111831. [CrossRef]
117. Du, J.; Jing, C. Anthropogenic PAHs in Lake Sediments: A Literature Review (2002–2018). *Environ. Sci. Process. Impacts* **2018**, *20*, 1649–1666. [CrossRef]
118. Yan, J.; Liu, J.; Shi, X.; You, X.; Cao, Z. Polycyclic Aromatic Hydrocarbons (PAHs) in Water from Three Estuaries of China: Distribution, Seasonal Variations and Ecological Risk Assessment. *Mar. Pollut. Bull.* **2016**, *109*, 471–479. [CrossRef]
119. Yang, J.; Yang, Y.; Chen, R.-S.; Meng, X.-Z.; Xu, J.; Qadeer, A.; Liu, M. Modeling and Evaluating Spatial Variation of Polycyclic Aromatic Hydrocarbons in Urban Lake Surface Sediments in Shanghai. *Environ. Pollut.* **2018**, *235*, 1–10. [CrossRef]
120. Golobokova, L.; Khodzher, T.; Khuriganova, O.; Marinayte, I.; Onishchuk, N.; Rusanova, P.; Potemkin, V. Variability of Chemical Properties of the Atmospheric Aerosol above Lake Baikal during Large Wildfires in Siberia. *Atmosphere* **2020**, *11*, 1230. [CrossRef]
121. Gorshkov, A.G.; Izosimova, O.N.; Kustova, O.V.; Marinaite, I.I.; Galachyants, Y.P.; Sinyukovich, V.N.; Khodzher, T.V. Wildfires as a Source of PAHs in Surface Waters of Background Areas (Lake Baikal, Russia). *Water* **2021**, *13*, 2636. [CrossRef]
122. Tao, Y.; Xue, B.; Feng, M. Spatial and Historical Occurrence, Sources, and Potential Toxicological Risk of Polycyclic Aromatic Hydrocarbons in Sediments of the Largest Chinese Deep Lake. *Arch. Environ. Contam. Toxicol.* **2019**, *77*, 501–513. [CrossRef]
123. Zhang, Y.; Cheng, D.; Lei, Y.; Song, J.; Xia, J. Spatiotemporal Distribution of Polycyclic Aromatic Hydrocarbons in Sediments of a Typical River Located in the Loess Plateau, China: Influence of Human Activities and Land-Use Changes. *J. Hazard. Mater.* **2022**, *424*, 127744. [CrossRef] [PubMed]

124. Zeng, S.; Ma, J.; Ren, Y.; Liu, G.-J.; Zhang, Q.; Chen, F. Assessing the Spatial Distribution of Soil PAHs and Their Relationship with Anthropogenic Activities at a National Scale. *Int. J. Environ. Res. Public Health* **2019**, *16*, 4928. [CrossRef] [PubMed]
125. Zhang, K.; Wang, J.-Z.; Liang, B.; Zeng, E.Y. Occurrence of Polycyclic Aromatic Hydrocarbons in Surface Sediments of a Highly Urbanized River System with Special Reference to Energy Consumption Patterns. *Environ. Pollut.* **2011**, *159*, 1510–1515. [CrossRef] [PubMed]
126. Wang, Y.; Shen, C.; Shen, Z.; Zhang, D.; Crittenden, J.C. Spatial Variation and Sources of Polycyclic Aromatic Hydrocarbons (PAHs) in Surface Sediments from the Yangtze Estuary, China. *Environ. Sci. Process. Impacts* **2015**, *17*, 1340–1347. [CrossRef]
127. Grmasha, R.A.; Stenger-Kovács, C.; Bedewy, B.A.H.; Al-sareji, O.J.; Al-Juboori, R.A.; Meiczinger, M.; Hashim, K.S. Ecological and Human Health Risk Assessment of Polycyclic Aromatic Hydrocarbons (PAH) in Tigris River near the Oil Refineries in Iraq. *Environ. Res.* **2023**, *227*, 115791. [CrossRef]
128. Adesina, O.B.; Paul, E.D.; Nuhu, A.A.; Onoyima, C.C.; Okibe, F.G. Spatiotemporal Variation and Health Risk Assessment of Selected Polycyclic Aromatic Hydrocarbons and Pesticides in Ogun River, Lagos, Nigeria. *J. Appl. Sci. Environ. Manag.* **2024**, *28*, 1501–1512. [CrossRef]
129. Qiao, M.; Wang, C.; Huang, S.; Wang, D.; Wang, Z. Composition, Sources, and Potential Toxicological Significance of PAHs in the Surface Sediments of the Meiliang Bay, Taihu Lake, China. *Environ. Int.* **2006**, *32*, 28–33. [CrossRef]
130. Ren, C.; Wu, Y.; Zhang, S.; Wu, L.-L.; Liang, X.-G.; Chen, T.-H.; Zhu, C.-Z.; Sojinu, S.O.; Wang, J.-Z. PAHs in Sediment Cores at Main River Estuaries of Chaohu Lake: Implication for the Change of Local Anthropogenic Activities. *Environ. Sci. Pollut. Res.* **2015**, *22*, 1687–1696. [CrossRef]
131. Garg, J.K. Wetland Assessment, Monitoring and Management in India Using Geospatial Techniques. *J. Environ. Manag.* **2015**, *148*, 112–123. [CrossRef]
132. Oyuela Leguizamo, M.A.; Fernández Gómez, W.D.; Sarmiento, M.C.G. Native Herbaceous Plant Species with Potential Use in Phytoremediation of Heavy Metals, Spotlight on Wetlands—A Review. *Chemosphere* **2017**, *168*, 1230–1247. [CrossRef]
133. Lettoof, D.C.; Bateman, P.W.; Aubret, F.; Gagnon, M.M. The Broad-Scale Analysis of Metals, Trace Elements, Organochlorine Pesticides and Polycyclic Aromatic Hydrocarbons in Wetlands Along an Urban Gradient, and the Use of a High Trophic Snake as a Bioindicator. *Arch. Environ. Contam. Toxicol.* **2020**, *78*, 631–645. [CrossRef] [PubMed]
134. Zheng, X.; Wang, H.; Tao, Y.; Kou, X.; He, C.; Wang, Z. Community Diversity of Soil Meso-Fauna Indicates the Impacts of Oil Exploitation on Wetlands. *Ecol. Indic.* **2022**, *144*, 109451. [CrossRef]
135. Yancheshmeh, R.A.; Bakhtiari, A.R.; Mortazavi, S.; Savabieasfahani, M. Sediment PAH: Contrasting Levels in the Caspian Sea and Anzali Wetland. *Mar. Pollut. Bull.* **2014**, *84*, 391–400. [CrossRef] [PubMed]
136. Rasta, M.; Sattari, M.; Taleshi, M.S.; Namin, J.I. Identification and Distribution of Microplastics in the Sediments and Surface Waters of Anzali Wetland in the Southwest Caspian Sea, Northern Iran. *Mar. Pollut. Bull.* **2020**, *160*, 111541. [CrossRef]
137. Amini-Birami, F.; Keshavarzi, B.; Esmaeili, H.R.; Moore, F.; Busquets, R.; Saemi-Komsari, M.; Zarei, M.; Zarandian, A. Microplastics in Aquatic Species of Anzali Wetland: An Important Freshwater Biodiversity Hotspot in Iran. *Environ. Pollut.* **2023**, *330*, 121762. [CrossRef]
138. Yang, W.; Cao, Z.; Lang, Y. Pollution Status of Polycyclic Aromatic Hydrocarbons (PAHs) in Northeastern China: A Review and Metanalysis. *Environ. Process.* **2021**, *8*, 429–454. [CrossRef]
139. Shi, C.; Qu, C.; Sun, W.; Zhou, J.; Zhang, J.; Cao, Y.; Zhang, Y.; Guo, J.; Zhang, J.; Qi, S. Multimedia Distribution of Polycyclic Aromatic Hydrocarbons in the Wang Lake Wetland, China. *Environ. Pollut.* **2022**, *306*, 119358. [CrossRef]
140. Chen, Z.; Ren, G.; Ma, X.; Zhou, B.; Yuan, D.; Liu, H.; Wei, Z. Presence of Polycyclic Aromatic Hydrocarbons among Multi-Media in a Typical Constructed Wetland Located in the Coastal Industrial Zone, Tianjin, China: Occurrence Characteristics, Source Apportionment and Model Simulation. *Sci. Total Environ.* **2021**, *800*, 149601. [CrossRef]
141. Cheshmvahm, H.; Keshavarzi, B.; Moore, F.; Zarei, M.; Esmaeili, H.R.; Hooda, P.S. Investigation of the Concentration, Origin and Health Effects of PAHs in the Anzali Wetland: The Most Important Coastal Freshwater Wetland of Iran. *Mar. Pollut. Bull.* **2023**, *193*, 115191. [CrossRef]
142. Rokhbar, M.; Keshavarzi, B.; Moore, F.; Zarei, M.; Hooda, P.S.; Risk, M.J. Occurrence and Source of PAHs in Miankaleh International Wetland in Iran. *Chemosphere* **2023**, *321*, 138140. [CrossRef]
143. Fakhradini, S.S.; Moore, F.; Keshavarzi, B.; Lahijanzadeh, A. Polycyclic Aromatic Hydrocarbons (PAHs) in Water and Sediment of Hoor Al-Azim Wetland, Iran: A Focus on Source Apportionment, Environmental Risk Assessment, and Sediment-Water Partitioning. *Environ. Monit. Assess.* **2019**, *191*, 233. [CrossRef] [PubMed]
144. Bemanikharanagh, A.; Riahi Bakhtiari, A.; Mohammadi, J.; Taghizadeh-Mehrjardi, R. Toxicity and Origins of PAHs in Sediments of Shadegan Wetland, in Khuzestan Province, Iran. *J. Maz. Univ. Med. Sci.* **2017**, *26*, 304–317.
145. Bao, K.; Shen, J.; Zhang, Y.; Wang, J.; Wang, G. A 200-Year Record of Polycyclic Aromatic Hydrocarbons Contamination in an Ombrotrophic Peatland in Great Hinggan Mountain, Northeast China. *J. Mt. Sci.* **2014**, *11*, 1085–1096. [CrossRef]
146. Russkikh, I.V.; Strel'nikova, E.B.; Serebrennikova, O.V.; Voistinova, E.S.; Kharanzhevskaya, Y.A. Identification of Hydrocarbons in the Waters of Raised Bogs in the Southern Taiga of Western Siberia. *Geochem. Int.* **2020**, *58*, 447–455. [CrossRef]

147. Salman, N.A.; Al-Saad, H.T.; Al-Imarah, F.J. The Status of Pollution in the Southern Marshes of Iraq: A Short Review. In *Southern Iraq's Marshes: Their Environment and Conservation*; Jawad, L.A., Ed.; Coastal Research Library; Springer International Publishing: Cham, Switzerland, 2021; pp. 505–516, ISBN 978-3-030-66238-7.
148. International Association of Hydrogeologists Groundwater—More About the Hidden Resource. Available online: https://iah.org/education/general-public/groundwater-hidden-resource (accessed on 16 January 2024).
149. Li, P.; Karunanidhi, D.; Subramani, T.; Srinivasamoorthy, K. Sources and Consequences of Groundwater Contamination. *Arch. Environ. Contam. Toxicol.* **2021**, *80*, 1–10. [CrossRef]
150. Li, J.; Li, F.; Liu, Q. PAHs Behavior in Surface Water and Groundwater of the Yellow River Estuary: Evidence from Isotopes and Hydrochemistry. *Chemosphere* **2017**, *178*, 143–153. [CrossRef]
151. Pan, Z.; Li, B.; Yang, J.; Zhang, D.; Yang, Y.; Zhang, S. Study on the Spatial and Temporal Distribution and Risk Assessment of PAHs between River and Groundwater—Take the Typical Section of Beijing North Canal as an Example. *J. Coast. Res.* **2020**, *115*, 361–366. [CrossRef]
152. Lan, J.; Sun, Y.; Yuan, D. Transport of Polycyclic Aromatic Hydrocarbons in a Highly Vulnerable Karst Underground River System of Southwest China. *Environ. Sci. Pollut. Res.* **2018**, *25*, 34519–34530. [CrossRef]
153. Chen, W.; Zhang, Z.; Zhu, Y.; Wang, X.; Wang, L.; Xiong, J.; Qian, Z.; Xiong, S.; Zhao, R.; Liu, W.; et al. Distribution, Sources and Transport of Polycyclic Aromatic Hydrocarbons (PAHs) in Karst Spring Systems from Western Hubei, Central China. *Chemosphere* **2022**, *300*, 134502. [CrossRef]
154. Lu, L.; Chen, Y.; Zou, S.; Wang, Z.; Fan, L. The Sources, Diffusion, and Health Risks of Polycyclic Aromatic Hydrocarbons in Water and Sediment of a Typical Underground River in South China. *Environ. Earth Sci.* **2024**, *83*, 100. [CrossRef]
155. Deka, J.; Sarma, K.P.; Gupta, N.; Sahbaz Ahmed, M.; Mazumder, M.A.J.; Hoque, R.R. Polycyclic Aromatic Hydrocarbons in Groundwater of Oil-Rich Regions of Upper Brahmaputra Valley, India: Linkages of Colloidal Transport. *Arab. J. Geosci.* **2023**, *16*, 66. [CrossRef]
156. Mansilha, C.; Melo, A.; Martins, Z.E.; Ferreira, I.M.P.L.V.O.; Pereira, A.M.; Espinha Marques, J. Wildfire Effects on Groundwater Quality from Springs Connected to Small Public Supply Systems in a Peri-Urban Forest Area (Braga Region, NW Portugal). *Water* **2020**, *12*, 1146. [CrossRef]
157. Ilić, P.; Nešković Markić, D.; Stojanović Bjelić, L. Evaluation of Sources and Ecological Risk of PAHs in Different Layers of Soil and Groundwater. *Preprints* **2020**. [CrossRef]
158. Edet, A.; Nyong, E.; Ukpong, A.; Edet, C. Evaluation and Risk Assessment of Polycyclic Aromatic Hydrocarbons in Groundwater and Soil near a Petroleum Distribution Pipeline Spill Site, Eleme, Nigeria. *Sustain. Water Resour. Manag.* **2021**, *7*, 50. [CrossRef]
159. Qi, X.; Lan, J.; Sun, Y.; Wang, S.; Liu, L.; Wang, J.; Long, Q.; Huang, M.; Yue, K. Linking PAHs Concentration, Risk to PAHs Source Shift in Soil and Water in Epikarst Spring Systems, Southwest China. *Ecotoxicol. Environ. Saf.* **2023**, *264*, 115465. [CrossRef]
160. Montuori, P.; De Rosa, E.; Cerino, P.; Pizzolante, A.; Nicodemo, F.; Gallo, A.; Rofrano, G.; De Vita, S.; Limone, A.; Triassi, M. Estimation of Polycyclic Aromatic Hydrocarbons in Groundwater from Campania Plain: Spatial Distribution, Source Attribution and Health Cancer Risk Evaluation. *Toxics* **2023**, *11*, 435. [CrossRef]
161. Liu, X.; Dong, Z.; Baccolo, G.; Gao, W.; Li, Q.; Wei, T.; Qin, X. Distribution, Composition and Risk Assessment of PAHs and PCBs in Cryospheric Watersheds of the Eastern Tibetan Plateau. *Sci. Total Environ.* **2023**, *890*, 164234. [CrossRef]
162. Pawlak, F.; Koziol, K.; Polkowska, Z. Chemical Hazard in Glacial Melt? The Glacial System as a Secondary Source of POPs (in the Northern Hemisphere). A Systematic Review. *Sci. Total Environ.* **2021**, *778*, 145244. [CrossRef]
163. Marchal, L.; Gateuille, D.; Naffrechoux, E.; Deline, P.; Baudin, F.; Clément, J.-C.; Poulenard, J. Polycyclic Aromatic Hydrocarbon Dynamics in Soils along Proglacial Chronosequences in the Alps. *Sci. Total Environ.* **2023**, *902*, 165998. [CrossRef]
164. Ademollo, N.; Spataro, F.; Rauseo, J.; Pescatore, T.; Fattorini, N.; Valsecchi, S.; Polesello, S.; Patrolecco, L. Occurrence, Distribution and Pollution Pattern of Legacy and Emerging Organic Pollutants in Surface Water of the Kongsfjorden (Svalbard, Norway): Environmental Contamination, Seasonal Trend and Climate Change. *Mar. Pollut. Bull.* **2021**, *163*, 111900. [CrossRef]
165. Li, R.; Gao, H.; Ji, Z.; Jin, S.; Ge, L.; Zong, H.; Jiao, L.; Zhang, Z.; Na, G. Distribution and Sources of Polycyclic Aromatic Hydrocarbons in the Water Column of Kongsfjorden, Arctic. *J. Environ. Sci.* **2020**, *97*, 186–193. [CrossRef] [PubMed]
166. Szopińska, M.; Szumińska, D.; Bialik, R.J.; Dymerski, T.; Rosenberg, E.; Polkowska, Ż. Determination of Polycyclic Aromatic Hydrocarbons (PAHs) and Other Organic Pollutants in Freshwaters on the Western Shore of Admiralty Bay (King George Island, Maritime Antarctica). *Environ. Sci. Pollut. Res.* **2019**, *26*, 18143–18161. [CrossRef] [PubMed]
167. Deelaman, W.; Pongpiachan, S.; Tipmanee, D.; Suttinun, O.; Choochuay, C.; Iadtem, N.; Charoenkalunyuta, T.; Promdee, K. Source Apportionment of Polycyclic Aromatic Hydrocarbons in the Terrestrial Soils of King George Island, Antarctica. *J. S. Am. Earth Sci.* **2020**, *104*, 102832. [CrossRef]
168. Pongpiachan, S.; Hattayanone, M.; Tipmanee, D.; Suttinun, O.; Khumsup, C.; Kittikoon, I.; Hirunyatrakul, P. Chemical Characterization of Polycyclic Aromatic Hydrocarbons (PAHs) in 2013 Rayong Oil Spill-Affected Coastal Areas of Thailand. *Environ. Pollut.* **2018**, *233*, 992–1002. [CrossRef]

169. Colby, G.A. Deposition of Polycyclic Aromatic Hydrocarbons (PAHs) into Northern Ontario Lake Sediments. *bioRxiv* **2019**, 786913. [CrossRef]
170. Miao, X.; Hao, Y.; Cai, J.; Xie, Y.; Zhang, J. The Distribution, Sources and Health Risk of Polycyclic Aromatic Hydrocarbons (PAHs) in Sediments of Liujiang River Basin: A Field Study in Typical Karstic River. *Mar. Pollut. Bull.* **2023**, *188*, 114666. [CrossRef]
171. Leizou, K.E.; Ashraf, M.A. Distribution, Compositional Pattern and Potential to Human Exposure of PAHs in Water, Amassoma Axis, Nun River, Bayelsa State, Nigeria. *Acta Chem. Malays.* **2019**, *3*, 16–20. [CrossRef]
172. Yuan, Z.; Shi, S.; Wu, X.; Wang, Q.; Wang, S.; Fan, Z. Polycyclic Aromatic Hydrocarbons (PAHs) in Riparian Soils of the Middle Reach of Huaihe River: A Typical Coal Mining Area in China. *Soil Sediment Contam. Int. J.* **2022**, *32*, 259–273. [CrossRef]
173. Nair, M.M.; Sreeraj, M.K.; Rakesh, P.S.; Thomas, J.K.; Kharat, P.Y.; Sukumaran, S. Distribution, Source and Potential Biological Impacts of Polycyclic Aromatic Hydrocarbons in the Core Sediments of a Networked Aquatic System in the Northwest Coast of India—A Special Focus on Thane Creek Flamingo Sanctuary (Ramsar Site). *Reg. Stud. Mar. Sci.* **2024**, *70*, 103377. [CrossRef]
174. Baldwin, A.K.; Corsi, S.R.; Oliver, S.K.; Lenaker, P.L.; Nott, M.A.; Mills, M.A.; Norris, G.A.; Paatero, P. Primary Sources of Polycyclic Aromatic Hydrocarbons to Streambed Sediment in Great Lakes Tributaries Using Multiple Lines of Evidence. *Environ. Toxicol. Chem.* **2020**, *39*, 1392–1408. [CrossRef]
175. Jabali, Y.; Iaaly, A.; Millet, M. Environmental Occurrence, Spatial Distribution, and Source Identification of PAHs in Surface and Groundwater Samples of Abou Ali River-North Lebanon. *Environ. Monit. Assess.* **2021**, *193*, 714. [CrossRef] [PubMed]
176. Qiao, X.; Zheng, B.; Li, X.; Zhao, X.; Dionysiou, D.D.; Liu, Y. Influencing Factors and Health Risk Assessment of Polycyclic Aromatic Hydrocarbons in Groundwater in China. *J. Hazard. Mater.* **2021**, *402*, 123419. [CrossRef] [PubMed]
177. Sreedevi, M.A.; Harikumar, P.S. Occurrence, Distribution, and Ecological Risk of Heavy Metals and Persistent Organic Pollutants (OCPs, PCBs, and PAHs) in Surface Sediments of the Ashtamudi Wetland, South-West Coast of India. *Reg. Stud. Mar. Sci.* **2023**, *64*, 103044. [CrossRef]
178. Xu, J.; Wang, H.; Sheng, L.; Liu, X.; Zheng, X. Distribution Characteristics and Risk Assessment of Polycyclic Aromatic Hydrocarbons in the Momoge Wetland, China. *Int. J. Environ. Res. Public Health* **2017**, *14*, 85. [CrossRef]
179. Singh, V.; Negi, R.; Jacob, M.; Gayathri, A.; Rokade, A.; Sarma, H.; Kalita, J.; Tasfia, S.T.; Bharti, R.; Wakid, A.; et al. Polycyclic Aromatic Hydrocarbons (PAHs) in Aquatic Ecosystem Exposed to the 2020 Baghjan Oil Spill in Upper Assam, India: Short-Term Toxicity and Ecological Risk Assessment. *PLoS ONE* **2023**, *18*, e0293601. [CrossRef]
180. Nim, N.; Morris, J.; Tekasakul, P.; Dejchanchaiwong, R. Fine and Ultrafine Particle Emission Factors and New Diagnostic Ratios of PAHs for Peat Swamp Forest Fires. *Environ. Pollut.* **2023**, *335*, 122237. [CrossRef]
181. Caumo, S.; Lázaro, W.L.; Sobreira Oliveira, E.; Beringui, K.; Gioda, A.; Massone, C.G.; Carreira, R.; de Freitas, D.S.; Ignacio, A.R.A.; Hacon, S. Human Risk Assessment of Ash Soil after 2020 Wildfires in Pantanal Biome (Brazil). *Air Qual Atmos Health* **2022**, *15*, 2239–2254. [CrossRef]
182. Hites, R.A. Polycyclic Aromatic Hydrocarbons in the Atmosphere near the Great Lakes: Why Do Their Concentrations Vary? *Environ. Sci. Technol.* **2021**, *55*, 9444–9449. [CrossRef]
183. Nguyen, T.N.T.; Park, M.-K.; Son, J.-M.; Choi, S.-D. Spatial Distribution and Temporal Variation of Polycyclic Aromatic Hydrocarbons in Runoff and Surface Water. *Sci. Total Environ.* **2021**, *793*, 148339. [CrossRef]
184. Wu, Y.; Salamova, A.; Venier, M. Using Diagnostic Ratios to Characterize Sources of Polycyclic Aromatic Hydrocarbons in the Great Lakes Atmosphere. *Sci. Total Environ.* **2021**, *761*, 143240. [CrossRef]
185. Yang, J.; Sun, P.; Zhang, X.; Wei, X.-Y.; Huang, Y.-P.; Du, W.-N.; Qadeer, A.; Liu, M.; Huang, Y. Source Apportionment of PAHs in Roadside Agricultural Soils of a Megacity Using Positive Matrix Factorization Receptor Model and Compound-Specific Carbon Isotope Analysis. *J. Hazard. Mater.* **2021**, *403*, 123592. [CrossRef] [PubMed]
186. Fei, W.; Ying, Z. Pollution Characteristics and Risk Assessment of PAHs in Agricultural Soil from Sewage Irrigation Area of Taiyuan City, Shanxi Province. *Ecol. Environ.* **2022**, *31*, 160. [CrossRef]
187. Meyers, R.A. (Ed.) *Encyclopedia of Physical Science and Technology*, 3rd ed.; Academic Press: Cambridge, MA, USA, 2001; ISBN 978-0-08-091795-5.
188. Cao, Y.; Lin, C.; Zhang, X.; Liu, X.; He, M.; Ouyang, W. Distribution, Source, and Ecological Risks of Polycyclic Aromatic Hydrocarbons in Lake Qinghai, China. *Environ. Pollut.* **2020**, *266*, 115401. [CrossRef] [PubMed]
189. Du, W.; Liu, M.; Li, Y.; Zhu, J.; Wei, X.; Yang, J.; Huang, Y.; Zhao, D.; Gao, D.; Qadeer, A. Cross-Interface Transfer of Polycyclic Aromatic Hydrocarbons (PAHs) in a Shallow Urban Lake in Shanghai, China Based on the Fugacity Model. *Sci. Total Environ.* **2020**, *736*, 139369. [CrossRef]
190. Nie, N.; Li, T.; Miao, Y.; Wei, X.; Zhao, D.; Liu, M. Environmental Fate and Health Risks of Polycyclic Aromatic Hydrocarbons in the Yangtze River Delta Urban Agglomeration during the 21st Century. *J. Hazard. Mater.* **2024**, *465*, 133407. [CrossRef]
191. Duttagupta, S.; Mukherjee, A.; Routh, J.; Devi, L.G.; Bhattacharya, A.; Bhattacharya, J. Role of Aquifer Media in Determining the Fate of Polycyclic Aromatic Hydrocarbons in the Natural Water and Sediments along the Lower Ganges River Basin. *J. Environ. Sci. Health Part A* **2020**, *55*, 354–373. [CrossRef]

192. Gregg, T.; Prahl, F.G.; Simoneit, B.R.T. Suspended Particulate Matter Transport of Polycyclic Aromatic Hydrocarbons in the Lower Columbia River and Its Estuary. *Limnol. Oceanogr.* **2015**, *60*, 1935–1949. [CrossRef]
193. Tucca, F.; Luarte, T.; Nimptsch, J.; Woelfl, S.; Pozo, K.; Casas, G.; Dachs, J.; Barra, R.; Chiang, G.; Galbán-Malagón, C. Sources and Diffusive Air–Water Exchange of Polycyclic Aromatic Hydrocarbons in an Oligotrophic North–Patagonian Lake. *Sci. Total Environ.* **2020**, *738*, 139838. [CrossRef]
194. Minick, D.J.; Anderson, K.A. Diffusive Flux of PAHs across Sediment–Water and Water–Air Interfaces at Urban Superfund Sites. *Environ. Toxicol. Chem.* **2017**, *36*, 2281–2289. [CrossRef]
195. Ma, X.; Kong, X.; Xue, B.; Mu, S.; Huang, C.; Huang, T.; Li, S.; Jiang, Q. Sediment Records and Multi-Media Transfer and Fate of Polycyclic Aromatic Hydrocarbons in Dianchi Lake over the Past 100 Years. *Ecol. Indic.* **2024**, *160*, 111774. [CrossRef]
196. Liu, C.; Huang, Z.; Qadeer, A.; Liu, Y.; Qiao, X.; Zheng, B.; Zhao, G.; Zhao, X. The Sediment-Water Diffusion and Risk Assessment of PAHs in Different Types of Drinking Water Sources in the Yangtze River Delta, China. *J. Clean. Prod.* **2021**, *309*, 127456. [CrossRef]
197. Zhang, Y.; Qu, C.; Qi, S.; Zhang, Y.; Mao, L.; Liu, J.; Qin, S.; Yang, D. Spatial-Temporal Variations and Transport Process of Polycyclic Aromatic Hydrocarbons in Poyang Lake: Implication for Dry–Wet Cycle Impacts. *J. Geochem. Explor.* **2021**, *226*, 106738. [CrossRef]
198. Zhang, J.; Liu, G.; Wang, R.; Huang, H. Polycyclic Aromatic Hydrocarbons in the Water-SPM-Sediment System from the Middle Reaches of Huai River, China: Distribution, Partitioning, Origin Tracing and Ecological Risk Assessment. *Environ. Pollut.* **2017**, *230*, 61–71. [CrossRef] [PubMed]
199. Hussain, T.; Athanasiou, D.; Rao, B.; Bejar, M.; Rakowska, M.; Drygiannaki, I.; Chadwick, D.B.; Colvin, M.A.; Hayman, N.T.; Rosen, G.H.; et al. Sediment Recontamination Potential and Biological Impacts of Hydrophobic Organics from Stormwater in a Mixed-Use Watershed. *Sci. Total Environ.* **2024**, *906*, 167444. [CrossRef]
200. Usanase, G.; Azema, N.; Bitouri, Y.E.; Souche, J.-C.; Gonzalez, C. Contribution of Settling Measurements to the Study of Polycyclic Aromatic Hydrocarbons' (PAHs) Mobilisation during Resuspension of PAHs-Associated Sediment. *Environ. Sci. Pollut. Res.* **2021**, *28*, 68349–68363. [CrossRef]
201. Sun, C.; Zhang, J.; Ma, Q.; Chen, Y.; Ju, H. Polycyclic Aromatic Hydrocarbons (PAHs) in Water and Sediment from a River Basin: Sediment–Water Partitioning, Source Identification and Environmental Health Risk Assessment. *Environ. Geochem. Health* **2017**, *39*, 63–74. [CrossRef]
202. Wan, N.; Gao, J.; Duan, D.; Yang, Y.; Ran, Y. Sedimentation and Resuspension Fluxes of Polycyclic Aromatic Hydrocarbons in Lake and Reservoir, South China 2022. *SSRN Electron. J.* **2022**. [CrossRef]
203. Hu, T.; Shi, M.; Mao, Y.; Liu, W.; Li, M.; Yu, Y.; Yu, H.; Cheng, C.; Zhang, Z.; Zhang, J.; et al. The Characteristics of Polycyclic Aromatic Hydrocarbons and Heavy Metals in Water and Sediment of Dajiuhu Subalpine Wetland, Shennongjia, Central China, 2018–2020: Insights for Sources, Sediment-Water Exchange, and Ecological Risk. *Chemosphere* **2022**, *309*, 136788. [CrossRef]
204. Mu, G.; Bian, D.; Zou, M.; Wang, X.; Chen, F. Pollution and Risk Assessment of Polycyclic Aromatic Hydrocarbons in Urban Rivers in a Northeastern Chinese City: Implications for Continuous Rainfall Events. *Sustainability* **2023**, *15*, 5777. [CrossRef]
205. Oyo-Ita, I.; Nkom, P.Y.; Ugim, S.U.; Bassey, F.I.; Oyo-Ita, O.E. Seasonal Changes of PAHs in Water and Suspended Particulate Matter from Cross River Estuary, SE Nigeria in Response to Human-Induced Activity and Hydrological Cycle. *Polycycl. Aromat. Compd.* **2022**, *42*, 5456–5473. [CrossRef]
206. Bacosa, H.P.; Steichen, J.; Kamalanathan, M.; Windham, R.; Lubguban, A.; Labonté, J.M.; Kaiser, K.; Hala, D.; Santschi, P.H.; Quigg, A. Polycyclic Aromatic Hydrocarbons (PAHs) and Putative PAH-Degrading Bacteria in Galveston Bay, TX (USA), Following Hurricane Harvey (2017). *Environ. Sci. Pollut. Res.* **2020**, *27*, 34987–34999. [CrossRef] [PubMed]
207. Shang, N.; Wang, C.; Kong, J.; Yu, H.; Li, J.; Hao, W.; Huang, T.; Yang, H.; He, H.; Huang, C. Dissolved Polycyclic Aromatic Hydrocarbons (PAHs-d) in Response to Hydrology Variation and Anthropogenic Activities in the Yangtze River, China. *J. Environ. Manag.* **2023**, *326*, 116673. [CrossRef] [PubMed]
208. Wu, Y.; Wang, X.; Ya, M.; Li, Y.; Liu, Y.; Chen, H. Spatial-Temporal Distribution and Transport Flux of Polycyclic Aromatic Hydrocarbons in a Large Hydropower Reservoir of Southeast China: Implication for Impoundment Impacts. *Environ. Pollut.* **2020**, *257*, 113603. [CrossRef] [PubMed]
209. Skic, K.; Boguta, P.; Klimkowicz-Pawlas, A.; Ukalska-Jaruga, A.; Baran, A. Effect of Sorption Properties on the Content, Ecotoxicity, and Bioaccumulation of Polycyclic Aromatic Hydrocarbons (PAHs) in Bottom Sediments. *J. Hazard. Mater.* **2023**, *442*, 130073. [CrossRef]
210. Soukarieh, B.; Hamieh, M.; Halloum, W.; Budzinski, H.; Jaber, F. The Effect of the Main Physicochemical Properties of Polycyclic Aromatic Hydrocarbons on Their Water/Sediments Distribution. *Int. J. Environ. Sci. Technol.* **2023**, *20*, 10261–10270. [CrossRef]
211. Huang, W.; Peng, P.; Yu, Z.; Fu, J. Effects of Organic Matter Heterogeneity on Sorption and Desorption of Organic Contaminants by Soils and Sediments. *Appl. Geochem.* **2003**, *18*, 955–972. [CrossRef]
212. Sun, Y.; Xie, Z.; Wu, K.; Lan, J.; Li, T.; Yuan, D. Speciation, Distribution and Migration Pathways of Polycyclic Aromatic Hydrocarbons in a Typical Underground River System in Southwest China. *J. Hydrol.* **2021**, *596*, 125690. [CrossRef]

213. Adeola, A.O.; Forbes, P.B.C. Influence of Natural Organic Matter Fractions on PAH Sorption by Stream Sediments and a Synthetic Graphene Wool Adsorbent. *Environ. Technol. Innov.* **2021**, *21*, 101202. [CrossRef]
214. Li, Z.; Zhang, W.; Shan, B. Effects of Organic Matter on Polycyclic Aromatic Hydrocarbons in Riverine Sediments Affected by Human Activities. *Sci. Total Environ.* **2022**, *815*, 152570. [CrossRef]
215. Lu, Z.; Tian, W.; Chu, M.; Zhang, S.; Zhao, J.; Liu, B.; Huo, B.; Chen, Z.; Zhang, R. A Novel and Thorough Research into Desorption Behavior of PAHs from Sediments to Seawater: Aging Process, Thermodynamics, Kinetics, Influencing Factors. *Chem. Eng. J.* **2024**, *480*, 148322. [CrossRef]
216. Zhang, S.; Zhao, J.; Zhu, L. New Insights into Thermal Desorption Remediation of Pyrene-Contaminated Soil Based on an Optimized Numerical Model. *J. Hazard. Mater.* **2024**, *461*, 132687. [CrossRef] [PubMed]
217. Song, X.; Wu, X.; Song, X.; Shi, C.; Zhang, Z. Sorption and Desorption of Petroleum Hydrocarbons on Biodegradable and Nondegradable Microplastics. *Chemosphere* **2021**, *273*, 128553. [CrossRef] [PubMed]
218. Almouallem, W.; Michel, J.; Dorge, S.; Joyeux, C.; Trouvé, G.; Le Nouen, D. A Comparative Study of the Sorption of O-PAHs and PAHs onto Soils to Understand Their Transport in Soils and Groundwater. *J. Environ. Sci.* **2023**, *124*, 61–75. [CrossRef] [PubMed]
219. He, Y.; Qin, N.; He, W.; Xu, F. The Impacts of Algae Biological Pump Effect on the Occurrence, Source Apportionment and Toxicity of SPM-Bound PAHs in Lake Environment. *Sci. Total Environ.* **2021**, *753*, 141980. [CrossRef]
220. Ding, Q.; Gong, X.; Jin, M.; Yao, X.; Zhang, L.; Zhao, Z. The Biological Pump Effects of Phytoplankton on the Occurrence and Benthic Bioaccumulation of Hydrophobic Organic Contaminants (HOCs) in a Hypereutrophic Lake. *Ecotoxicol. Environ. Saf.* **2021**, *213*, 112017. [CrossRef]
221. Dong, Y.; Yan, Z.; Wu, H.; Zhang, G.; Zhang, H.; Yang, M. Polycyclic Aromatic Hydrocarbons in Sediments from Typical Algae, Macrophyte Lake Bay and Adjoining River of Taihu Lake, China: Distribution, Sources, and Risk Assessment. *Water* **2021**, *13*, 470. [CrossRef]
222. Ololade, I.A.; Oladoja, N.A.; Ololade, O.O.; Saliu, T.D.; Alabi, A.B.; Obadawo, S.B.; Anifowose, M.M. Bioaccumulation and Toxic Potencies of Polycyclic Aromatic Hydrocarbons in Freshwater Biota from the Ogbese River, Nigeria. *Environ. Monit. Assess.* **2020**, *193*, 8. [CrossRef]
223. Pouch, A.; Zaborska, A.; Dąbrowska, A.M.; Pazdro, K. Bioaccumulation of PCBs, HCB and PAHs in the Summer Plankton from West Spitsbergen Fjords. *Mar. Pollut. Bull.* **2022**, *177*, 113488. [CrossRef]
224. Pastorino, P.; Nocita, A.; Ciccotelli, V.; Zaccaroni, A.; Anselmi, S.; Giugliano, R.; Tomasoni, M.; Silvi, M.; Menconi, V.; Vivaldi, B.; et al. Health Risk Assessment of Potentially Toxic Elements, Persistence of NDL-PCB, PAHs, and Microplastics in the Translocated Edible Freshwater Sinotaia Quadrata (Gasteropoda, Viviparidae): A Case Study from the Arno River Basin (Central Italy). *Expo. Health* **2021**, *13*, 583–596. [CrossRef]
225. Huang, Y.; Wang, B.; Yang, Y.; Yang, S.; Dong, M.; Xu, M. Microbial Carriers Promote and Guide Pyrene Migration in Sediments. *J. Hazard. Mater.* **2022**, *424*, 127188. [CrossRef]
226. Ou, D.; Liu, M.; Xu, S.; Cheng, S.; Hou, L.; Wang, L. Distribution and ecological risk assessment of polycyclic aromatic hydrocarbons in overlying waters and surface sediments from the Yangtze estuarine and coastal areas. *Huan Jing Ke Xue* **2009**, *30*, 3043–3049. [PubMed]
227. Hung, H.; Halsall, C.; Ball, H.; Bidleman, T.; Dachs, J.; Silva, A.D.; Hermanson, M.; Kallenborn, R.; Muir, D.; Sühring, R.; et al. Climate Change Influence on the Levels and Trends of Persistent Organic Pollutants (POPs) and Chemicals of Emerging Arctic Concern (CEACs) in the Arctic Physical Environment—A Review. *Environ. Sci. Process. Impacts* **2022**, *24*, 1577–1615. [CrossRef] [PubMed]
228. Na, M.; Zhao, Y.; Rina, S.; Wang, R.; Liu, X.; Tong, Z.; Zhang, J. Residues, Potential Source and Ecological Risk Assessment of Polycyclic Aromatic Hydrocarbons (PAHs) in Surface Water of the East Liao River, Jilin Province, China. *Sci. Total Environ.* **2023**, *886*, 163977. [CrossRef] [PubMed]
229. Cao, Y.; Wang, J.; Xin, M.; Wang, B.; Lin, C. Spatial Distribution and Partition of Polycyclic Aromatic Hydrocarbons (PAHs) in the Water and Sediment of the Southern Bohai Sea: Yellow River and PAH Property Influences. *Water Res.* **2024**, *248*, 120873. [CrossRef]
230. Cai, T.; Ding, Y.; Zhang, Z.; Wang, X.; Wang, T.; Ren, Y.; Dong, Y. Effects of Total Organic Carbon Content and Leaching Water Volume on Migration Behavior of Polycyclic Aromatic Hydrocarbons in Soils by Column Leaching Tests. *Environ. Pollut.* **2019**, *254*, 112981. [CrossRef]
231. Garcia, M.R.; Martins, C.C. A Systematic Evaluation of Polycyclic Aromatic Hydrocarbons in South Atlantic Subtropical Mangrove Wetlands under a Coastal Zone Development Scenario. *J. Environ. Manag.* **2021**, *277*, 111421. [CrossRef]
232. Lin, B.; Qi, F.; An, X.; Zhao, C.; Gao, Y.; Liu, Y.; Zhong, Y.; Qiu, B.; Wang, Z.; Hu, Q.; et al. Review: The Application of Source Analysis Methods in Tracing Urban Non-Point Source Pollution: Categorization, Hotspots, and Future Prospects. *Environ. Sci. Pollut. Res.* **2024**, *31*, 23482–23504. [CrossRef]
233. Li, Y.; Wang, X.; Gong, P. Combined Risk Assessment Method Based on Spatial Interaction: A Case for Polycyclic Aromatic Hydrocarbons and Heavy Metals in Taihu Lake Sediments. *J. Clean. Prod.* **2021**, *328*, 129590. [CrossRef]

234. Tien, C.-J.; Wang, Z.-X.; Chen, C.S. Microplastics in Water, Sediment and Fish from the Fengshan River System: Relationship to Aquatic Factors and Accumulation of Polycyclic Aromatic Hydrocarbons by Fish. *Environ. Pollut.* **2020**, *265*, 114962. [CrossRef]
235. José, S.; Jordao, L. Exploring the Interaction between Microplastics, Polycyclic Aromatic Hydrocarbons and Biofilms in Freshwater. *Polycycl. Aromat. Compd.* **2022**, *42*, 2210–2221. [CrossRef]
236. Ali, M.; Xu, D.; Yang, X.; Hu, J. Microplastics and PAHs Mixed Contamination: An in-Depth Review on the Sources, Co-Occurrence, and Fate in Marine Ecosystems. *Water Res.* **2024**, *257*, 121622. [CrossRef] [PubMed]
237. Li, Y.; Liu, M.; Hou, L.; Li, X.; Yin, G.; Sun, P.; Yang, J.; Wei, X.; He, Y.; Zheng, D. Geographical Distribution of Polycyclic Aromatic Hydrocarbons in Estuarine Sediments over China: Human Impacts and Source Apportionment. *Sci. Total Environ.* **2021**, *768*, 145279. [CrossRef] [PubMed]
238. Huang, Y.; Li, K.; Liu, H.; Yuan, X.; Li, M.; Xiong, B.; Du, R.; Johnson, D.M.; Xi, Y. Distribution, Sources and Risk Assessment of PAHs in Soil from the Water Level Fluctuation Zone of Xiangxi Bay, Three Gorges Reservoir. *Environ. Geochem. Health* **2022**, *44*, 2615–2628. [CrossRef] [PubMed]
239. Łyszczarz, S.; Lasota, J.; Szuszkiewicz, M.M.; Błońska, E. Soil Texture as a Key Driver of Polycyclic Aromatic Hydrocarbons (PAHs) Distribution in Forest Topsoils. *Sci. Rep.* **2021**, *11*, 14708. [CrossRef]
240. Pillay, V.; Moodley, B. Assessment of the Impact of Reforestation on Soil, Riparian Sediment and River Water Quality Based on Polyaromatic Hydrocarbon Pollutants. *J. Environ. Manag.* **2022**, *324*, 116331. [CrossRef]
241. Domínguez, C.; Sarkar, S.K.; Bhattacharya, A.; Chatterjee, M.; Bhattacharya, B.D.; Jover, E.; Albaigés, J.; Bayona, J.M.; Alam, M.A.; Satpathy, K.K. Quantification and Source Identification of Polycyclic Aromatic Hydrocarbons in Core Sediments from Sundarban Mangrove Wetland, India. *Arch. Environ. Contam. Toxicol.* **2010**, *59*, 49–61. [CrossRef]
242. Stephansen, D.A.; Arias, C.A.; Brix, H.; Fejerskov, M.L.; Nielsen, A.H. Relationship between Polycyclic Aromatic Hydrocarbons in Sediments and Invertebrates of Natural and Artificial Stormwater Retention Ponds. *Water* **2020**, *12*, 2020. [CrossRef]
243. Rasheed, R.O. Seasonal Variations of Polycyclic Aromatic Hydrocarbons in the Muscle Tissue of *Silurus Triostegus* Heckel, 1843 from Derbendikhan Reservoir. *Polycycl. Aromat. Compd.* **2023**, *43*, 2144–2151. [CrossRef]
244. Lécrivain, N.; Duparc, A.; Clément, B.; Naffrechoux, E.; Frossard, V. Tracking Sources and Transfer of Contamination According to Pollutants Variety at the Sediment-Biota Interface Using a Clam as Bioindicator in Peri-Alpine Lakes. *Chemosphere* **2020**, *238*, 124569. [CrossRef]
245. Sheng, Y.; Yan, C.; Nie, M.; Ju, M.; Ding, M.; Huang, X.; Chen, J. The Partitioning Behavior of PAHs between Settled Dust and Its Extracted Water Phase: Coefficients and Effects of the Fluorescent Organic Matter. *Ecotoxicol. Environ. Saf.* **2021**, *223*, 112573. [CrossRef]
246. Li, R.; Hua, P.; Zhang, J.; Krebs, P. Effect of Anthropogenic Activities on the Occurrence of Polycyclic Aromatic Hydrocarbons in Aquatic Suspended Particulate Matter: Evidence from Rhine and Elbe Rivers. *Water Res.* **2020**, *179*, 115901. [CrossRef] [PubMed]
247. Nybom, I.; van Grimbergen, J.; Forsell, M.; Mustajärvi, L.; Martens, J.; Sobek, A. Water Column Organic Carbon Composition as Driver for Water-Sediment Fluxes of Hazardous Pollutants in a Coastal Environment. *J. Hazard. Mater.* **2024**, *465*, 133393. [CrossRef] [PubMed]
248. Ge, X.; Ren, J.; Li, S.; Rene, E.R.; Zhou, D.; Zhang, P.; Hu, Q.; Ma, W. Prediction of the Impact of Benzo[a]Pyrene on Shallow Groundwater during Natural Infiltration of Reclaimed Water-Receiving Rivers: A Case Study of Liangshui, China. *J. Environ. Manag.* **2022**, *323*, 116070. [CrossRef] [PubMed]
249. Wang, Z.; Li, J.; Lu, L.; Cao, J.; Zhao, L.; Luan, S. Source, Partition and Ecological Risk of Polycyclic Aromatic Hydrocarbons in Karst Underground River Environment, Southern China. *Water* **2021**, *13*, 2655. [CrossRef]
250. Wu, Z.; Lin, T.; Hu, L.; Guo, T.; Guo, Z. Polycyclic Aromatic Hydrocarbons in Sediment–Porewater System from the East China Sea: Occurrence, Partitioning, and Diffusion. *Environ. Res.* **2022**, *209*, 112755. [CrossRef]
251. Hu, S.-Y.; Hsieh, C.-Y.; Dahms, H.-U.; Tseng, Y.-H.; Chen, J.; Wu, M.-C.; Kim, J.-H.; Liu, C.-H. Toxic Effects of Heavy Metals and Organic Polycyclic Aromatic Hydrocarbons in Sediment Porewater on the Amphipod Hyalella Azteca and Zebrafish Brachydanio Rerio Embryos from Different Rivers in Taiwan. *Appl. Sci.* **2021**, *11*, 8021. [CrossRef]
252. MacKay, C.E.; Knock, G.A. Control of Vascular Smooth Muscle Function by Src-Family Kinases and Reactive Oxygen Species in Health and Disease. *J. Physiol.* **2015**, *593*, 3815–3828. [CrossRef]
253. Vondráček, J.; Machala, M. The Role of Metabolism in Toxicity of Polycyclic Aromatic Hydrocarbons and Their Non-Genotoxic Modes of Action. *Curr. Drug Metab.* **2021**, *22*, 584–595. [CrossRef]
254. Cheng, X.-M.; Hu, Y.-Y.; Yang, T.; Wu, N.; Wang, X.-N. Reactive Oxygen Species and Oxidative Stress in Vascular-Related Diseases. *Oxid. Med. Cell. Longev.* **2022**, *2022*, 7906091. [CrossRef]
255. Harvey, R.G.; Dai, Q.; Ran, C.; Lim, K.; Blair, I.; Penning, T.M. Syntheses of adducts of active metabolites of carcinogenic polycyclic aromatic hydrocarbons with 2′-deoxyribonucleosides. *Polycycl. Aromat. Compd.* **2005**, *25*, 371–391. [CrossRef]
256. Hrdina, A.I.H.; Kohale, I.N.; Kaushal, S.; Kelly, J.; Selin, N.E.; Engelward, B.P.; Kroll, J.H. The Parallel Transformations of Polycyclic Aromatic Hydrocarbons in the Body and in the Atmosphere. *Environ. Health Perspect.* **2022**, *130*, 025004. [CrossRef] [PubMed]

257. Barnes, J.L.; Zubair, M.; John, K.; Poirier, M.C.; Martin, F.L. Carcinogens and DNA Damage. *Biochem. Soc. Trans.* **2018**, *46*, 1213–1224. [CrossRef] [PubMed]
258. Luo, K.; Hochalter, J.B.; Carmella, S.G.; Hecht, S.S. Quantitation of Phenanthrene Dihydrodiols in the Urine of Smokers and Non-Smokers by Gas Chromatography-Negative Ion Chemical Ionization-Tandem Mass Spectrometry. *J. Chromatogr. B* **2020**, *1141*, 122023. [CrossRef] [PubMed]
259. Cheng, T.; Lam, A.K.; Gopalan, V. Diet Derived Polycyclic Aromatic Hydrocarbons and Its Pathogenic Roles in Colorectal Carcinogenesis. *Crit. Rev. Oncol./Hematol.* **2021**, *168*, 103522. [CrossRef]
260. Peng, B.; Dong, Q.; Li, F.; Wang, T.; Qiu, X.; Zhu, T. A Systematic Review of Polycyclic Aromatic Hydrocarbon Derivatives: Occurrences, Levels, Biotransformation, Exposure Biomarkers, and Toxicity. *Environ. Sci. Technol.* **2023**, *57*, 15314–15335. [CrossRef]
261. Penning, T.M.; Burczynski, M.E.; Hung, C.-F.; McCoull, K.D.; Palackal, N.T.; Tsuruda, L.S. Dihydrodiol Dehydrogenases and Polycyclic Aromatic Hydrocarbon Activation: Generation of Reactive and Redox Active o-Quinones. *Chem. Res. Toxicol.* **1999**, *12*, 1–18. [CrossRef]
262. Shiraiwa, M.; Ueda, K.; Pozzer, A.; Lammel, G.; Kampf, C.J.; Fushimi, A.; Enami, S.; Arangio, A.M.; Fröhlich-Nowoisky, J.; Fujitani, Y.; et al. Aerosol Health Effects from Molecular to Global Scales. *Environ. Sci. Technol.* **2017**, *51*, 13545–13567. [CrossRef]
263. Liu, J.; Hanzhong, J.; Kecheng, Z.; Zhao, S.; Eric, L. Formation of Environmentally Persistent Free Radicals and Reactive Oxygen Species during the Thermal Treatment of Soils Contaminated by Polycyclic Aromatic Hydrocarbons. *Environ. Chem. Lett.* **2020**, *18*, 1329–1336. [CrossRef]
264. Sun, N.; Li, M.; Liu, G.; Jing, M.; He, F.; Cao, Z.; Zong, W.; Tang, J.; Gao, C.; Liu, R. Toxic Mechanism of Pyrene to Catalase and Protective Effects of Vitamin C: Studies at the Molecular and Cell Levels. *Int. J. Biol. Macromol.* **2021**, *171*, 225–233. [CrossRef]
265. Jomova, K.; Raptova, R.; Alomar, S.Y.; Alwasel, S.H.; Nepovimova, E.; Kuca, K.; Valko, M. Reactive Oxygen Species, Toxicity, Oxidative Stress, and Antioxidants: Chronic Diseases and Aging. *Arch. Toxicol.* **2023**, *97*, 2499–2574. [CrossRef]
266. Aranda-Rivera, A.K.; Cruz-Gregorio, A.; Arancibia-Hernández, Y.L.; Hernández-Cruz, E.Y.; Pedraza-Chaverri, J. RONS and Oxidative Stress: An Overview of Basic Concepts. *Oxygen* **2022**, *2*, 437–478. [CrossRef]
267. Madamanchi, N.R.; Vendrov, A.; Runge, M.S. Oxidative Stress and Vascular Disease. *Arterioscler. Thromb. Vasc. Biol.* **2005**, *25*, 29–38. [CrossRef] [PubMed]
268. He, J.; Pang, Q.; Huang, C.; Xie, J.; Hu, J.; Wang, L.; Wang, C.; Meng, L.; Fan, R. Environmental Dose of 16 Priority-Controlled PAHs Mixture Induce Damages of Vascular Endothelial Cells Involved in Oxidative Stress and Inflammation. *Toxicol. Vitr.* **2022**, *79*, 105296. [CrossRef] [PubMed]
269. Malik, S.; Prasad, S.; Kishore, S.; Kumar, A.; Upadhyay, V. A Perspective Review on Impact and Molecular Mechanism of Environmental Carcinogens on Human Health. *Biotechnol. Genet. Eng. Rev.* **2021**, *37*, 178–207. [CrossRef]
270. Sun, Y.; Shi, Z.; Lin, Y.; Zhang, M.; Liu, J.; Zhu, L.; Chen, Q.; Bi, J.; Li, S.; Ni, Z.; et al. Benzo(a)Pyrene Induces MUC5AC Expression through the AhR/Mitochondrial ROS/ERK Pathway in Airway Epithelial Cells. *Ecotoxicol. Environ. Saf.* **2021**, *210*, 111857. [CrossRef]
271. Basu, A.K.; Essigmann, J.M. Establishing Linkages Among DNA Damage, Mutagenesis, and Genetic Diseases. *Chem. Res. Toxicol.* **2022**, *35*, 1655–1675. [CrossRef]
272. Qin, C.; Hu, X.; Yang, B.; Liu, J.; Gao, Y. Amino, Nitro, Chloro, Hydroxyl and Methyl Substitutions May Inhibit the Binding of PAHs with DNA. *Environ. Pollut.* **2021**, *268*, 115798. [CrossRef]
273. Zeng, J.; Li, Y.; Dai, Y.; Wu, Y.; Lin, X. Effects of Polycyclic Aromatic Hydrocarbon Structure on PAH Mineralization and Toxicity to Soil Microorganisms after Oxidative Bioremediation by Laccase. *Environ. Pollut.* **2021**, *287*, 117581. [CrossRef]
274. Shukla, S.; Khan, R.; Bhattacharya, P.; Devanesan, S.; AlSalhi, M.S. Concentration, Source Apportionment and Potential Carcinogenic Risks of Polycyclic Aromatic Hydrocarbons (PAHs) in Roadside Soils. *Chemosphere* **2022**, *292*, 133413. [CrossRef]
275. Bukowska, B.; Sicińska, P. Influence of Benzo(a)Pyrene on Different Epigenetic Processes. *Int. J. Mol. Sci.* **2021**, *22*, 13453. [CrossRef]
276. Xie, S.; Zhou, A.; Feng, Y.; Zhang, Y.; Li, J.; Sun, Z.; Fan, L.; Zou, J. Cytochrome P450 1A mRNA in the Gambusia Affinis and Response to Several PAHs. *Biochem. Genet.* **2020**, *58*, 551–565. [CrossRef] [PubMed]
277. Vallée, A.; Ceccaldi, P.-F.; Carbonnel, M.; Feki, A.; Ayoubi, J.-M. Pollution and Endometriosis: A Deep Dive into the Environmental Impacts on Women's Health. *BJOG Int. J. Obstet. Gynaecol.* **2024**, *131*, 401–414. [CrossRef] [PubMed]
278. Doan, T.; Connolly, L.; Igout, A.; Nott, K.; Muller, M.; Scippo, M. In Vitro Profiling of the Potential Endocrine Disrupting Activities Affecting Steroid and Aryl Hydrocarbon Receptors of Compounds and Mixtures Prevalent in Human Drinking Water Resources. *Chemosphere* **2020**, *258*, 127332. [CrossRef]
279. Bozinovic, G.; Shea, D.; Feng, Z.; Hinton, D.; Sit, T.; Oleksiak, M.F. PAH-Pollution Effects on Sensitive and Resistant Embryos: Integrating Structure and Function with Gene Expression. *PLoS ONE* **2021**, *16*, e0249432. [CrossRef] [PubMed]
280. Pironti, C.; Ricciardi, M.; Proto, A.; Bianco, P.M.; Montano, L.; Motta, O. Endocrine-Disrupting Compounds: An Overview on Their Occurrence in the Aquatic Environment and Human Exposure. *Water* **2021**, *13*, 1347. [CrossRef]

281. Alaekwe, I.O.; Abba, O. Polycyclic Aromatic Hydrocarbons in Water: A Review of the Sources, Properties, Exposure Pathways, Bionetwork and Strategies for Remediation. *J. Geosci. Environ. Prot.* **2022**, *10*, 137–144. [CrossRef]
282. Ramesh, A.; Harris, K.J.; Archibong, A.E. Chapter 38—Reproductive Toxicity of Polycyclic Aromatic Hydrocarbons. In *Reproductive and Developmental Toxicology*, 3rd ed.; Gupta, R.C., Ed.; Academic Press: Cambridge, MA, USA, 2022; pp. 759–778, ISBN 978-0-323-89773-0.
283. Šimečková, P.; Pěnčíková, K.; Kováč, O.; Slavík, J.; Pařenicová, M.; Vondráček, J.; Machala, M. In Vitro Profiling of Toxic Effects of Environmental Polycyclic Aromatic Hydrocarbons on Nuclear Receptor Signaling, Disruption of Endogenous Metabolism and Induction of Cellular Stress. *Sci. Total Environ.* **2022**, *815*, 151967. [CrossRef]
284. Peng, F.-J.; Palazzi, P.; Viguié, C.; Appenzeller, B.M.R. Measurement of Hair Thyroid and Steroid Hormone Concentrations in the Rat Evidence Endocrine Disrupting Potential of a Low Dose Mixture of Polycyclic Aromatic Hydrocarbons. *Environ. Pollut.* **2022**, *313*, 120179. [CrossRef]
285. Vieira, L.R.; Guilhermino, L. Multiple Stress Effects on Marine Planktonic Organisms: Influence of Temperature on the Toxicity of Polycyclic Aromatic Hydrocarbons to *Tetraselmis chuii*. *J. Sea Res.* **2012**, *72*, 94–98. [CrossRef]
286. Marlatt, V.L.; Bayen, S.; Castaneda-Cortès, D.; Delbès, G.; Grigorova, P.; Langlois, V.S.; Martyniuk, C.J.; Metcalfe, C.D.; Parent, L.; Rwigemera, A.; et al. Impacts of Endocrine Disrupting Chemicals on Reproduction in Wildlife and Humans. *Environ. Res.* **2022**, *208*, 112584. [CrossRef]
287. Esmaeilbeigi, M.; Kalbassi, M.R.; Seyedi, J.; Tayemeh, M.B.; Moghaddam, J.A. Intra and Extracellular Effects of Benzo [α] Pyrene on Liver, Gill and Blood of Caspian White Fish (Rutilus Frissi Kutum): Cyto-Genotoxicity and Histopathology Approach. *Mar. Pollut. Bull.* **2021**, *163*, 111942. [CrossRef] [PubMed]
288. Black, T.A.; White, M.S.; Blais, J.M.; Hollebone, B.; Orihel, D.M.; Palace, V.P.; Rodriguez-Gil, J.L.; Hanson, M.L. Surface Oil Is the Primary Driver of Macroinvertebrate Impacts Following Spills of Diluted Bitumen in Freshwater. *Environ. Pollut.* **2021**, *290*, 117929. [CrossRef] [PubMed]
289. Indiketi, N.; Lhoste, E.; Grenon, M.C.; Gagnon, M.; Veilleux, É.; Triffault-Bouchet, G.; Couture, P. Toxicity and Risk Management of Oil-Spiked Sediments by Diluted Bitumen for Two Freshwater Benthic Invertebrates. *Environ. Pollut.* **2023**, *327*, 121497. [CrossRef] [PubMed]
290. Canadian Council of Ministers of the Environment. Canadian Sediment Quality Guidelines for the Protection of Aquatic Life—Polycyclic Aromatic Hydrocarbons. Available online: https://ccme.ca/en/res/polycyclic-aromatic-hydrocarbons-pahs-canadian-sediment-quality-guidelines-for-the-protection-of-aquatic-life-en.pdf (accessed on 9 June 2024).
291. Asl, A.G.; Nabavi, S.M.B.; Rouzbahani, M.M.; Alipour, S.S.; Monavari, S.M. Persistent Organic Pollutants Influence the Marine Benthic Macroinvertebrate Assemblages in Surface Sediments of Nayband National Park and Bay, Northern Persian Gulf, Iran. *Environ. Sci. Pollut. Res.* **2023**, *30*, 30254–30270. [CrossRef]
292. Yan, Z.; Song, N.; Wang, C.; Jiang, H. Functional Potential and Assembly of Microbes from Sediments in a Lake Bay and Adjoining River Ecosystem for Polycyclic Aromatic Hydrocarbon Biodegradation. *Environ. Microbiol.* **2021**, *23*, 628–640. [CrossRef]
293. Idan, F.S.; Jazza, S.H. Bioaccumulation of Hydrocarbon Compounds in the Muscle of Three Aquatic Birds in Um Alnaaj Marsh, Iraq. *Int. J. Aquat. Biol.* **2022**, *10*, 234–241.
294. Li, H.; Duan, D.; Beckingham, B.; Yang, Y.; Ran, Y.; Grathwohl, P. Impact of Trophic Levels on Partitioning and Bioaccumulation of Polycyclic Aromatic Hydrocarbons in Particulate Organic Matter and Plankton. *Mar. Pollut. Bull.* **2020**, *160*, 111527. [CrossRef]
295. Liu, B.; Gao, L.; Ding, L.; Lv, L.; Yu, Y. Trophodynamics and Bioaccumulation of Polycyclic Aromatic Hydrocarbons (PAHs) in Marine Food Web from Laizhou Bay, China. *Mar. Pollut. Bull.* **2023**, *194*, 115307. [CrossRef]
296. Qin, N.; He, W.; Liu, W.; Kong, X.; Xu, F.; Giesy, J.P. Tissue Distribution, Bioaccumulation, and Carcinogenic Risk of Polycyclic Aromatic Hydrocarbons in Aquatic Organisms from Lake Chaohu, China. *Sci. Total Environ.* **2020**, *749*, 141577. [CrossRef]
297. Mai, Y.; Wang, Y.; Geng, T.; Peng, S.; Lai, Z.; Wang, X.; Li, H. A Systematic Toxicologic Study of Polycyclic Aromatic Hydrocarbons on Aquatic Organisms via Food-Web Bioaccumulation. *Sci. Total Environ.* **2024**, *929*, 172362. [CrossRef]
298. Bukowska, B.; Mokra, K.; Michałowicz, J. Benzo[a]Pyrene—Environmental Occurrence, Human Exposure, and Mechanisms of Toxicity. *Int. J. Mol. Sci.* **2022**, *23*, 6348. [CrossRef] [PubMed]
299. Bignell, J.P.; Barber, J.; Bateman, K.S.; Etherton, M.; Feist, S.W.; Galloway, T.S.; Katsiadaki, I.; Sebire, M.; Scott, A.P.; Stentiford, G.D.; et al. Insights into the Development of Hepatocellular Fibrillar Inclusions in European Flounder (*Platichthys flesus*) from UK Estuaries. *Chemosphere* **2020**, *256*, 126946. [CrossRef]
300. Brette, F.; Machado, B.; Cros, C.; Incardona, J.P.; Scholz, N.L.; Block, B.A. Crude Oil Impairs Cardiac Excitation-Contraction Coupling in Fish. *Science* **2014**, *343*, 772–776. [CrossRef] [PubMed]
301. Cherr, G.N.; Fairbairn, E.; Whitehead, A. Impacts of Petroleum-Derived Pollutants on Fish Development. *Annu. Rev. Anim. Biosci.* **2017**, *5*, 185–203. [CrossRef] [PubMed]
302. Yamaguchi, A.; Uchida, M.; Ishibashi, H.; Hirano, M.; Ichikawa, N.; Arizono, K.; Koyama, J.; Tominaga, N. Potential Mechanisms Underlying Embryonic Developmental Toxicity Caused by Benzo[a]Pyrene in Japanese Medaka (*Oryzias latipes*). *Chemosphere* **2020**, *242*, 125243. [CrossRef] [PubMed]

303. Nayak, S.; Patnaik, L. Acetylcholinesterase, as a Potential Biomarker of Naphthalene Toxicity in Different Tissues of Freshwater Teleost, Anabas Testudineus. *J. Environ. Eng. Landsc. Manag.* **2021**, *29*, 403–409. [CrossRef]
304. Haverinen, J.; Badr, A.; Korajoki, H.; Hassinen, M.; Vornanen, M. Dual Effect of Polyaromatic Hydrocarbons on Sarco(Endo)Plasmic Reticulum Calcium ATPase (SERCA) Activity of a Teleost Fish (*Oncorhynchus mykiss*). *Comp. Biochem. Physiol. Part C Toxicol. Pharmacol.* **2024**, *276*, 109785. [CrossRef]
305. Bramatti, I.; Matos, B.; Figueiredo, N.; Pousão-Ferreira, P.; Branco, V.; Martins, M. Interaction of Polycyclic Aromatic Hydrocarbon Compounds in Fish Primary Hepatocytes: From Molecular Mechanisms to Genotoxic Effects. *Sci. Total Environ.* **2023**, *855*, 158783. [CrossRef]
306. Khan, S.; Qamar, Z.; Khan, A.; Waqas, M.; Nawab, J.; Khisroon, M.; Khan, A. Genotoxic Effects of Polycyclic Aromatic Hydrocarbons (PAHs) Present in Vehicle-Wash Wastewater on Grass Carp (*Ctenopharyngodon idella*) and Freshwater Mussels (*Anodonta cygnea*). *Environ. Pollut.* **2023**, *327*, 121513. [CrossRef]
307. Ho, K.T.; Konovets, I.M.; Terletskaya, A.V.; Milyukin, M.V.; Lyashenko, A.V.; Shitikova, L.I.; Shevchuk, L.I.; Afanasyev, S.A.; Krot, Y.G.; Zorina-Sakharova, K.Y.; et al. Contaminants, Mutagenicity and Toxicity in the Surface Waters of Kyiv, Ukraine. *Mar. Pollut. Bull.* **2020**, *155*, 111153. [CrossRef]

toxics

MDPI

Article

Concentrations, Compositions and Human Exposure Risks to Organophosphate Esters in Indoor Air from Various Microenvironments in Guangzhou, China

Yunmei Cai [1,†], Maoyuan Xu [2,†], Minghui Ouyang [2], Yusheng Wu [2], Ruijie Wang [2], Kewen Zheng [2] and Guofa Ren [2,*]

1 School of Environmental Monitoring, Guangdong Polytechnic of Environmental Protection Engineering, Foshan 528216, China; 18702030877@163.com

2 Institute of Environmental Pollution and Health, School of Environmental and Chemical Engineering, Shanghai University, Shanghai 200444, China

* Correspondence: rgf2008@shu.edu.cn; Tel./Fax: +86-21-66137753

† These authors contributed equally to this work.

Abstract: Limited research has characterized the occurrence of organophosphate esters (OPEs) in indoor microenvironment air. To address this gap, ten OPE congeners were measured in air samples collected from 46 homes, 12 offices, 6 student dormitories, and 60 private cars in Guangzhou, China. Among the four microenvironments, private vehicles exhibited the highest total OPE concentrations (ΣOPEs), with an average of 264.89 ng/m^3—statistically significantly higher than the other three environments ($p < 0.05$). This finding underscores the need for increased attention to OPE environmental fate in vehicles and associated human exposure risks. Distinct compositional profiles of OPEs were observed across microenvironments. In homes, offices, and student dormitories, tris(2-chloroethyl) phosphate (TCEP) and tris(2-chloropropyl) phosphate (TCPP) dominated the OPE mixture, accounting for 56% and 34% of ΣOPEs, respectively. By contrast, private cars were characterized by elevated levels of TCPP (68% of ΣOPEs) and tris(1,3-dichloro-2-propyl) phosphate (TDCP, 12%), reflecting source-specific emission patterns related to automotive materials. Significant correlations existed in most of the OPEs in the private cars, indicating that there are many potential sources of OPEs in private cars, and one source may release multiple OPEs. Human inhalation exposure to OPEs was estimated based on measured air concentrations. Daily respiratory exposure doses ranged from 9.1 to 30.85 ng/kg/d across different populations, with all values falling below established thresholds for non-carcinogenic and carcinogenic risks. These results indicate that current indoor air OPE levels in the studied microenvironments do not pose significant health hazards via inhalation pathways under typical exposure scenarios.

Keywords: organophosphate esters; private car microenvironments; indoor air; human exposure

1. Introduction

Organophosphate esters (OPEs) represent a class of synthetic industrial additives widely employed as flame retardants and plasticizers in diverse applications, including furniture, electrical/electronic devices, textiles, and automotive components. After the restriction and phase-out of brominated flame retardants such as polybrominated diphenyl ethers (PBDEs) and hexabromocyclododecane (HBCD), OPEs are the main alternative flame retardants [1]. However, accumulating evidence indicates that certain OPE congeners have the characteristics of persistent organic pollutants (POPs): high toxicity, persistence, bioaccumulation, and long-range transport [2]. Similar to PBDEs and HBCD, OPEs are used additively in the materials, which means that they are not fixed in the polymer product through chemical binding and could easily leak into the environment via self-volatilization and product wear during the use of materials, causing potential hazards to the ecological environment and human health. Currently, OPEs have been detected in a broad spectrum of environmental matrices worldwide, including air [3], water [4,5], soil [6], sediment [7], aquatic biota (e.g., fish) [8], and human biomonitoring samples (e.g., breast milk) [9]. Regarding health effects, a recent review has concluded that organophosphate esters (OPEs) are rapidly absorbed after oral intake and inhalation, and can induce hazards such as oxidative stress, inflammatory responses, and endocrine disruption in organisms [10].

Human exposure risk to OPEs is significantly elevated in indoor microenvironments compared to outdoor settings, attributable to two primary factors. Firstly, modern urban residents typically spend over 70% of their time in indoor spaces, increasing prolonged contact with OPE-laden materials. Secondly, analytical data consistently demonstrate substantially higher OPE concentrations in indoor air (often orders of magnitude greater than outdoor air), driven by continuous emissions from building materials and consumer products. OPEs are intentionally incorporated into up to 5% of common indoor materials—such as polyurethane foams, plastic composites, and textile coatings—where they exist as non-covalently bound additives. These material matrices serve as significant reservoirs and continuous sources of releases of OPEs in the indoor environment. Wang et al. [11] reported that target OPEs were ubiquitously detected in the indoor air of 16 Beijing households, with total OPE concentrations (ΣOPEs) exceeding ΣPBDEs by at least an order of magnitude. In a Norwegian residential study, Kucharska et al. [12] observed distinct distribution patterns between air and dust matrices: tris(2-chloropropyl) phosphate (TCPP) dominated indoor air with a median concentration of 128 ng/m^3, while tris(2-butoxyethyl) phosphate (TBEP) exhibited higher abundance in dust samples, reaching a maximum of 8100 ng/g. Additionally, Sun et al. [13] monitored the change in OPE concentration in the indoor air of a household in Harbin, China, over a year, and found that the concentrations of OPEs were higher in the spring and summer than in the autumn and winter.

Indoor contamination of OPEs in public spaces warrants equal attention. Kim et al. [14] analyzed OPEs in ten U.S. indoor microenvironments and recorded the highest average OPE concentration of 258 ng/m^3 in auto parts stores, followed by electronic stores, nail salons, and furniture stores (in descending order of contamination). In a Swedish study, Bergh et al. [15] investigated air and dust samples from three Stockholm microenvironments—residential homes, day-care centers, and offices—where nine OPE congeners were ubiquitously detected. Day-care centers exhibited the highest air pollution levels, surpassing offices and homes, indicating elevated exposure risks for young children in these high-activity environments. Wang et al. [11] reported varying OPE concentrations across Beijing microenvironments: activity rooms (14.4 ng/m^3) and student dormitories (19.4 ng/m^3) had lower levels compared to offices (29 ng/m^3) and family homes (24 ng/m^3), reflecting source contributions from office equipment and consumer products. In addition, some studies have shown that OPE concentrations in indoor environments correlate with internal exposure levels in humans, e.g., positive correlations have been identified between air/dust OPE concentrations and their presence in human hair [12], while urinary metabolites of OPEs demonstrate strong associations with parent compound levels in indoor dust or hand towel samples [16]. These findings underscore the role of indoor microenvironments as critical pathways for OPE uptake.

The automobile interior represents another critical indoor microenvironment with pronounced OPE contamination. Although daily exposure duration in vehicles is generally shorter than in residential or office settings, OPE concentrations measured within car compartments frequently exceed those of other indoor environments. He et al. [17] analyzed dust samples from offices, public spaces, and automobiles in Nanjing, reporting the highest OPE levels in vehicle dust—16-fold and 6-fold greater than office and public microenvironment samples, respectively. Ali et al. [18] observed similar trends in Kuwaiti and Pakistani vehicles, where OPE concentrations in automobile dust were approximately five times higher than in residential or workplace dust matrices. Brommer et al. [19] compared OPE levels in dust from UK automobiles, school classrooms, homes, and offices, identifying tris(1,3-dichloro-2-propyl) phosphate (TDCIPP) as the dominant congener in vehicles with significantly elevated concentrations relative to other microenvironments. Brandsma et al. [20] further reported TDCIPP in car seat dust at 1100 μg/g—the highest concentration documented in the literature to date—highlighting automotive interiors as hotspots for OPE accumulation. While existing research confirms elevated OPE pollution in vehicle compartments, current investigations are overwhelmingly focused on dust matrices, with limited data available on airborne concentrations and associated human exposure risks. A recent screening assessment demonstrated that partial dust removal interventions in vehicles fail to reduce human contact with OPEs, implying the existence of additional exposure pathways beyond dust-mediated routes [21]. These findings underscore the urgent need for a comprehensive evaluation of airborne OPE levels in private vehicles. As the average person spends over 200 h per year in vehicles, understanding the exposure risks in this confined microenvironment is crucial for developing effective mitigation strategies. In this study, passive air sampling was employed to collect air samples from four distinct indoor microenvironments—residential homes, offices, student dormitories, and private vehicles. This method offers several advantages over traditional active sampling techniques, including lower cost, longer sampling duration, and reduced disturbance to the indoor environment. The research aims to systematically investigate the concentration profiles and pollution characteristics of OPEs across these environments, characterize their potential emission sources, and assess the human exposure risk.

2. Materials and Methods

2.1. Sampling Collection

From September to October 2020, a total of 124 indoor air sampling sites were selected in Guangzhou, China, including 46 homes, 12 offices, 6 student dormitories, and 60 private cars (gasoline-powered, 3 to 6 years old), to characterize the occurrence and compositional profiles of OPEs in diverse microenvironments. At each sampling site, at least one polyurethane foam-based passive air sampler (PUF-PAS) was deployed at 0.5–1.5 m height for about 30 consecutive days. The PUF disks (14 cm diameter, 13.5 cm thickness, 0.017 g/cm^3 density, 200 cm^3 volume) followed the standardized design described in Li et al. [22], with detailed sampling protocols available in these references and Supporting Information (Text S1). Post-sampling, PUF disks were immediately sealed in brown glass bottles and stored at −20 °C to prevent analyte degradation until chemical analysis. Sampling rates for target OPEs were derived from a prior Northern China regional study [23], with individual compound-specific rates ranging from 0.59 to 2.01 m^3/d. These values account for environmental variability in temperature, humidity, and air turbulence, ensuring reliable exposure metric calculations.

2.2. Chemicals and Materials

Ten organophosphate ester (OPE) reference standards were acquired from AccuStandard (New Haven, CT, USA), including tri-n-butyl phosphate (TBP, 99%), tris(2-chloroethyl) phosphate (TCEP), tris(1-chloro-2-propyl) phosphate (TCPP), tris(1,3-dichloro-

2-propyl) phosphate (TDCPP), triphenyl phosphate (TPEP), 2-ethylhexyl diphenyl phosphate (DPEHP), triphenyl phosphate (TPHP), tris(2-butoxyethyl) phosphate (TBEP), tris(2-ethylhexyl) phosphate (TEHP), and tricresyl phosphate (TCP). Surrogate standards (D15-TPHP and d27-TBEP) and an internal standard (d27-TnBP) were sourced from C/D/N Isotopes Inc. (Quebec, QC, Canada) and Cambridge Isotope Laboratories (Cambridge, MA, USA), respectively. All solvents—n-hexane, dichloromethane, acetone, methanol, acetonitrile, and ethyl acetate—were of pesticide-grade purity (Sigma-Aldrich, Taufkirchen, Germany). Analytical-grade silica gel (60–200 μm) and sodium sulfate were purchased from Sinopharm Group Co. Ltd. (Beijing, China). Prior to use, silica gel and sodium sulfate were heated at 450 °C for 4 h to remove organic contaminants and activated under anhydrous conditions.

2.3. Sample Preparation and Analysis

Detailed protocols for air sample extraction and cleanup have been described previously [24]. Briefly, PUF disks were spiked with known quantities of surrogate standards (D15-TPHP and d27-TBEP) and subjected to Soxhlet extraction for 48 h using 300 mL of a n-hexane/dichloromethane (1:1, v/v) mixture. Extracts were concentrated to 1 mL via rotary evaporation and solvent-exchanged to hexane. The concentrated samples were purified using silica gel columns: 25 mL of ethyl acetate was employed as the eluant to fractionate target analytes. Eluates containing OPEs were further reduced to 200 μL under a gentle nitrogen stream, after which a defined amount of internal standard (d27-TnBP) was added prior to instrumental analysis. Sample quantification was performed using an Agilent 6890N gas chromatograph coupled to a 5973 mass spectrometer (GC-MS), with high-purity helium as the carrier gas. Separation was achieved on a DB-5MS capillary column (30 m × 0.25 mm i.d., 0.25 μm film thickness) under electron ionization (EI+) selected ion monitoring (SIM) mode. Detailed information on quantification ions, verification ions, and instrument parameters is provided in the Supplementary Information (SI).

2.4. Quality Assurance/Quality Control

To minimize blank contamination, a series of analytical precautions was implemented. All glassware was baked at 450 °C for 4 h and rinsed with n-hexane and ethyl acetate prior to use. Additionally, all samples and glassware were covered with aluminum foil to prevent dust contamination. A procedural blank was analyzed alongside each batch of 5 samples to evaluate potential contamination. Instrumental Detection Limit (IDL) was determined by injecting low-concentration target analytes with a signal-to-noise ratio of no less than 3:1. The Method Detection Limit (MDL) was calculated as the mean blank sample concentration plus three times the standard deviation. When target compounds were not detected in blanks, half of the IDL was used as the MDL. In this study, the MDLs of OPEs in indoor air samples ranged from 0.022 to 1.21 ng/m^3. The recovery efficiency of the procedure was verified by analyzing uncontaminated PUF spiked with OPE standards. The overall recovery was generally greater than 68%. The recovery rates of the surrogate standards ranged from 76% to 104% for D_{27}—TBEP, and from 63% to 109% for D_{15}—TPHP, respectively. No sample was corrected for recovery.

2.5. Exposure and Risk Assessment

The estimated daily intake (EDI) by inhalation of OPEs was estimated using Equation (1):

$$EDI = C \times IR \times IEF/BW \quad (1)$$

where EDI is the estimated daily intake of each OPE through atmosphere inhalation (ng/kg/day); C is the concentration of OPEs in indoor ambient air (ng/m^3); IR is the respiration rate (m^3/day); IEF is the fraction of individual's daily exposure; BW is the body weight of human (kg). We assumed 100% absorption of inhalation in the absence of experimental data for this parameter. IR and BW for Chinese were obtained from the report "Risk Assessment Guidance for Industrial Contaminated Sites in China". The exposure assessment parameters for different age groups can be found in Table S2. The EDI of each OPE values was calculated based on a Monte Carlo simulation, which was performed using Crystal Ball software (Oracle, Austin, TX, USA.version:11.1.3.0.0) with independent runs of 1000 trials.

The non-carcinogenic hazard quotient (HQ) and carcinogenic risk (CR) index of inhaling OPEs were estimated using Equations (2) and (3), respectively:

$$HQ = EDI/RfD \quad (2)$$

$$CR = EDI \times CSF \quad (3)$$

where HQ is the hazard quotient, EDI is the estimated daily intake of each OPE (ng/kg/day), RfD is the reference doses values for each OPEs (ng/kg bw/day), and CSF is the carcinogenic effect slope factor for each OPEs (mg/kg bw/day). Assuming that residents of Guangzhou are exposed for 24 h per day. HQ > 1 suggests that there is a potential health risk, whereas CR > 1×10^{-6} indicates potential adverse effects.

2.6. Data Analysis

Statistical analyses of air sample data were performed using SPSS version 19.0 (SPSS Inc., Chicago, IL, USA). Differences in airborne organophosphate ester (OPE) concentrations across microenvironments were evaluated using independent-samples *t*-tests to assess statistical significance. Spearman's rank correlation analysis was employed to quantify the strength of associations between individual OPE congeners, with statistical significance defined as a p value < 0.05. Estimated Daily Intakes (EDIs) for individual OPEs were derived from Monte Carlo simulations implemented in Crystal Ball software, which generated probabilistic exposure distributions by incorporating input parameters such as air concentrations, breathing rates, and exposure durations. This approach accounts for variability in human activity patterns and microenvironmental conditions, providing robust estimates of potential human exposure.

3. Results and Discussion

Concentrations of ten OPEs were measured in indoor air samples collected from 60 private cars, 46 homes, 12 offices, and 6 student dormitories in Guangzhou, China. OPEs were ubiquitously detected across all four microenvironments. Average total OPE concentrations (ΣOPEs) in private car air (264.89 ng/m^3) were significantly higher than those in homes (42.12 ng/m^3), offices (36.02 ng/m^3), and student dormitories (18.14 ng/m^3), as determined by one-way ANOVA ($p < 0.05$; Figure 1). No statistically significant differences were observed in ΣOPEs levels among homes, offices, and dormitories. Given this pronounced disparity, the unique contamination profile of private car interiors is discussed separately in the subsequent analysis.

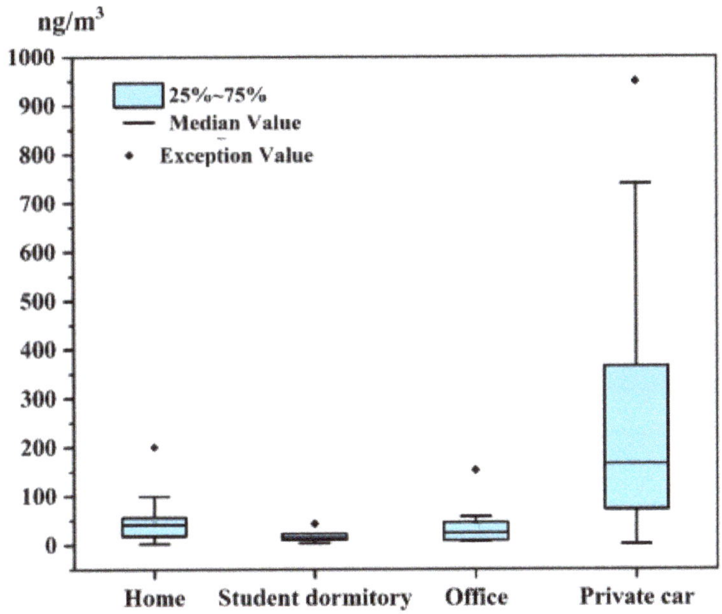

Figure 1. Total concentrations of OPEs in indoor air from different microenvironments.

3.1. Total Concentrations and Profiles of OPEs in Indoor Air from China Homes, Offices, and Student Dormitories

All the target flame retardants (FRs) were detected in at least one of the indoor air samples, which implies the extensive use of a wide variety of OPEs in the study area. Table 1 presents the summary statistics and detection frequencies (DF) of OPEs in indoor air collected from homes, offices, and student dormitories in Guangzhou, China. TCEP, TCPP, DPEHP, TPHP, and TDCPP were the five most prevalent contaminants, with detection rates of 100%, 100%, 98.4%, 98.4%, and 88.7%, respectively. TEHP, TBP, and TBEP followed, with detection rates of 62.9%, 45.2%, and 25.8%, respectively. Given that the detection rates of TCP and TPEP were less than 3%, these two compounds were not further discussed. The total concentration of eight OPEs (Σ_8OPEs) in indoor microenvironment air ranged from 2.99 to 200.76 ng/m^3, with an average of 38.62 ng/m^3 and a median of 32.78 ng/m^3. Generally, the OPE concentrations in the indoor air of Guangzhou were higher than those previously reported for outdoor air [25]. This suggests that indoor environments serve as the primary source of OPEs to the outdoor environment, due to emissions from household products and materials used indoors. These results are consistent with previous research findings. In homes, offices, and student dormitories, the overall OPE concentration trend followed home > office > student dormitory, although statistical analysis revealed no significant differences among the three ($p > 0.05$). This finding aligns with previous observations of similar concentration patterns in Dalian residential microenvironments [26]. The relatively low Σ_8OPE levels in student dormitories can be attributed to their simpler interior finishes and fewer consumer products, which likely reduce OPE emission sources. Available data on indoor air OPE concentrations from global studies are summarized in Table S3. Compared to these datasets, the average OPE concentration in Guangzhou indoor air was lower than reported values from Germany (81.89 ng/m^3) [27], the United States (101 ng/m^3) [13], and Sweden (160 ng/m^3) [28]. Conversely, it was significantly higher than levels measured in India (0.483 ng/m^3) [29], Egypt (0.007–0.064 ng/m^3) [30], and Dalian, China (14.9 ng/m^3) [26], while comparable to results from Australia (44 ng/m^3) [31]

and Canada (46 ng/m^3) [32]. These regional disparities suggest a strong correlation between indoor OPE levels and the degree of urbanization/industrialization, with more developed areas generally exhibiting higher contamination, likely due to intensive use of OPE-containing materials and products.

Table 1. Summary statistics and detection frequencies (DF) of OPEs in indoor air collected from homes, offices, and student dormitories in Guangzhou, China (ng/m^3).

Compounds	DF (%)	Mean	Percentile					
			Min	25th	50th	75th	90th	Max
TBP	45.2	0.95	ND	<MDL	<MDL	1.02	2.59	16.15
TCEP	100	18.01	0.27	2.16	11.71	30.92	42.65	92.20
TCPP	100	14.52	0.10	3.74	7.26	13.90	36.26	108.36
TDCPP	88.7	0.96	ND	0.67	0.85	1.11	1.61	6.86
TBEP	25.8	0.29	ND	<MDL	<MDL	0.50	1.01	3.39
TPHP	98.4	2.61	ND	0.31	1.26	2.51	6.06	43.18
DPEHP	98.4	0.58	ND	0.26	0.33	0.40	0.94	11.51
TEHP	62.9	0.70	ND	<MDL	0.36	0.52	1.84	8.79
Σ_8OPEs	100	38.62	2.99	14.88	32.78	52.11	71.12	200.76

DF: detectable frequencies; ND: not detected. MDL: Method Detection Limit.

Among the analyzed compounds, tris(2-chloroethyl) phosphate (TCEP) was the most abundant OPE congener, with a mean concentration of 18.01 ng/m^3, followed by tris(2-chloropropyl) phosphate (TCPP, 14.52 ng/m^3) and triphenyl phosphate (TPHP, 2.61 ng/m^3). The compositional profile of indoor air OPEs followed the order: TCEP (46.6%) > TCPP (37.6%) > TPHP (6.8%) > TDCPP (2.5%) > TBP (2.4%) > TEHP (1.8%) > DPEHP (1.5%) > TBEP (0.8%) (Figure 2, Left). Chlorinated OPEs dominated the mixture, accounting for 84.2% of total concentrations, with TCEP and TCPP as the primary homologs—consistent with findings from Lai et al. in Northern Chinese atmospheric fine particles [33]. Notable compositional differences were observed compared to global studies: TCPP and triethyl phosphate (TEP) accounted for 43% and 33% of U.S. air samples [14], while TCPP, triisobutyl phosphate (TiBP), and tri-n-butyl phosphate (TnBP) represented 45%, 28%, and 12% in German air [27]. In Dalian, China, TCPP (74%), TBP (12%), and TDCPP (5%) dominated the profile [26]. These disparities reflect regional regulatory and usage patterns: since 2016, the U.S. and EU have restricted TCEP due to its neurotoxic and carcinogenic properties, leading to increased reliance on TCPP as a substitute—explaining its elevated presence in those regions. In this study, the high abundance of TCEP and TCPP can be attributed to multiple factors: their relatively high saturated vapor pressures facilitate volatilization from materials, combined with environmental persistence due to slow degradation rates. Additionally, their historical widespread use in China for flame retardancy in plastics, foams, and textiles likely contributes to their dominant presence in indoor air.

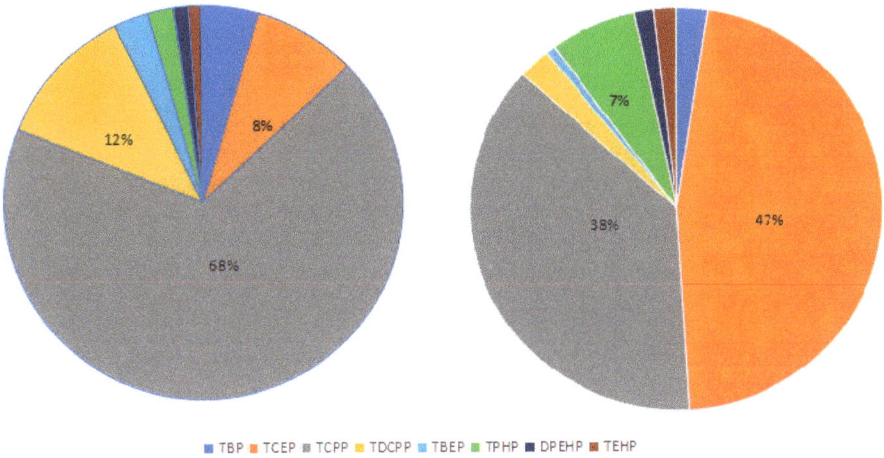

Figure 2. Distribution characteristics of OPEs in indoor air from different microenvironments. (**Left**): homes, offices, and student dormitories as a whole; (**Right**): Private car.

3.2. Total Concentrations and Profiles of OPEs in Indoor Air of Private Cars from China

Table 2 presents the concentration profiles of OPEs in private vehicle air. All target OPEs were detected in at least one sample, demonstrating their universal presence in car interiors. Tris(2-chloropropyl) phosphate (TCPP) was the only congener with a 100% detection rate, followed by TCEP, TDCPP, and TPHP (detection rates > 90%). Moderate detection rates were observed for TBP (75%), DPEHP (73.7%), and TEHP (71.7%), while tBEP (21.7%) and TPEP (11.7%) exhibited lower detection frequencies. TCP was detected in only 3 out of 60 vehicles and was excluded from further analysis. Total OPE concentrations (Σ_9OPEs) in private car air exhibited substantial variability, ranging from 1.22 to 1288.81 ng/m^3—a 1000-fold difference between minimum and maximum values. TCPP dominated the OPE profile with a mean concentration of 180.33 ng/m^3, significantly higher than all other congeners ($p < 0.001$). The highest individual TCPP concentration reached 1087.38 ng/m^3, reflecting diverse emission sources in vehicle interiors. This dominance is likely attributed to TCPP's widespread use as a flame retardant in polyvinyl chloride (PVC) and polyurethane foam materials commonly found in automotive upholstery and dashboard components [34]. Global comparisons (Table S4) highlight limited airborne OPE research in vehicle microenvironments. Swedish vehicle air reported an average ΣOPEs of 2065 ng/m^3—approximately eight times higher than this study [28]. In Japan's Yokohama region, although most OPEs were undetected or below limits, TCPP reached a maximum of 1500 ng/m^3, comparable to the highest TCPP concentration observed here (1087 ng/m^3) [35]. Kim et al. found U.S. private car air averaged 59 ng/m^3, with auto parts stores exhibiting higher levels (258 ng/m^3), consistent with OPE use in automotive plastics and flame retardants [14]. German studies in the Rhine region reported ΣOPEs ranging from 26.18 to 751 ng/m^3 (mean: 265.9 ng/m^3) [27], while Swiss measurements in Zurich recorded a TCPP mean of 260 ng/m^3 [36]. Collectively, these data indicate that OPE concentrations in this study fall within the moderate range compared to global vehicle microenvironments.

The compositional profile of OPEs in private vehicle air (Figure 2, Right) followed the order: TCPP (68.1%) > TDCPP (11.6%) > TCEP (8.4%) > TBP (4.6%) > TBEP (2.8%) > TPHP (2.1%) > DPEHP (1.2%) > TEHP (1.0%) > TPEP (0.1%). Chlorinated OPEs—TCPP, TDCPP, and TCEP—dominated the mixture, accounting for 88.2% of total concentrations, with TCPP as the primary contaminant. Non-chlorinated alkyl esters (TBP, TBEP, TEHP, TPEP) contributed 8.5% of the total OPEs load. This composition closely resembled that reported for German Rhine-region private cars, where chlorinated OPEs accounted for 60% of total concentrations [27]. Notably, TCPP represented 86% of total chlorinated OPEs in this study, with TDCPP comprising 14%, while TCEP was undetected in German samples—likely due to regional restrictions on TCEP production and use. Consistent with findings from Sweden, the U.S., Switzerland, and Japan, TCPP emerged as the dominant congener in transportation microenvironments globally [14,28,35,36]. This compositional similarity suggests shared emission sources across different regions, primarily the use of chlorinated OPEs (e.g., TCPP, TDCPP) as flame retardants in flexible polyurethane foams and polyvinyl chloride materials commonly used in automotive interiors. Additional influencing factors include vehicle service life, manufacturing origins, and regional regulatory policies on specific OPE congeners. Correlation analysis among seven OPEs (excluding TBEP and TPEP due to low detection rates) revealed significant associations, see Table S5. TPHP correlated strongly with five congeners ($p < 0.05$), except for a weak relationship with TBP ($r = 0.188$, $p > 0.05$), likely due to shared physicochemical properties with DPEHP/TEHP, indicating common sources. TCPP exhibited significant positive correlations with TCEP ($r = 0.401$, $p < 0.01$) and TDCPP ($r = 0.546$, $p < 0.001$), consistent with co-use in industrial formulations. While most OPE pairs showed significant correlations, moderate correlation coefficients ($r < 0.6$) suggest contributions from multiple emission pathways—such as in-car air exchange, occupant-introduced contaminants, or differential environmental behaviors (e.g., volatility, degradation rates) of individual congeners. These findings highlight the complex interplay between source inputs and microenvironmental processes in shaping OPE distribution patterns.

Table 2. Summary statistics and detection frequencies (DF) of OPEs in indoor air collected from private cars in Guangzhou, China (ng/m^3).

Compounds	DF (%)	Mean	Percentile					
			Min	25th	50th	75th	90th	Max
TBP	75%	12.12	ND	0.0076	3.68	12.95	37.73	89.6
TCEP	95%	22.36	ND	9.25	14.04	33.24	47.82	105.99
TCPP	100%	180.33	0.06	17.78	75.52	242.96	540.17	1087.38
TDCPP	91.7%	30.84	ND	1.19	3.66	7.05	15.00	827.49
TBEP	21.7%	7.36	ND	<MDL	<MDL	<MDL	38.60	72.86
TPHP	93.3%	5.65	ND	0.64	1.85	5.07	10.03	108.8
DPEHP	73.7%	3.30	ND	<MDL	0.83	8.60	8.87	22.94
TEHP	71.7%	2.61	ND	<MDL	0.87	2.99	4.16	70.21
TPEP	11.7%	0.31	ND	<MDL	<MDL	<MDL	1.25	3.81
Σ_9OPEs	100%	264.89	1.22	71.79	165.58	369.65	702.25	1288.81

DF: detectable frequencies; ND: not detected. MDL: Method Detection Limit.

3.3. Exposure Levels and Health Risks of OPEs in Four Indoor Environments for Different Populations

In this study, the Monte Carlo algorithm was employed to calculate the daily respiratory exposure doses of different populations, using the total indoor air concentrations of OPEs measured in homes, offices, student dormitories, and private cars in Guangzhou. The results are presented in Table 3. Infants exhibited the highest exposure level, with a mean daily exposure of 30.85 ng/kg/day, followed by toddlers and children, whose mean daily exposures were 15.89 ng/kg/day and 12.41 ng·kg^{-1}·d^{-1}, respectively. Adolescents and adults had lower daily respiratory exposures compared to infants, toddlers, and children. A general trend of decreasing daily OPE exposure levels in indoor ambient air with increasing age was observed among different populations. When compared with other studies, the mean daily exposures of young children and adults to indoor air OPEs in homes in Beijing were 2.1 ng/kg/day and 1.2 ng/kg/day, respectively [37]. In Australia, the daily respiratory exposures of young children and adults to indoor air OPEs were 7 ng/kg/day and 7.9 ng/kg/day [31], both of which were lower than those of the corresponding age groups in this study. The relatively high exposure levels of OPEs in the indoor ambient air observed in this study might be attributed to the elevated OPE concentrations detected in the microenvironment of private cars.

Table 3. Daily respiratory exposure dose (ng/kg/day) of OPEs in indoor air of different age groups.

Infants	TBP	TCEP	TCPP	TDCPP	TPHP	DPEHP	TEHP	ΣOPEs
Mean	0.73	15.13	11.26	0.76	0.29	2.35	0.5	30.85
P5	0.02	2.78	1.15	0.16	0	0.75	0.02	7.42
P50	0.26	10.66	6.32	0.51	0.01	1.97	0.18	23.15
P95	2.71	42.13	37.08	1.62	0.74	5.17	1.96	79.18
Toddlers	TBP	TCEP	TCPP	TDCPP	TPHP	DPEHP	TEHP	ΣOPEs
Mean	0.39	7.77	5.61	0.38	0.15	1.19	0.27	15.89
P5	0.01	1.54	0.63	0.09	0	0.44	0.01	4.25
P50	0.14	5.61	3.32	0.27	0.01	1.03	0.09	12.05
P95	1.45	21.09	17.85	0.77	0.41	2.48	1.02	39.91
Children	TBP	TCEP	TCPP	TDCPP	TPHP	DPEHP	TEHP	ΣOPEs
Mean	0.45	2.69	7.55	0.94	0.44	0.17	0.13	12.41
P5	0.09	0.97	2.08	0.23	0.15	0.08	0.02	4.74
P50	0.31	2.24	5.79	0.39	0.35	0.13	0.08	10.34
P95	1.25	5.94	18.92	1.66	0.96	0.36	0.35	27.39
Adolescents	TBP	TCEP	TCPP	TDCPP	TPHP	DPEHP	TEHP	ΣOPEs
Mean	0.32	1.94	5.55	0.66	0.31	0.12	0.09	8.86
P5	0.06	0.71	1.53	0.17	0.11	0.06	0.02	3.43
P50	0.22	1.6	4.2	0.28	0.25	0.1	0.06	7.37
P95	0.86	4.26	13.6	1.18	0.68	0.26	0.25	19.12
Adults	TBP	TCEP	TCPP	TDCPP	TPHP	DPEHP	TEHP	ΣOPEs
Mean	0.29	3.37	4.27	0.48	0.18	0.39	0.13	9.1
P5	0.06	1.16	1.09	0.1	0.04	0.16	0.02	3.63
P50	0.2	2.78	3.28	0.24	0.12	0.34	0.08	7.9
P95	0.8	7.51	10.69	0.81	0.46	0.76	0.39	18.59

Sensitivity analyses were conducted to investigate the impacts of parameters, including the concentration of OPEs in various indoor environments (C), respiration rate (IR), and body weight (BW), on the respiratory exposure doses of different populations. The contribution of each factor to the estimated daily intake (EDI) was calculated, and the results are detailed in Table S6 and Figure 3. The contribution ratios of IR and BW to the infant exposure dose (approximately 10%) were higher than those for other populations. This indicates that infants are more sensitive to OPE exposure, likely due to their unique crawling posture and lower body weight compared with the other four populations. For infants and toddlers, only their exposure in the home environment was considered. The concentration of OPEs in home air contributed 93–99.5% and 75.3–98.6% to the exposure of toddlers and infants, respectively. For children and adolescents, the OPE concentration in private cars was the primary contributor to their daily respiratory exposure doses (children: 60.2%; adolescents: 61.6%), followed by the influence of the study office environment (children: 22.6%; adolescents: 23.1%). For adults, the OPE concentration in homes was the main contributor to their daily respiratory exposure doses (50.7%), with private cars following as the second major contributor (30.3%).

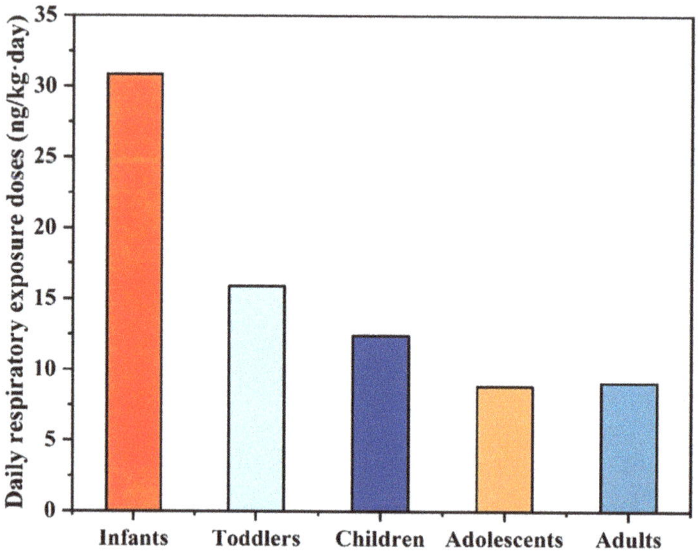

Figure 3. Comparison of the average daily respiratory exposure of OPEs to different populations.

The non-carcinogenic risk quotient (HQ) and carcinogenic risk index (CR) were employed to assess the respiratory exposure risks of OPEs in different indoor air environments for residents of Guangzhou City, as presented in Table 4. The table shows that the mean HQ values for different populations in this region were much less than one, suggesting that the current exposure levels of OPEs in various indoor environments in Guangzhou do not pose non-carcinogenic risks to humans. Carcinogenic risk assessment results indicated that the CR values for infants, toddlers, children, adolescents, and adults were all below 1×10^{-6}, indicating that the OPE exposure levels in different indoor environments in Guangzhou City did not present carcinogenic risks in this study. However, it is noteworthy that the mean carcinogenic effect value of OPEs for infants was 0.33×10^{-6}, the highest among the five age groups. Considering that OPEs can enter the human body not only through respiratory exposure but also via dermal contact, dust ingestion, and diet, and that infants have more frequent dermal contact and dust ingestion routes compared with the other four age groups, potential health risks for infants may still persist.

Table 4. Health risks of OPEs in indoor air of Guangzhou to different populations.

Compounds	Non-Carcinogenic Risk Quotient [a] ($\times 10^{-4}$)					Carcinogenic Risk Index [b] ($\times 10^{-6}$)				
	Infants	Toddlers	Children	Adolescents	Adults	Infants	Toddlers	Children	Adolescents	Adults
TBP	0.73	0.39	0.45	0.32	0.29	0.0066	0.0035	0.0041	0.0029	0.0026
TCEP	21.61	11.10	3.84	2.77	4.81	0.30	0.16	0.054	0.039	0.067
TCPP	11.26	5.61	7.71	5.55	4.27	-	-	-	-	-
TDCPP	0.38	0.19	0.47	0.33	0.24	0.024	0.012	0.029	0.020	0.015
TPHP	0.041	0.021	0.063	0.044	0.026	-	-	-	-	-
TEHP	0.050	0.027	0.013	0.009	0.013	0.002	0.001	0.000	0.000	0.000
Σ_6OPEs	34.08	17.34	12.55	9.02	9.65	0.33	0.17	0.087	0.062	0.085

[a] Reference dose (ng·kg^{-1} bw day^{-1}), data from United States Environmental Protection Agency. The values of TBP, TCEP, TCPP, TDCPP, TPHP, and TEHP are 10,000, 7000, 10,000, 20,000, 70,000, and 100,000, respectively. [b] Cancer slope factor ((mg·kg^{-1} bw day^{-1})$^{-1}$), data from United States Environmental Protection Agency. The values of TBP, TCEP, TDCPP, and TEHP are 0.009, 0.02, 0.031, and 0.0032, respectively.

4. Conclusions

This study investigated the concentration levels and compositional characteristics of OPEs in 124 air samples from four indoor microenvironments in Guangzhou, China—households, offices, student dormitories, and private cars—and evaluated human respiratory exposure risks. Results showed that organophosphorus flame retardants are ubiquitous organic pollutants in Guangzhou's indoor air, with concentrations falling within the global low-to-medium pollution range. Significantly higher ∑OPEs (sum of OPEs) concentrations were observed in private car indoor air compared to the other three microenvironments, highlighting the need for increased attention to OPE environmental behavior in vehicles and associated human exposure risks. Respiratory exposure risk assessments indicated low inhalation-based exposure levels of OPEs for all populations in Guangzhou, with no significant overall non-carcinogenic or carcinogenic health risks identified. Nevertheless, this study also has certain limitations. In terms of spatial and temporal scales, the research only selected four types of micro-environments in Guangzhou, without considering climate, ventilation design, and industrial differences, and did not cover seasonal variations. Therefore, it is difficult to reflect the temporal and spatial dynamics of OPE concentrations. The pollution source analysis is not in-depth, there is no direct traceability of materials, and no detection of emerging OPEs. In terms of exposure pathways, only respiratory exposure was evaluated, without considering dermal contact, dietary intake, and dust contact, especially the multi-pathway exposure situations of special groups such as infants, which may underestimate the actual risk. Future research can be conducted in the directions of multi-media sampling, cross-regional seasonal studies, and material traceability.

Supplementary Materials: The following supporting information can be downloaded at: https://www.mdpi.com/article/10.3390/toxics13070531/s1, Text S1: Passive Sampling Procedure; Text S2: Instrumental Analysis; Table S1: Retention time and qualitative and quantitative ions of OPEs; Table S2: Exposure assessment parameters of different age groups; Table S3: Comparison of ΣOPE concentrations (ng/m^3) in indoor air around different areas in the world range (mean or mean range); Table S4: Concentration level of OPEs in the air of private cars (ng/m^3); Table S5: Correlations between various OPEs in air of private cars; Table S6. Contribution rate (%) of different exposure parameters (C, IR, and BW) to the daily respiratory exposure of OPEs in different populations [14,26–31,35,36].

Author Contributions: Methodology, M.X.; Software, Y.W.; Resources, Y.C.; Data curation, K.Z.; Writing—original draft, M.O.; Visualization, R.W.; Writing—original draft, Funding acquisition; Supervision, G.R. All authors have read and agreed to the published version of the manuscript.

Funding: We are grateful for the support from the following Funding Projects: The National Natural Science Foundation of China (20107073, 4247071761), the National Key R&D Plan (2019YFC1805501), and the Innovative Research Team in University (No. IRT13078).

Institutional Review Board Statement: Not Applicable.

Informed Consent Statement: Not Applicable.

Data Availability Statement: The data that support the findings of this study are available from the corresponding author upon reasonable request.

Conflicts of Interest: The authors declare no conflict of interest.

References

1. Huang, J.; Ye, L.; Fang, M.; Su, G. Industrial Production of Organophosphate Flame Retardants (OPFRs): Big Knowledge Gaps Need to Be Filled? *Bull. Environ. Contam. Toxicol.* **2022**, *108*, 809–818. [CrossRef] [PubMed]
2. Bekele, T.G.; Zhao, H.; Yang, J.; Chegen, R.G.; Chen, J.; Mekonen, S.; Qadeer, A. A review of environmental occurrence, analysis, bioaccumulation, and toxicity of organophosphate esters. *Environ. Sci. Pollut. Res.* **2021**, *28*, 49507–49528. [CrossRef]

3. Fu, J.; Fu, K.; Chen, Y.; Li, X.; Ye, T.; Gao, K.; Pan, W.; Zhang, A.; Fu, J. Long-Range Transport, Trophic Transfer, and Ecological Risks of Organophosphate Esters in Remote Areas. *Environ. Sci. Technol.* **2021**, *55*, 10192–10209. [CrossRef]
4. Lv, J.; Guo, C.; Luo, Y.; Liu, Y.; Deng, Y.; Sun, S.; Xu, J. Spatial distribution, receptor modelling and risk assessment of organophosphate esters in surface water from the largest freshwater lake in China. *Ecotoxicol. Environ. Saf.* **2022**, *238*, 113618. [CrossRef]
5. Ai, S.; Chen, X.; Zhou, Y. Critical review on organophosphate esters in water environment: Occurrence, health hazards and removal technologies. *Environ. Pollut.* **2024**, *343*, 123218. [CrossRef] [PubMed]
6. Zhou, G.; Zhang, Y.; Wang, Z.; Li, M.; Li, H.; Shen, C. Distribution Characteristics and Ecological Risk Assessment of Organophosphate Esters in Surface Soils of China. *Toxics* **2024**, *12*, 686. [CrossRef]
7. Castro-Jimenez, J.; Cuny, P.; Militon, C.; Sylvi, L.; Royer, F.; Papillon, L.; Sempere, R. Effective degradation of organophosphate ester flame retardants and plasticizers in coastal sediments under high urban pressure. *Sci. Rep.* **2022**, *12*, 20228. [CrossRef] [PubMed]
8. Xie, Z.; Wang, P.; Wang, X.; Castro-Jimenez, J.; Kallenborn, R.; Liao, C.; Mi, W.; Lohmann, R.; Vila-Costa, M.; Dachs, J. Organophosphate ester pollution in the oceans. *Nat. Rev. Earth Environ.* **2022**, *3*, 309–322. [CrossRef]
9. Shahin, S.; Medley, E.A.; Naidu, M.; Trasande, L.; Ghassabian, A. Exposure to organophosphate esters and maternal-child health. *Environ. Res.* **2024**, *252*, 118955. [CrossRef]
10. Alharbi, A.; Alhujaily, M. Molecular Mechanism of Indoor Exposure to Airborne Halogenated Flame Retardants TCIPP (Tris(1,3-Dichloro-2-Propyl) Phosphate) and TCEP Tris(2-chloroethyl) Phosphate and Their Hazardous Effects on Biological Systems. *Metabolites* **2024**, *14*, 697. [CrossRef]
11. Wang, D.; Wang, P.; Wang, Y.; Zhang, W.; Zhu, C.; Sun, H.; Matsiko, J.; Zhu, Y.; Li, Y.; Meng, W.; et al. Temporal variations of $PM_{2.5}$-bound organophosphate flame retardants in different microenvironments in Beijing, China, and implications for human exposure. *Sci. Total Environ.* **2019**, *666*, 226–234. [CrossRef] [PubMed]
12. Kucharska, A.; Cequier, E.; Thomsen, C.; Becher, G.; Covaci, A.; Voorspoels, S. Assessment of human hair as an indicator of exposure to organophosphate flame retardants. Case study on a Norwegian mother-child cohort. *Environ. Int.* **2015**, *83*, 50–57. [CrossRef] [PubMed]
13. Sun, Y.; Guo, J.-Q.; Liu, L.-Y.; Sverko, E.; Zhang, Z.; Tian, C.-G.; Huo, C.-Y.; Li, H.-L.; Ma, W.-L.; Zhang, Z.-F.; et al. Seasonal variation and influence factors of organophosphate esters in air particulate matter of a northeastern Chinese test home. *Sci. Total Environ.* **2020**, *740*, 140048. [CrossRef] [PubMed]
14. Kim, U.-J.; Wang, Y.; Li, W.; Kannan, K. Occurrence of and human exposure to organophosphate flame retardants/plasticizers in indoor air and dust from various microenvironments in the United States. *Environ. Int.* **2019**, *125*, 342–349. [CrossRef] [PubMed]
15. Bergh, C.; Torgrip, R.; Emenius, G.; Ostman, C. Organophosphate and phthalate esters in air and settled dust—A multi-location indoor study. *Indoor Air* **2011**, *21*, 67–76. [CrossRef]
16. Dodson, R.E.; Van den Eede, N.; Covaci, A.; Perovich, L.J.; Brody, J.G.; Rudel, R.A. Urinary Biomonitoring of Phosphate Flame Retardants: Levels in California Adults and Recommendations for Future Studies. *Environ. Sci. Technol.* **2014**, *48*, 13625–13633. [CrossRef]
17. He, R.-W.; Li, Y.-Z.; Xiang, P.; Li, C.; Cui, X.-Y.; Ma, L.Q. Impact of particle size on distribution and human exposure of flame retardants in indoor dust. *Environ. Res.* **2018**, *162*, 166–172. [CrossRef]
18. Ali, N.; Van den Eede, N.; Dirtu, A.C.; Neels, H.; Covaci, A. Assessment of human exposure to indoor organic contaminants via dust ingestion in Pakistan. *Indoor Air* **2012**, *22*, 200–211. [CrossRef]
19. Brommer, S.; Harrad, S. Sources and human exposure implications of concentrations of organophosphate flame retardants in dust from UK cars, classrooms, living rooms, and offices. *Environ. Int.* **2015**, *83*, 202–207. [CrossRef]
20. Brandsma, S.H.; de Boer, J.; van Velzen, M.J.M.; Leonards, P.E.G. Organophosphorus flame retardants (PFRs) and plasticizers in house and car dust and the influence of electronic equipment. *Chemosphere* **2014**, *116*, 3–9. [CrossRef]
21. Reddam, A.; Herkert, N.; Stapleton, H.M.; Volz, D.C. Partial dust removal in vehicles does not mitigate human exposure to organophosphate esters. *Environ. Res.* **2022**, *205*, 112525. [CrossRef] [PubMed]
22. Li, Q.; Yang, K.; Li, J.; Zeng, X.; Yu, Z.; Zhang, G. An assessment of polyurethane foam passive samplers for atmospheric metals compared with active samplers. *Environ. Pollut.* **2018**, *236*, 498–504. [CrossRef] [PubMed]
23. Liu, J. Theory and Concentration Calculations of PUF Atmospheric Passive Sampling Technique for Persistent Organic Pollutants. *Environ. Monit. China* **2012**, *28*, 107–112.
24. Ren, G.; Chen, Z.; Feng, J.; Ji, W.; Zhang, J.; Zheng, K.; Yu, Z.; Zeng, X. Organophosphate esters in total suspended particulates of an urban city in East China. *Chemosphere* **2016**, *164*, 75–83. [CrossRef] [PubMed]
25. Ma, S.; Yue, C.; Tang, J.; Lin, M.; Zhuo, M.; Yang, Y.; Li, G.; An, T. Occurrence and distribution of typical semi-volatile organic chemicals (SVOCs) in paired indoor and outdoor atmospheric fine particle samples from cities in southern China. *Environ. Pollut.* **2021**, *269*, 116123. [CrossRef]

26. Wang, Y.; Zhang, Z.; Tan, F.; Rodgers, T.F.M.; Hou, M.; Yang, Y.; Li, X. Ornamental houseplants as potential biosamplers for indoor pollution of organophosphorus flame retardants. *Sci. Total Environ.* **2021**, *767*, 144433. [CrossRef]
27. Zhou, L.; Hiltscher, M.; Gruber, D.; Puettmann, W. Organophosphate flame retardants (OPFRs) in indoor and outdoor air in the Rhine/Main area, Germany: Comparison of concentrations and distribution profiles in different microenvironments. *Environ. Sci. Pollut. Res.* **2017**, *24*, 10992–11005. [CrossRef]
28. Staaf, T.; Ostman, C. Organophosphate triesters in indoor environments. *J. Environ. Monit.* **2005**, *7*, 883–887. [CrossRef]
29. Yadav, I.C.; Devi, N.L.; Kumar, A.; Li, J.; Zhang, G. Airborne brominated, chlorinated and organophosphate ester flame retardants inside the buildings of the Indian state of Bihar: Exploration of tag source and human exposure. *Ecotoxicol. Environ. Saf.* **2020**, *191*, 110212. [CrossRef]
30. Khairy, M.A.; Lohmann, R. Organophosphate flame retardants in the indoor and outdoor dust and gas-phase of Alexandria, Egypt. *Chemosphere* **2019**, *220*, 275–285. [CrossRef]
31. He, C.; Wang, X.; Phong, T.; Baduel, C.; Gallen, C.; Banks, A.; Bainton, P.; English, K.; Mueller, J.F. Organophosphate and brominated flame retardants in Australian indoor environments: Levels, sources, and preliminary assessment of human exposure. *Environ. Pollut.* **2018**, *235*, 670–679. [CrossRef] [PubMed]
32. Okeme, J.O.; Yang, C.; Abdollahi, A.; Dhal, S.; Harris, S.A.; Jantunen, L.M.; Tsirlin, D.; Diamond, M.L. Passive air sampling of flame retardants and plasticizers in Canadian homes using PDMS, XAD-coated PDMS and PUF samplers. *Environ. Pollut.* **2018**, *239*, 109–117. [CrossRef] [PubMed]
33. Lai, S.; Xie, Z.; Song, T.; Tang, J.; Zhang, Y.; Mi, W.; Peng, J.; Zhao, Y.; Zou, S.; Ebinghaus, R. Occurrence and dry deposition of organophosphate esters in atmospheric particles over the northern South China Sea. *Chemosphere* **2015**, *127*, 195–200. [CrossRef] [PubMed]
34. Wang, Y.; Yang, Y.; Zhang, Y.; Tan, F.; Li, Q.; Zhao, H.; Xie, Q.; Chen, J. Polyurethane heat preservation materials: The significant sources of organophosphorus flame retardants. *Chemosphere* **2019**, *227*, 409–415. [CrossRef]
35. Tokumura, M.; Hatayama, R.; Tatsu, K.; Naito, T.; Takeda, T.; Raknuzzaman, M.; Habibullah-Al-Mamun, M.; Masunaga, S. Organophosphate flame retardants in the indoor air and dust in cars in Japan. *Environ. Monit. Assess.* **2017**, *189*, 48. [CrossRef]
36. Hartmann, P.C.; Bürgi, D.; Giger, W. Organophosphate flame retardants and plasticizers in indoor air. *Chemosphere* **2004**, *57*, 781–787. [CrossRef]
37. Cao, D.; Lv, K.; Gao, W.; Fu, J.; Wu, J.; Fu, J.; Wang, Y.; Jiang, G. Presence and human exposure assessment of organophosphate flame retardants (OPEs) in indoor dust and air in Beijing, China. *Ecotoxicol. Environ. Saf.* **2019**, *169*, 383–391. [CrossRef]

toxics

MDPI

Article

Migration Behavior of Technetium-99 in Granite, Clay Rock, and Shale: Insights into Anionic Exclusion Effects

Yunfeng Shi, Song Yang, Wenjie Chen, Aiming Zhang, Zhou Li, Longjiang Wang * and Bing Lian *

Department of Nuclear Environmental Science, China Institute for Radiation Protection (CIRP), Taiyuan 030006, China; shiyunfeng@cirp.org.cn (Y.S.)
* Correspondence: wanglongjiang@cirp.org.cn (L.W.); lianbing@cirp.org.cn (B.L.); Tel.: +86-15103418769 (L.W.)

Abstract: One of the key tasks in the geological disposal of radioactive waste is to investigate the blocking ability of different host rocks on nuclide migration in the disposal site. This study conducted experimental and numerical methods to the adsorption, diffusion, and advection–dispersion behavior of ^{99}Tc in three types of rocks: granite, clay rock, and mudstone shale, with a focus on the influence of anion exclusion during migration. The research results found that the three types of rocks have no significant adsorption effect on ^{99}Tc, and the anion exclusion during diffusion and advection–dispersion processes can block small "channels", causing some nuclide migration to lag, and accelerate the nuclide migration rate in larger "channels". In addition, parameters characterizing the size of anion exclusion in different migration behaviors, such as effective diffusion coefficient (De) and immobile liquid region porosity (θ_{im}), were fitted and obtained.

Keywords: granite; clay rock; mudstone shale; anion exclusion; ^{99}Tc

1. Introduction

Radioactive waste is a unique byproduct generated during the use of atomic energy, and its safe disposal has always been a focal point of public concern. According to the IAEA safety standards, radioactive wastes are categorized into six groups based on their radioactivity levels, ranging from highest to lowest: high-level waste (HLW), intermediate-level waste (ILW), low-level waste (LLW), very low-level waste (VLLW), very short-lived waste (VSLW), and exempt waste (EW) [1]. With the exception of short-lived waste and EW, all other types of waste must undergo various disposal methods, such as near-surface disposal, intermediate-depth disposal, and deep geological repository, for long-term management. The migration of radioactive nuclides from the waste to the surrounding environment must be effectively impeded and isolated through a "multi-barrier system" [2]. Therefore, evaluating the barrier capabilities of different rock types against nuclides is a crucial task in the selection and construction of radioactive waste disposal facilities [3].

Currently, numerous studies have conducted experiments and numerical simulations on the migration of radioactive nuclides through adsorption, diffusion, and convection–dispersion in various types of rocks, yielding a substantial amount of data [4–7]. In recent years, the research focus has transitioned from determining key parameters of nuclide migration behavior to comprehending the primary factors and mechanisms involved in these processes. For instance, Lee investigated the diffusion behavior of Se (IV) in Tamusu clay rocks, emphasizing the impact of compaction density on diffusion behavior [8]. Sun examined the adsorption and diffusion behavior of Se (IV) and Se (VI) in clay rocks, exploring the influence of different valence states of the same nuclide on their migration

https://doi.org/10.3390/toxics13090760

behavior [9]. Qi delved into the migration and adsorption behavior of Se (IV) and Sr (II) in granite minerals and fracture fillers, and scrutinized the influence of different chemical forms on the migration behavior of nuclides [10].

In the past, many scholars previously categorized anionic nuclides such as ^{99}Tc (TcO_4^-), ^{129}I (IO_3^-), ^{36}Cl (Cl^-), and ^{3}H (H^+) as non-adsorbed nuclides. Consequently, in numerous hydrogeological experiments, researchers tend to employ anionic elements as tracers to derive geological parameters [11,12], overlooking the "accelerated" migration of nuclides due to anionic exclusion. While some studies have addressed the impact of anionic exclusion on the diffusion and convection–dispersion migration behavior of elements like Cl^- and IO_3^- [13–15], there has been no comprehensive investigation into the effects of anionic exclusion on the adsorption, diffusion, and convection–dispersion of a specific anion nuclide. In 2022, Shi examined the convection–dispersion migration behavior of various iodine ion forms in granite, emphasizing the influence of different chemical forms (I^-/I_3^-) on anionic exclusion [16]. This study offers insights into understanding the "accelerated migration" behavior of anionic nuclides.

^{99}Tc is a significant nuclide found in the waste produced by nuclear power plants, boasting a high fission yield (6.13%) in reactors. Compounds of Tc can display oxidation states ranging from -I to VII, typically existing as TcO_4^- in aqueous solutions [17]. Research on the migration behavior of ^{99}Tc has primarily centered on the impact of changes in nuclide valence states on solubility [18,19] and anionic exclusion [20,21], driven by its unique characteristics of "multiple chemical valence states" and "non-adsorption." The former elucidates how redox conditions in the environment influence the migration of ^{99}Tc in complex settings, aiding in the separation and elimination of ^{99}Tc from multi-nuclide mixed solutions. The latter explores the effects of anionic exclusion on ^{99}Tc migration by comparing its behavior with that of other nuclides in various solid-phase media [22].

Diffusion and advection–dispersion are the primary mechanisms through which radioactive nuclides migrate in the rock medium of disposal sites, influenced by physical and chemical factors. Physical factors pertain to the characteristics of nuclide movement channels within the solid phase medium, defined by parameters like porosity, effective diffusion coefficient, and dispersibility [23]. Chemical factors involve the interaction between nuclides and the surrounding media during migration, particularly in terms of adsorption, with characteristic parameters including distribution coefficient and adsorption rate [24]. The migration of nuclides is concurrently affected by physical and chemical factors. To elucidate the specific role of an influencing factor in nuclide migration, it is typically necessary to initially conduct diffusion and advection–dispersion tests using non-adsorbed nuclides (e.g., HTO) and subsequently introduce designated nuclides to investigate the influencing factors of nuclide migration behavior [25,26].

This study utilized batch experiments [27], diffusion experiments [28], and column experiments [29] to replicate the adsorption, diffusion, and advection–dispersion characteristics of ^{99}Tc in granite, clay rock, and shale. The essential parameters governing the diffusion and advection–dispersion behavior of ^{99}Tc were determined through numerical model fitting [30]. A comparison with the diffusion and advection–dispersion parameters of HTO in the same rock medium facilitated a discussion on the impact of anion repulsion on the migration behavior of ^{99}Tc.

2. Theory of Adsorption, Diffusion, and Advection–Dispersion

2.1. Distribution Coefficient Model

The distribution coefficient is a parameter that characterizes the distribution of nuclides in the solid–liquid phase. According to the IAEA definition, the distribution coefficient is determined by the ratio of nuclide activity concentration in the solid phase to that in the liquid phase [30]. The expression formula is as follows:

$$K_d = \frac{(c_0 - c_e) \times V}{c_e \times m} \tag{1}$$

where c_0 is the initial activity concentration of nuclides in the liquid phase (Bq/L), c_e is the activity concentration of nuclides in the liquid phase at adsorption equilibrium (Bq/L), V is the total volume of the solution (mL), and M is the mass of the solid phase medium (g).

2.2. Diffusion Model

According to the theory of porous media, rock samples are generalized as isotropic homogeneous porous media, and the diffusion behavior of nuclides in the sample is equivalent to one-dimensional. The diffusion equation is established based on the principle of conservation of mass and the concept of typical unit cells, and the obtained diffusion curve is fitted to the solution of the equation to obtain the diffusion coefficient [31].

$$\frac{\partial C}{\partial t} = \frac{D_a}{R}\frac{\partial^2 C}{\partial x^2} \tag{2}$$

$$D_a = \frac{D_e}{\alpha} \tag{3}$$

$$\alpha = \theta + \rho_b \times k_d \tag{4}$$

$$\theta = 1 - \frac{\rho_b}{\rho_s} \tag{5}$$

The initial conditions and boundary conditions are as follows:
Boundary conditions: $C(x,0) = 0$, $x > 0$
Initial conditions: $C(0,t) = C_0$, $t > 0$

$$C(\infty, t) = 0,\ t \geq 0 \tag{6}$$

where $C(x,t)$ is the concentration distribution of tracer in the vertical direction of the thin slice (Bq/mL), D_e is the effective diffusion coefficient of the tracer in the vertical direction of the thin film (m^2/s), D_a is the apparent diffusion coefficient of the tracer in the vertical direction of the thin film (m^2/s), α is the capacity factor, x is the one-dimensional diffusion distance (cm), t is the diffusion time (d), C_0 is the initial concentration value of each tracer in the source solution tank (Bq/mL), K_d is the adsorption distribution coefficient of the nuclide (mL/g), θ is the total porosity of the porous medium, which is the ratio of the pore volume to the total volume of the medium, ρ_b is the dry density of the solid phase medium (kg/m^3), and ρ_s is the particle density (kg/m^3).

2.3. Advection–Dispersion Model

2.3.1. Equilibrium Transport (E-T)

The one-dimensional equation is as follows [32]:

$$\frac{\partial}{\partial t}(\theta c_r + \rho_b s) = \frac{\partial}{\partial x}\left(\theta D\frac{\partial c_r}{\partial x} - J_w c\right) - \theta\mu_l c_r - \rho_b\mu_s s + \theta r_l(x) + \rho_b r_s(x) \tag{7}$$

where c_r is the volume-averaged or resident concentration of the liquid phase (ML^{-3}); s is the concentration of the adsorbed phase (MM^{-1}); D is the dispersion coefficient (L^2T^{-1});

θ is the volumetric water content (L^3L^{-3}); J_w is the volumetric water flux density (LT^{-1}); ρ_b is the soil bulk density (ML^{-3}); μ_l and μ_s are the first-order decay coefficients for the degradation of the solute in the liquid and adsorbed phases, respectively (T^{-1}); r_l (ML^{-3} T^{-1}) and r_s (MM^{-1} T^{-1}) are the zero-order production terms for the liquid and adsorbed phases, respectively; x is the distance (L); and t is the time (T).

The solute adsorption by the solid phase is expressed using a linear isotherm as follows:

$$s = K_d c_r \tag{8}$$

where K_d is an empirical distribution constant ($M^{-1}L^3$). Using (8) and assuming a steady-state flow in a homogeneous soil, (7) may be rewritten as follows:

$$R\frac{\partial c_r}{\partial t} = D\frac{\partial^2 c_r}{\partial x^2} - v\frac{\partial c_r}{\partial x} - \mu c_r + r(x) \tag{9}$$

where $v(=\frac{J_w}{\theta})$ is the average pore-water velocity; R is the retardation factor given by $R = 1 + \frac{\rho_b K_d}{\theta}$; and μ and r are combined first- and zero-order rate coefficients: $\mu = \mu_l + \frac{\rho_b K_d \mu_s}{\theta}$, $r(x) = r_l(x) + \frac{\rho_b r_s(x)}{\theta}$, respectively.

2.3.2. Two-Region Nonequilibrium Transport (T-N)

A two-region transport model was proposed, and it was assumed that the liquid phase can be partitioned into mobile (flowing) and immobile (stagnant) regions (which are produced by anion exclusion). The solute exchange between the two liquid regions is simulated as a first-order kinetic process. The two-region solute transport model is given by [33].

$$(\theta_m + f\rho_b K_d)\frac{\partial c_m}{\partial t} = \theta_m D_m \frac{\partial^2 c_m}{\partial x^2} - J_w\frac{\partial c_m}{\partial x} - a(c_m - c_{im}) - (\theta_m \mu_{l,m} + f\rho_b K_d \mu_{s,m})c_m + \theta_m r_{l,m}(x) + f\rho_b r_{s,m}(x) \tag{10}$$

$$(\theta_{im} + (1-f)\rho_b K_d)\frac{\partial c_{im}}{\partial t} = a(c_m - c_{im}) - (\theta_{im}\mu_{l,im} + (1-f)\rho_b K_d \mu_{s,im})c_{im} + \theta_{im} r_{l,im}(x) + (1-f)\rho_b r_{s,im}(x) \tag{11}$$

where the subscripts m and im refer to the mobile and immobile liquid regions, respectively; $J_w = v \times \theta = v_m \times \theta_m$ is the volumetric water flux density; f represents the fraction of adsorption sites that equilibrates with the mobile liquid phase; and a is the first-order mass transfer coefficient governing the rate of solute exchange between the mobile and immobile liquid regions. θ is equal to $\theta_m + \theta_{im}$; $\mu_{l,m}$ and $\mu_{l,im}$ are first-order decay coefficients for the mobile and immobile liquid phases, respectively; $\mu_{s,m}$ and $\mu_{s,im}$ are first-order decay coefficients for the mobile and immobile adsorbed phases, respectively; $r_{l,m}$ and $r_{l,im}$ are zero-order production for the mobile and immobile liquid phases, respectively; and $r_{s,m}$. and $r_{s,im}$ are zero-order production terms for the mobile and immobile adsorbed phases, respectively.

3. Experiments

3.1. Rock

The granite, clay rock, and mudstone utilized in this study were sourced from radioactive waste disposal sites under construction, planned construction, and existing sites in northwest, northern, and southwestern China, respectively. Following retrieval, all the rock samples were crushed, washed thrice with deionized water, and dried at 100 ± 10 °C for 24 h before being stored. Both batch and column experiments were conducted. Batch experiments and diffusion experiments were conducted using rock samples with a particle size of 0.075 mm or less. In the column experiment, the particle size of the granite was less than 0.075 mm, while the particle size of the clay rock and mudstone shale ranged from 0.425 mm to 0.25 mm.

3.2. Rock Sample Analysis

X-ray Powder diffractometer (XRD,D-Power, Anhui Guoke Instrument Technology Co., Ltd., Hefei, China), X-ray Fluorescence Spectrometer (XRF, Zetium, Malvern Panalytical, Malvern, UK), and Scanning Electron Microscopy–Energy-Dispersive Spectrometer (SEM-EDS,TH-F120, CIQTEK Co., Ltd., Hefei, China) were utilized to analyze the mineral and chemical composition of the granite, clay rock, and mudstone shale, respectively.

3.3. Batch Experiment

The adsorption behavior was investigated using a batch experiment method. A 1.0 g treated rock sample was weighed and placed in a 15 mL centrifuge tube. Subsequently, a 10 mL nuclide solution (initial activity concentrations of 800 Bq, 1600 Bq, 2400 Bq, 3200 Bq, 4000 Bq) mixed with ultrapure water was added, maintaining a solid–liquid ratio of 1:10 g/mL and a pH value of 7.3–8.0. The mixture was shaken continuously at 25 °C for 48 h to ensure full contact of the sample. Following this, a centrifuge was used to separate the solid and liquid phases. The water sample was then extracted to measure the nuclide's activity and calculate the activity concentration. The ^{99}Tc sample (China Tongfu Co., Ltd., Nantong, China) was analyzed and measured using an ultra-low background liquid scintillation spectrometer (SIM-MAX LSA3000, Shanghai Xinman Sensing Technology Research and Development Co., Ltd., Shanghai, China).

3.4. Diffusion Experiment

3.4.1. Diffusion Experimental Device

Figure 1 illustrates the equipment utilized in the diffusion experiment, comprising three main components: the source liquid tank, the liquid collection tank, and the rock slice. Introduce nuclides into the source liquid tank, fill the liquid tank with ultrapure water devoid of solutes, secure the rock sheet using a porous filter, and position it at the center of the diffusion device using two "O" rings. Additionally, to maintain the long-term integrity of the sample medium and prevent damage, secure the sample with two polyethylene filters on both sides and insert it into the experimental apparatus. The experimental setup parameters are detailed in Table 1.

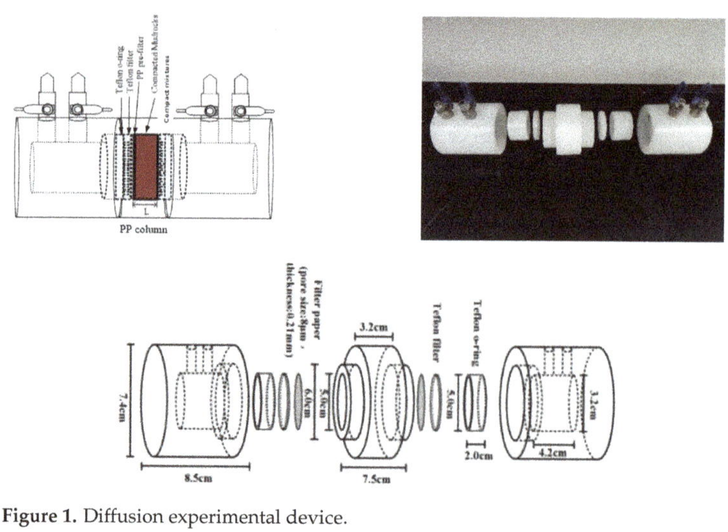

Figure 1. Diffusion experimental device.

Table 1. Diffusion experimental device parameters.

Device	Length (cm)	Diameter (cm)
Diffusion device	20.2 cm	7.4 cm
Source liquid tank (liquid collection tank)	4.2 cm	3.2 cm
rock slice	0.3 cm	5 cm

3.4.2. Diffusion Experimental Procedure

a. Granite, clay rock, and mudstone with a particle size less than 0.075 mm are compacted to obtain experimental samples with a density of 2.0 g/cm^3, and fixed through filters before being placed in a diffusion tank.

b. Water saturation treatment is performed on rock fragments by filling the source liquid tank with ultrapure water. The water is allowed to permeate into the extraction tank under the influence of a water level difference, saturating the rock fragments with water.

c. After saturating the rock fragments with water, the nuclide HTO (20 Bq/mL) is initially introduced into the source liquid tank. This tank is linked to an external storage tank via a peristaltic pump (30–60 min per cycle, with a water flow rate of 500–1000 mL). The solubility of nuclides in the source liquid tank is maintained at a constant level by continuously monitoring the HTO content in the storage tank. Simultaneously, ultrapure water is added to the extraction tank to ensure a consistent water level on both sides. Periodically, all the solutions in the extraction tank are extracted for HTO activity measurement. The diffusion experiment of HTO is concluded once analysis and calculations confirm the attainment of equilibrium. An ultra-low background liquid scintillation spectrometer (SIM-MAX LSA3000, Shanghai Xinman Sensing Technology Research and Development Co., Ltd.) is employed for the analysis and quantification of HTO (China Tongfu Co., Ltd.).

d. After rinsing the HTO clean, the ^{99}Tc (15 Bq/mL) solution is placed into the source tank and connected to an external storage tank using a peristaltic pump (30–60 min per cycle, with a water flow rate of 500–1000 mL). The ^{99}Tc content in the storage tank is continuously measured to ensure constant solubility of nuclides in the source tank. All the solutions are regularly taken out from the tank, and the activity of nuclides is measured. After confirming diffusion equilibrium through analysis and calculation, all diffusion tests are stopped.

3.5. Column Experiment

3.5.1. Column Experimental Device

The device is shown in Figure 2 and consists of three parts: the injection end, column, and collection end. The injection end includes bottle 1 (for storing nuclide solution) and bottle 2 (for storing ultrapure water). The injection of nuclides is achieved by controlling a three-way valve connected to a peristaltic pump. The column is filled with rock and soil samples, and filters are installed at the upper and lower ends to prevent sample loss. The collection end is used for collecting the effluent from the column, and the automatic collector completes the sample sampling at regular intervals. The device parameters are shown in Table 2.

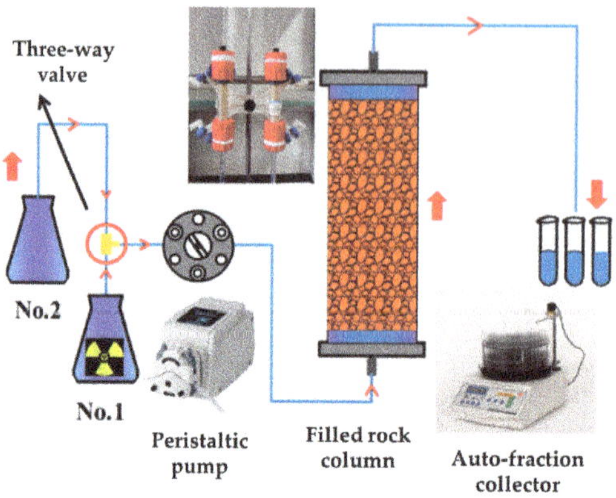

Figure 2. Column experimental device.

Table 2. Column experimental device parameters.

Device	Length	Diameter
Granite column (Particle size < 0.075 mm)	10 cm	1.6 cm
Clay rock column (0.425–0.25 mm)	10 cm	1.6 cm
Mudstone shale column (0.425–0.25 mm)	10 cm	1.6 cm

3.5.2. Column Experimental Procedure

a. Fill the processed granite, clay rock, and mudstone samples into glass columns with a density of 1.25 g/cm^3. Install filter screens at the upper and lower ends of the column to prevent the loss of soil samples during water flow erosion.

b. To saturate the rock column with water, adjust the three-way valve and open the peristaltic pump to slowly inject ultrapure water from the bottom of the column in bottle 2. Stop when water flows out from the outlet to fully saturate the rock and soil column.

c. Inject HTO solution into bottle 1, adjust the three-way valve, and open the peristaltic pump to inject HTO (5000 Bq) into the soil column instantly. Then, regulate the three-way valve to introduce ultrapure water into the column at a flow rate of 0.25 mL/min while keeping the pH within the range of 7.3–8.0. Periodically collect and analyze samples at the outflow end. Cease the HTO penetration test once the HTO content in the outflow falls below the lower limit of detection.

d. After rinsing the HTO in bottle 1, replace it with ^{99}Tc solution and inject ^{99}Tc (5000 Bq) into the soil column instantly following the same steps. Collect samples at the outflow end and analyze them periodically. Terminate the test when the nuclide content in the outflow is below the detection lower limit.

3.6. Mathematical Model and Parameter Estimations

Use 1stOpt 6.0 (First Optimization) software and STANMOD 2.10 (STudio of ANalytic MODels), respectively, to fit the experimental data obtained from diffusion experiments and convection–dispersion experiments. 1stOpt is a comprehensive mathematical optimization analysis tool software package that has achieved good results in handling nonlinear regression, curve fitting, and other problems through the use of the Universal Global Optimization (UGO) algorithm. STANMOD is a Windows-based computer software package used to evaluate solute transport in porous media using analytical solutions of the advection–dispersion solute transport equation. The software includes modified and updated versions of the CXTFIT code for estimating solute transport parameters using nonlinear least squares parameter optimization methods. Both software use root mean square error (RMSE) to represent the degree of fitting, which is defined as follows:

$$RMSE = \sqrt{\frac{\sum_i^N (C_p - C_e)^2}{N}} \tag{12}$$

where C_p denotes the results of the numerical fitting and C_e denotes the experimental data.

4. Results and Discussion

4.1. Mineral Composition and Chemical Composition

The mineral and chemical composition analysis of the granite, clay rock, and mudstone shale is depicted in Figures 3 and 4, and Table 3.

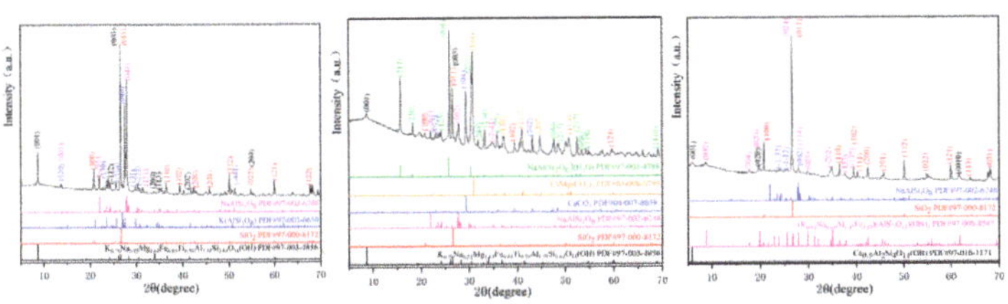

(**a**) Granite (**b**) Clay rock (**c**) Mudstone shale

Figure 3. XRD analysis results of rock samples.

Table 3. Mineral composition and chemical analysis results of rock samples (%).

Rock	Mica	Quartz	Potassium Feldspar	Plagioclase	Illite	Montmorillonite	Zeolite
Granite	10.8	20.5	30.2	38.5	/	/	/
Clay rock	11.1	4.0	/	16.9	/	/	23.6
Mudstone shale	/	46.7	/	9.7	24.6	19.0	/
Rock	**Calcite**	**Dolomite**	**SiO_2**	**Al_2O_3**	**Na_2O**	**K_2O**	**CaO**
Granite	/	/	66.81	15.88	3.83	4.82	3.43
Clay rock	9.6	34.8	30.61	9.72	3.08	2.14	27.95
Mudstone shale	/	/	61.15	20.81	0.33	5.15	0.67
Rock	**Fe_2O_3**	**MgO**	**TiO_2**	**P_2O_5**	**SO_3**	**MnO**	**SrO**
Granite	3.04	1.24	0.469	0.146	0.10	0.05	0.05
Clay rock	6.74	14.72	0.621	0.080	3.74	0.20	0.27
Mudstone shale	8.47	2.14	0.88	0.16	0.03	0.08	0.01
Rock	**BaO**	**Rb_2O**	**ZrO_2**	**ZnO**	**La_2O_3**	**V_2O_5**	**Cr_2O_3**
Granite	0.05	0.02	0.01	0.01	0.01	0.01	0.004
Clay rock	/	0.02	0.02	0.01	/	0.02	0.021
Mudstone shale	0.02	0.03	0.03	/	0.01	0.02	0.020
Rock	**Ga_2O_3**	**PbO**	**Co_3O_4**	**Cl**	**CeO_2**	**CuO**	**Er_2O_3**
Granite	0.003	0.003	0.002	0.032	/	/	/
Clay rock	0.002	/	0.003	0.083	0.006	0.005	0.004
Mudstone shale	0.003	/	0.003	0.023	/	/	/

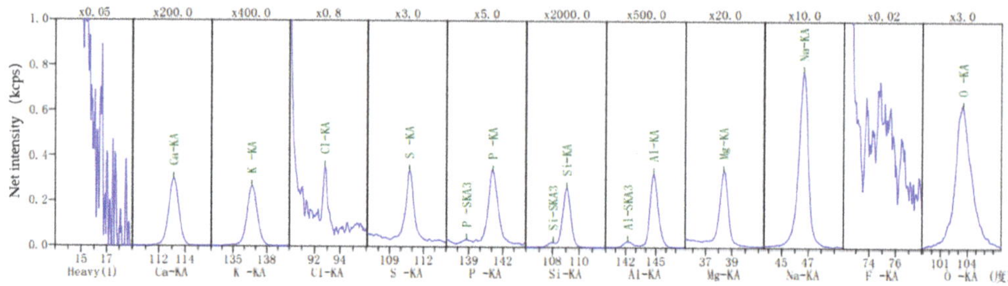

(**a**) Granite

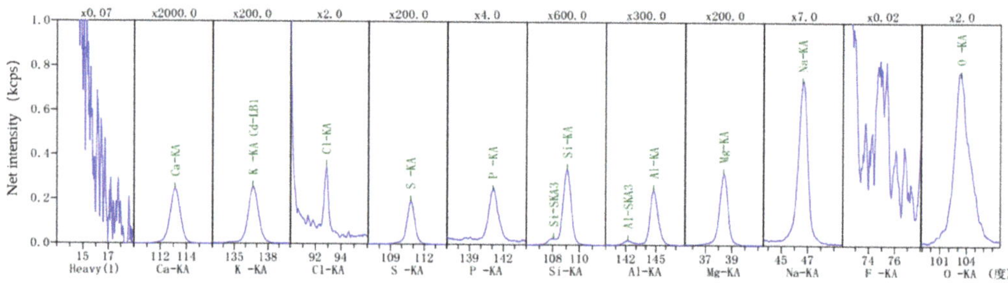

(**b**) Clay rock

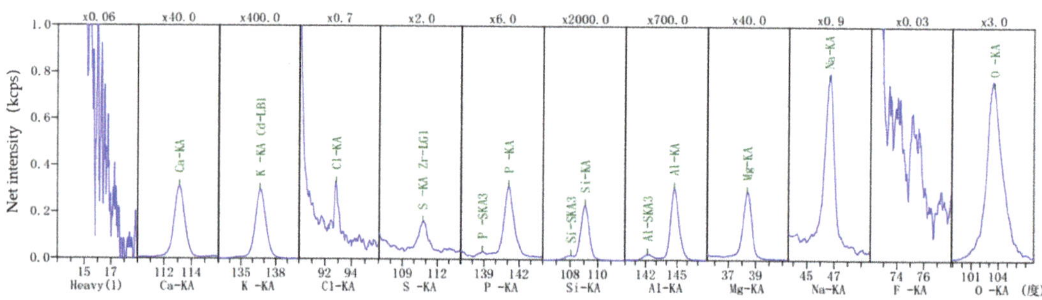

(**c**) Mudstone shale

Figure 4. XRF analysis results of rock samples.

4.2. Batch Experiment Results

Batch experiments were carried out to investigate the adsorption behavior of ^{99}Tc in the granite, clay rock, and mudstone shale. The results are illustrated in Figure 5. The distribution coefficients of ^{99}Tc for the granite, clay rock, and mudstone shale are 0.85 mL/g, 0.78 mL/g, and 0.30 mL/g, respectively. The solid-phase adsorption capacity in this study was determined by subtracting the liquid phase content post-adsorption from the total input amount, considering measurement errors in the liquid phase sample and solute losses during operations. Subsequent diffusion and convection–dispersion experiments indicated that the granite, clay rock, and mudstone shale exhibited negligible adsorption capacity for ^{99}Tc in this investigation.

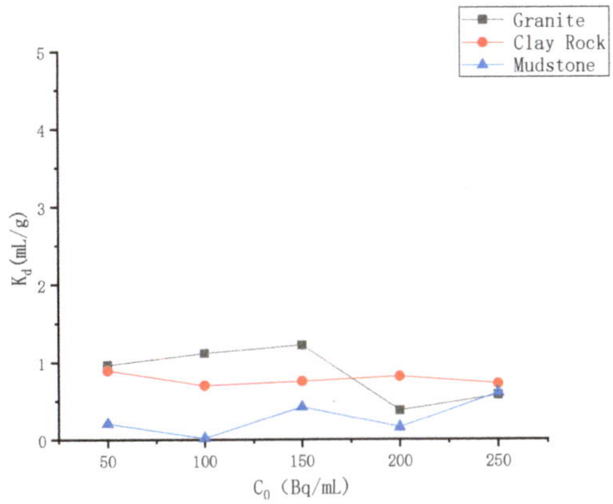

Figure 5. Adsorption capacity of different rocks for ^{99}Tc.

4.3. Diffusion Experiment Results

4.3.1. Diffusion Experiment Results of HTO

To examine the diffusion pathways within the granite, clay rock, and mudstone shale, experiments were conducted to analyze the diffusion characteristics of tritiated water (HTO) within them. The experimental findings are shown in Figure 6. Each material was tested in triplicate, with each test lasting 588 h. Following the initiation of the experiments, the diffusion profiles of HTO in all the substrates exhibited a swift escalation. The diffusion capacity of HTO in the three rock types follows the order granite > clay rock > mudstone shale. The diffusion parameters were determined through numerical simulation methods, and the fitting effect, along with related parameters, is presented in Figure 7 and Table 4. The experimental results demonstrate variations in porosity among different rock types: granite (Ø = 0.26) > mudstone shale (Ø = 0.20) > claystone (Ø = 0.18). Regarding the effective diffusion coefficient, a notable discrepancy exists within the granite samples, with values ranging from 2.48×10^{-12} to 4.05×10^{-11} m^2/s. In contrast, the disparity between clay rock and mudstone shale is insignificant, with diffusion coefficients ranging from 2.60×10^{-11} to 1.99×10^{-11} m^2/s and 2.03×10^{-11} to 2.50×10^{-11} m^2/s, respectively.

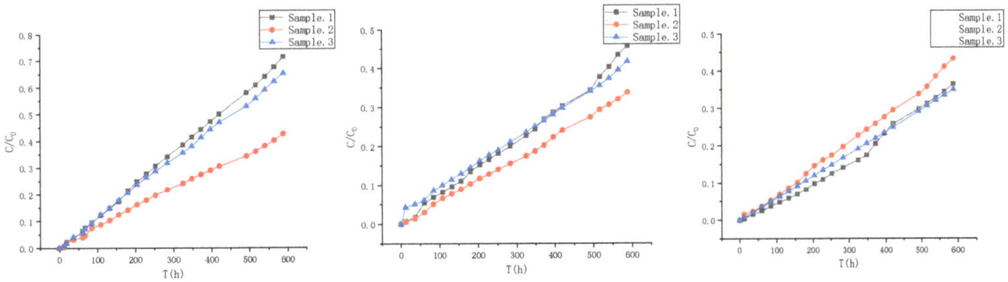

(**a**) Granite (**b**) Clay rock (**c**) Mudstone shale

Figure 6. Diffusion curves of HTO in different rocks.

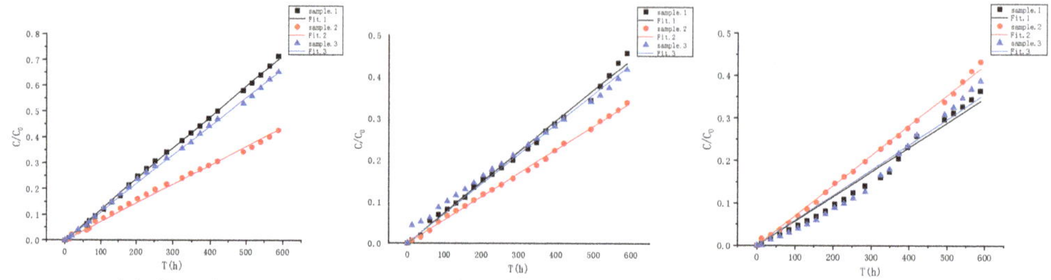

(**a**) Granite (**b**) Clay rock (**c**) Mudstone shale

Figure 7. Fitting of diffusion curves of HTO in different rock media.

Table 4. Parameters for fitting diffusion curves of HTO in different rock media.

Rock	Ø	α	D_a	D_e	R^2	*RMSE*
Granite-1	0.26	0.26	1.56×10^{-10} m²/s	4.05×10^{-11} m²/s	0.999	0.060
Granite-2	0.26	0.26	9.54×10^{-11} m²/s	2.48×10^{-12} m²/s	0.998	0.094
Granite-3	0.26	0.26	1.45×10^{-10} m²/s	3.76×10^{-11} m²/s	0.999	0.067
Clay rock-1	0.18	0.18	1.42×10^{-10} m²/s	2.60×10^{-11} m²/s	0.998	0.096
Clay rock-2	0.18	0.18	1.09×10^{-10} m²/s	1.99×10^{-11} m²/s	0.999	0.042
Clay rock-3	0.18	0.18	1.38×10^{-10} m²/s	2.53×10^{-11} m²/s	0.998	0.161
Mudstone shale-1	0.20	0.20	1.02×10^{-10} m²/s	2.03×10^{-11} m²/s	0.994	0.172
Mudstone shale-2	0.20	0.20	1.25×10^{-10} m²/s	2.50×10^{-11} m²/s	0.999	0.063
Mudstone shale-3	0.20	0.20	1.05×10^{-10} m²/s	2.10×10^{-11} m²/s	0.989	0.260

4.3.2. Diffusion Experiment of ^{99}Tc

The ^{99}Tc diffusion curve is depicted in Figure 8. Generally, the diffusion pattern of ^{99}Tc resembles that of HTO, suggesting negligible adsorption impact of the medium on ^{99}Tc.

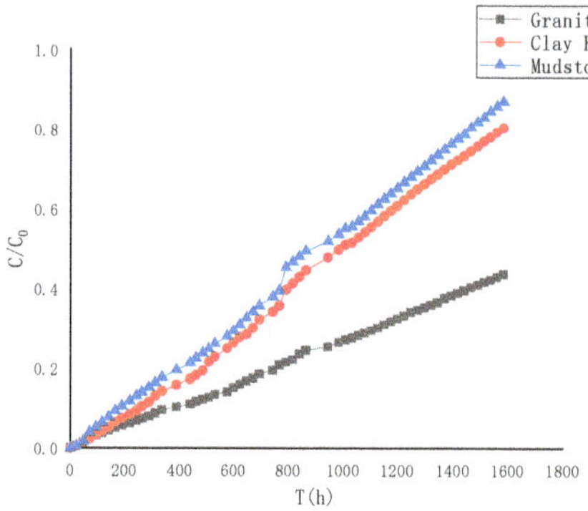

Figure 8. Diffusion curves of ^{99}Tc in different rocks.

The diffusion experiment lasted for 1584 h, and the cumulative concentration ratio of ^{99}Tc in different rock media was granite < clay rock < mudstone shale, which contrasts significantly with the diffusion behavior of HTO. This difference is attributed to ^{99}Tc existing in an anionic state (TcO_4^-) in aqueous solution, leading to repulsion by anions in the rock pores. This repulsion results in the blockage of small pores while accelerating the migration of nuclides in larger pores. Moreover, a higher clay mineral content intensifies the repulsive effect, indicating that ^{99}Tc diffuses most rapidly in mudstone shale and slowest in granite.

For the fitting of the diffusion curve of ^{99}Tc, the "anion repulsion" effect, not explicitly defined in current diffusion models, influences both the distribution coefficient (represented by negative values) and the effective diffusion coefficient (leading to a significant decrease). These parameters directly impact the capacity factor α and porosity Ø in the fitting process. To enhance the precision of the fitting parameters, the primary approach involves initially determining the apparent diffusion coefficient *Da* and capacity factor α through curve fitting. Subsequently, the effective diffusion coefficient *De* is calculated, with the fixed porosity Ø remaining constant at its original value and the distribution coefficient set at 0. This method consolidates all the anion repulsion effects within the effective diffusion coefficient.

The results of the fitting are displayed in Figure 9 and Table 5. Upon comparing the parameters across various rock types, it was observed that the apparent diffusion coefficient follows the order: mudstone > clay rock > granite. Similarly, the order for mud shale > clay rock > granite remains consistent. Regarding the effective diffusion coefficient, the sequence mudstone shale > clay rock > granite was noted, suggesting that the "anion repulsion" of ^{99}Tc leads to an overall "accelerated" diffusion of nuclides. Notably, a stronger "anion repulsion" corresponds to a smaller effective diffusion coefficient in these parameters.

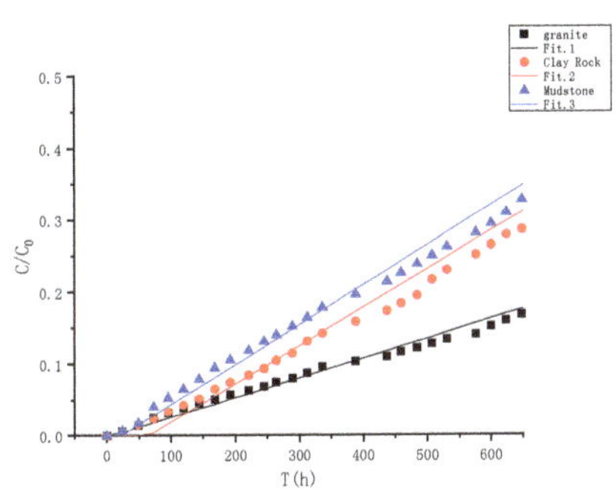

Figure 9. Fitting of diffusion curves of ^{99}Tc in different rock media.

Table 5. Parameters for fitting diffusion curves of ^{99}Tc in different rock media.

Rock	Ø	α	D_a (m^2/s)	D_e (m^2/s)	K_d (mL/g)	R^2	*RMSE*
Granite	0.26	0.26	3.59×10^{-11}	9.34×10^{-12}	0	0.999	0.006
Clay rock	0.18	0.18	8.97×10^{-11}	1.71×10^{-11}	0	0.998	0.014
Mudstone shale	0.20	0.20	9.28×10^{-11}	1.86×10^{-11}	0	0.997	0.014

4.4. Advection–Dispersion Experiment of ^{99}Tc

4.4.1. Advection–Dispersion Experiment Results of HTO

The experimental results are depicted in Figure 10. Overall, the advection–dispersion migration curves of HTO in different rocks exhibit an "S" shape, consistent with the advection–dispersion curve under the instantaneous input mode of non-adsorbed solutes in homogeneous media. To ensure future accuracy in determining relevant physical migration parameters of the medium, three parallel samples are analyzed for each case. The HTO experiment in granite columns lasted 255 min, with peak occurrences at 30, 35, and 30 min, and corresponding peak values of 0.094, 0.193, and 0.159, respectively. For clay rock columns, the experimental durations were 96 min, 103 min, and 97 min, with peak values observed at 24, 28, and 32 min, with corresponding peak values of 0.163, 0.100, and 0.100, respectively. In shale columns, the experimental duration of HTO was 102 min, with peak values recorded at 27, 36, and 27 min, with corresponding peak values of 0.175, 0.235, and 0.190, respectively.

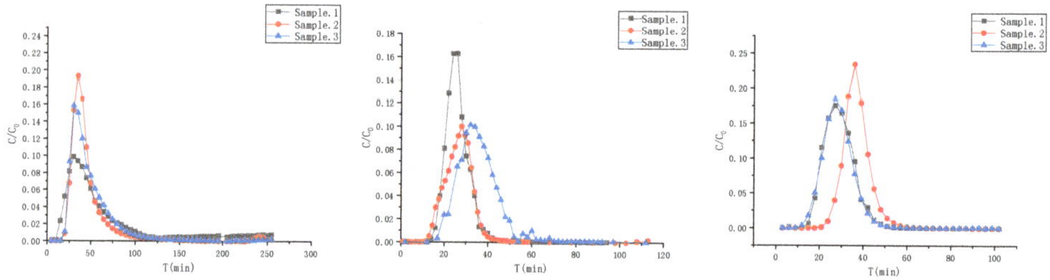

(**a**) Granite (**b**) Clay rock (**c**) Mudstone shale

Figure 10. Advection–dispersion curves of HTO in different rocks.

The advection–dispersion parameters were determined through a balanced model fitting. The fitting effect and associated parameters are presented in Figure 11 and Table 6, respectively. The dispersion values for granite range from 0.631 to 0.640 cm, for clay rock from 0.248 to 0.285 cm, and for mudstone from 0.263 to 0.323 cm.

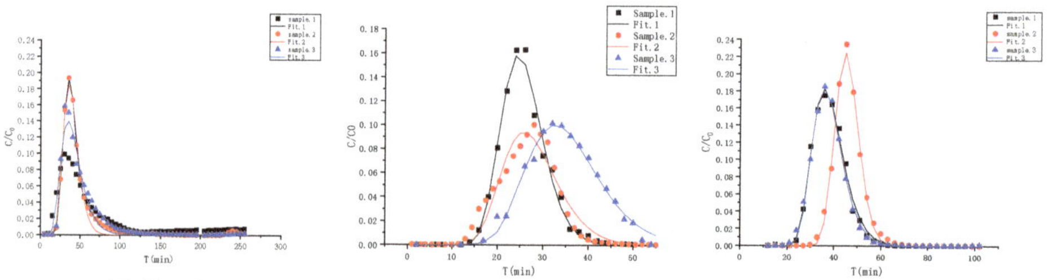

(**a**) Granite (**b**) Clay rock (**c**) Mudstone shale

Figure 11. Fitting of advection–dispersion curves of HTO in different rock media.

Table 6. Parameters for fitting advection–dispersion curves of HTO in different rock media.

Rock	V	D	D_L	R	$RMSE$
Granite-1	0.241 cm/min	0.152 cm^2/min	0.631 cm	1.00	5.25×10^{-6}
Granite-2	0.242 cm/min	0.155 cm^2/min	0.640 cm	1.00	3.89×10^{-6}
Granite-3	0.242 cm/min	0.154 cm^2/min	0.636 cm	1.00	9.45×10^{-5}
Clay rock-1	0.383 cm/min	0.095 cm^2/min	0.248 cm	1.00	1.57×10^{-5}
Clay rock-2	0.358 cm/min	0.095 cm^2/min	0.269 cm	1.00	3.46×10^{-10}
Clay rock-3	0.322 cm/min	0.092 cm^2/min	0.285 cm	1.00	9.26×10^{-6}
Mudstone shale-1	0.342 cm/min	0.098 cm^2/min	0.287 cm	1.00	1.04×10^{-5}
Mudstone shale-2	0.312 cm/min	0.101 cm^2/min	0.323 cm	1.00	1.24×10^{-5}
Mudstone shale-3	0.346 cm/min	0.091 cm^2/min	0.263 cm	1.00	9.14×10^{-6}

4.4.2. Advection–Dispersion Experiment Results of ^{99}Tc

The advection–dispersion behavior of ^{99}Tc in the granite, clay rock, and mudstone shale is illustrated in Figure 12. The migration curve of ^{99}Tc in granite and clay rock exhibits an "S" shape, while the migration curve of mudstone shows a distinct "tailing" phenomenon. This observation is attributed to the "anion repulsion" of ^{99}Tc during the convection–dispersion process in mudstone. This phenomenon causes a more pronounced "blockage" of small pores in mudstone compared to granite and clay rock, leading to delayed outflow of certain nuclides and the emergence of a "tailing" effect.

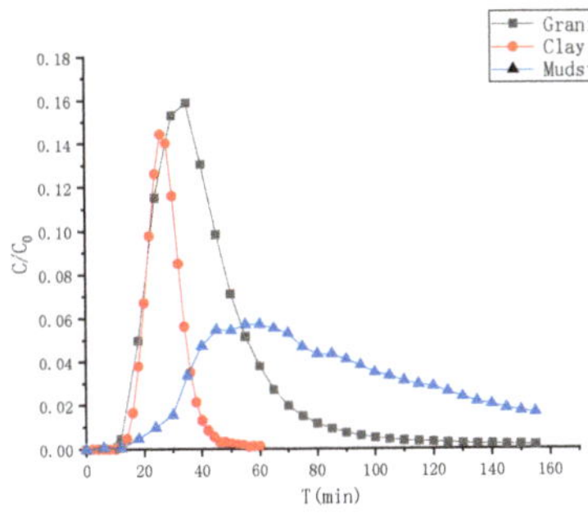

Figure 12. Advection–dispersion curves of ^{99}Tc in different rocks.

Different numerical models were utilized to fit the ^{99}Tc advection–dispersion migration curves in granite, clay rock, and mudstone shale. The fitting outcomes and associated parameters are presented in Figure 13 and Table 7. Initially, the equilibrium transport (E-T) model was applied to fit the advection–dispersion migration curves of ^{99}Tc in various rocks, demonstrating a satisfactory fitting effect. However, the calculated distribution coefficients were consistently negative, which contradicted the actual conditions. This discrepancy suggests that ^{99}Tc experiences "anion repulsion" during migration, leading to some channels being "blocked", resulting in negative distribution coefficients derived from the equilibrium transport model. Subsequently, the two-region nonequilibrium transport (T-N) model was employed to fit the curves and derive the relevant transfer parameters. The dispersion, movable zone porosity, and immovable zone porosity of ^{99}Tc in granite migration are 0.630 cm, 0.26 cm, and 0.11 cm, respectively; in clay rock, they are 0.215 cm, 0.16 cm, and 0.15 cm, respectively; and in mudstone, they are 0.320 cm, 0.17 cm, and 0.18 cm, respectively. Comparing the immovable zone porosity proportions, it was observed that the "anion repulsion" effect influences the ^{99}Tc migration behavior in mudstone in the following order: (51%) > clay rock (48%) > granite (30%).

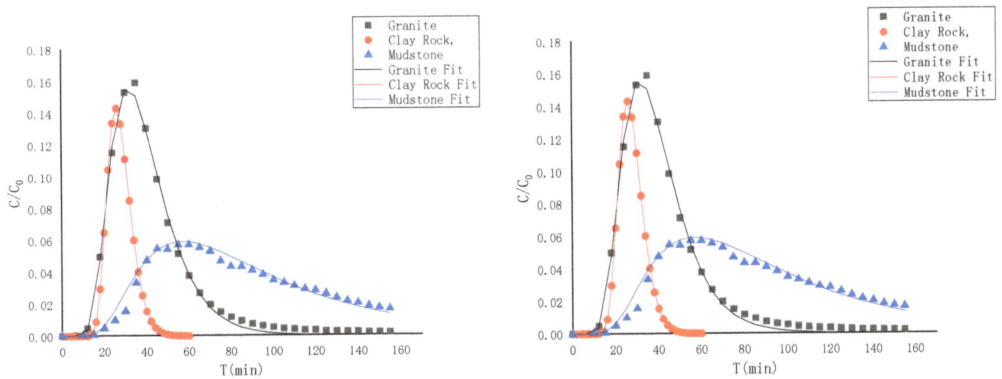

(**a**) E-T model (**b**) T-N model

Figure 13. Fitting of advection–dispersion curves of ^{99}Tc in different rock media.

Table 7. Parameters for fitting advection–dispersion curves of ^{99}Tc in different rock media.

Model	Granite Column					
E-T	V 0.254 cm/min	D 0.16 cm^2/min	D_L 0.630 cm	R 0.981	K_d <0	RMSE 1.97×10^{-5}
T-N	V 0.254 cm/min	D 0.16 cm^2/min	D_L 0.630 cm	θ_m 0.26	θ_{im} 0.11	RMSE 1.10×10^{-5}
Model	Clay rock column					
E-T	V 0.256 cm/min	D 0.055 cm^2/min	D_L 0.215 cm	R 0.707	K_d <0	RMSE 1.60×10^{-5}
T-N	V 0.256 cm/min	D 0.055 cm^2/min	D_L 0.215 cm	θ_m 0.16	θ_{im} 0.15	RMSE 1.45×10^{-5}
Model	Mudstone shale column					
E-T	V 0.286 cm/min	D 0.092 cm^2/min	D_L 0.320 cm	R 0.954	K_d <0	RMSE 1.14×10^{-5}
T-N	V 0.286 cm/min	D 0.092 cm^2/min	D_L 0.320 cm	θ_m 0.17	θ_{im} 0.18	RMSE 1.07×10^{-5}

5. Conclusions

A systematic study was conducted on the adsorption, diffusion, and advection–dispersion behavior of ^{99}Tc in granite, clay rock, and mudstone shale, with a focus on the influence of "anion repulsion" on the migration process of ^{99}Tc. The key parameters were obtained through numerical fitting to quantify their degree of influence. The research conclusions are as follows:

1. Batch experiments were conducted, revealing that granite, clay rock, and mudstone shale exhibited negligible adsorption of ^{99}Tc.

2. The anion exclusion observed in the diffusion of ^{99}Tc can impede certain narrow channels, demonstrating that a higher anion exclusion leads to a reduced effective diffusion coefficient. Conversely, the anion exclusion can enhance the diffusion rate of ^{99}Tc in wider water channels. Moreover, the anion exclusion encountered during the diffusion of ^{99}Tc is more pronounced in shale than in clay or granite.

3. ^{99}Tc is influenced by anion exclusion in the advection–dispersion process, making it challenging to achieve satisfactory fitting results using the equilibrium transport model. Therefore, employing a two-region nonequilibrium transport model becomes essential to accurately determine parameters such as porosity in the immobile zone due to anion exclusion across various media.

Author Contributions: Conceptualization, Y.S. and A.Z.; methodology, Y.S. and S.Y.; software, Y.S. and W.C.; validation, Y.S. and B.L.; formal analysis, Y.S. and Z.L.; investigation, A.Z. and S.Y.; resources, B.L. and L.W.; data curation, Y.S. and W.C.; writing—original draft preparation, Y.S. and S.Y.; writing—review and editing, Y.S. and Z.L.; visualization, Y.S. and S.Y.; supervision, L.W. and B.L.; project administration, Y.S. and B.L.; funding acquisition, Z.L. and B.L. All authors have read and agreed to the published version of the manuscript.

Funding: This project was mainly supported by the "Study on the Mechanism and Model of Multi-nuclide Transport in Granite Fracture under Competitive Adsorption". (No. 42402242), Natural Science Foundation of China, Shanxi Province Basic Research Project (No. SN22010301), and Start-up funds for doctoral research of East China University of Technology (No. 1410000434).

Institutional Review Board Statement: Not applicable for studies not involving humans or animals.

Informed Consent Statement: Not applicable.

Data Availability Statement: The raw data supporting the conclusions of this article wil be made available by the authors on request.

Acknowledgments: The experimentation conducted in this study was assisted by Chuan-Pin Lee, members of the Chemistry and Analysis Division of the Institute of Nuclear Energy Research, Taiwan. Assistance in conducting the experimental and instrumental analyses of this study was provided by the East China University of Technology, Nanchang, Jiangxi, China.

Conflicts of Interest: The authors declare no conflicts of interest.

References

1. IAEA. *IAEA Safety Standards Series No. GSG-1: Classification of Radioactive Waste*; IAEA: Vienna, Austria, 2009.
2. Sharif, A.D.; Hossam, A.G.; Vahid, D. A Comprehensive Review on Radioactive Waste Cycle from Generation to Disposal. *J. Radioanal. Nucl. Chem.* **2021**, *329*, 15–31.
3. Bruno, J.; Delos, A. *Safety Assessment of Nuclear Waste Repositories: A Radionuclide Migration Perspective*; A volume in Woodhead Publishing Series in Energy; Woodhead Publishing: Cambridge, UK, 2012; pp. 646–692.
4. Wang, X.; Liu, C.; Wang, C.; Li, C.; Chen, T. Adsorption and Diffusion of Some Important Radionuclides in Beishan Granites and Gaomiaozi Bentonites. *Sci. Sin. Chim.* **2020**, *50*, 1585–1599. [CrossRef]
5. Daniel, R. Sorption of Neptunium on Clays and Clay Minerals—A Review. *Clays Clay Miner.* **2015**, *64*, 262–276.
6. Marsily, G.D.; Julio, G.; Sophie, V. Migration mechanisms of radionuclides from a clay repository toward adjacent aquifers and the surfaceMécanismes de migration des radionucléides d'un stockage de déchets radioactifs dans l'argile vers la surface. *Comptes Rendus Phys.* **2002**, *3*, 945–959. [CrossRef]
7. IAEA. *Handbook of Parameter Values for the Prediction of Radionuclide Transfer in Terrestrial and Freshwater Environments*; IAEA: Vienna, Austria, 2010.
8. Lee, C.P.; Hu, Y.Q.; Tien, N.C.; Tsai, S.-C.; Shi, Y.; Liu, W.; Kong, J.; Sun, Y. Molecule Diffusion Behavior of Tritium and Selenium in Mongolia Clay Rock by Numerical Analysis of the Spatial and Temporal Variation. *Minerals* **2021**, *11*, 875. [CrossRef]
9. Sun, Y.Z. Study on the Adsorption and Diffusion of Nuclide ^{79}Se(IV)/^{79}Se(IV) in Tamusu Clay Rock. *East China University of Technology* **2023**.
10. Qi, L.L. The Sorption and Transport of Se(IV) and Sr(II) in Granite and the Impact Mechanism of Fracture Filling Materials. 2023.
11. Wang, K.; Zhang, R.D. Heterogeneous Soil Water Flow and Macropores Described with Combined Tracers of Dye and Iodine. *J. Hydrol.* **2011**, *397*, 105–117. [CrossRef]
12. Zaheer, M.; Zhang, W.; Zhan, H.B. An Experimental Study on Solute Transport in One-Dimensional Clay Soil Columns. *Geofluids.* **2017**, *2017*, 6390607. [CrossRef]
13. McCarter, C.P.R.; Rezanezhad, F.; Gharedaghloo, B. Transport of Chloride and Deuterated Water in Peat: The Role of Anion Exclusion, Diffusion, and Anion Adsorption in A Dual Porosity Organic Media. *J. Contam. Hydrol.* **2019**, *225*, 103497. [CrossRef]
14. Mäder, U. Advective Displacement Method for the Characterisation of Pore Water Chemistry and Transport Properties in Claystone. *Geofluids* **2018**, *2018*, 8198762. [CrossRef]
15. Van Loon, L.R.; Soler, J.M.; Jakob, A. Effect of Confining Pressure on the Diffusion of HTO, $^{36}Cl^-$ and $^{125}I^-$ in A Layered Argillaceous Rock (Opalinus Clay): Diffusion Perpendicular to the Fabric. *Geochem* **2003**, *18*, 1653–1662. [CrossRef]
16. Shi, Y.F.; Yang, S.; Chen, W.J. Study on Tritium and Iodine Species Transport Through Porous Granite: A Non-Sorption Effect by Anion Exclusion. *Toxics* **2022**, *10*, 540. [CrossRef]
17. Steigman, J.; Meinken, G.; Richards, P. The Reduction of Pertechnetate-99 by stannous Chloride-II. The Stoichiometry of the Reaction in Aqueous Solutions of Several Phosphorus(V) Compounds. *Int. J. Appl. Radiat. Isot.* **1978**, *29*, 653–660. [CrossRef]
18. He, A.D.; Fu, Z.H.; Yin, Z. The Coordination of Low-valent Re/Tc with Glutarimide Dioxime and the Fate of Tc in Aqueous Solution: Spectroscopy, ESI-MS and EXAFS. *J. Radioanal. Nucl. Chem.* **2021**, *328*, 1279–1289. [CrossRef]
19. Sheng, G.D.; Guo, Z.Q.; Yang, S.T. Application of XAFS Technique in Interface Interaction Study of Radionuclides in Environment. *Prog. Chem.* **2011**, *23*, 1455–1468.
20. Liu, C.L.; Wang, X.Y.; Gao, H.C. Diffusion Behavior of Some Weakly Absorbed Nuclide Species in Granite. *J. Nucl. Radiochem.* **2003**, *25*, 204–209.
21. Lee, C.P.; Hu, Y.Q.; Chen, D.Y. A Statistical Evaluation to Compare and Analyze Estimations of the Diffusion Coefficient of Pertechnetate ($^{99}TcO_4^-$) in Compacted Bentonite. *Minerals* **2021**, *11*, 1075. [CrossRef]
22. Tsai, T.L.; Tsai, S.C.; Shih, Y.H. Diffusion Characteristics of HTO and $^{99}TcO_4^-$ in Compacted Gaomiaozi (GMZ) Bentonite. *Nucl. Sci. Tech.* **2017**, *28*, 67. [CrossRef]
23. Poinssot, C.; Geckeis, H. *Radionuclide Behaviour in the Natural Environment. Implications and Lessons for the Nuclear Industry*; Woodhead Publishing: Cambridge, UK, 2012.

24. Šimůnek, J.; Šejna, M.; Sait, H. *The HYDRUS-1D Software Package for Simulating the One-Dimensional Movement of Water, Heat, and Multiple Solutes in Variably-Saturated Media*; Department of Environmental Sciences University of California Riverside Riverside: Riverside, CA, USA, 2018.
25. Kong, J.; Sun, Y.Z.; Hua, R. Anion Exclusion and Sorption Effect for Compacted Bentonite: The Dependency of Diffusion Coefficients and Capacity of HTO and Se(IV). *J. Radioanal. Nucl. Chem.* **2021**, *4*, 1588–2780. [CrossRef]
26. Shi, Y.F.; Lee, C.P.; Yu, H.Q. Study on Advection-dispersion Behavior for Simulation of HTO and Se Transport in Crushed Granite. *J. Radioanal. Nucl. Chem.* **2021**, *328*, 1329–1338. [CrossRef]
27. *ASTM C1733-17a*; Standard Test Method for Distribution Coefficients of Inorganic Species by the Batch Method. ASTM: West Conshohocken, PA, USA, 2017.
28. Zhang, X.Y.; Ma, F.; Dai, Z.X. Radionuclide Transport in Multi-scale Fractured Rocks: A Review. *J. Hazard. Mater.* **2021**, *424*, 1016. [CrossRef] [PubMed]
29. Shi, Y.F.; Yang, S.; Wu, E.H. Advection-Dispersion Behavior for Simulation of H-3 and Pu-238 Transport in Undisturbed Argillaceous Shale of a Near-Surface Repository. *Toxics* **2023**, *11*, 124. [CrossRef] [PubMed]
30. IAEA. *Generic Models for Use in Assessing the Impact of Discharges of Radioactive Substances to the Environment*; IAEA: Vienna, Austria, 2001.
31. Grathwohl, P. *Diffusion in Natural Porous Media: Contaminant Transport, Sorption/Desorption and Dissolution Kinetics*; Springer Science & Business Media: Berlin/Heidelberg, Germany, 1998.
32. Hemond, H.F.; Fechner, E.J. *Chemical Fate and Transport in the Environment*; Academic Press: New York, NY, USA, 1994.
33. Krupp, H.K.; Biggar, J.W.; Nielsen, D.R. Relative Fow Rates of Salt and Water in Soil. *Soil Sci. Soc. Am. Proc.* **1972**, *36*, 412–417. [CrossRef]

toxics

MDPI

Article

Occurrence and Distribution of Three Low Molecular Weight PAHs in Caño La Malaria, Cucharillas Marsh (Cataño, Puerto Rico): Spatial and Seasonal Variability, Sources, and Ecological Risk

Pedro J. Berríos-Rolón, Francisco Márquez * and María C. Cotto *

Nanomaterials Research Group, Department of Physics, Chemistry, and Mathematics, School of Natural Sciences and Technology, Universidad Ana G. Méndez-Gurabo Campus, Gurabo, PR 00778, USA; berriospl@uagm.edu

* Correspondence: fmarquez@uagm.edu (F.M.); mcotto48@uagm.edu (M.C.C.); Tel.: +1-787-743-7979 (ext. 4250) (F.M.); +1-787-743-7979 (ext. 4491) (M.C.C.)

Abstract: Polycyclic aromatic hydrocarbons (PAHs) are persistent organic pollutants with significant ecological and public health implications, particularly in urban wetlands exposed to chronic anthropogenic stress. This study evaluates the occurrence, spatial distribution, seasonal variability, and ecological risk of three low molecular weight PAHs—naphthalene (NAP), phenanthrene (PHEN), and anthracene (ANT)—in surface waters of Caño La Malaria, the main freshwater source of Cucharillas Marsh, Puerto Rico's largest urban wetland. Surface water samples were collected at four locations during both wet- and dry-season campaigns. Samples were extracted and quantified by GC-MS. NAP was the dominant compound, Σ3PAHs concentrations ranging from 7.4 to 2198.8 ng/L, with higher wet-season levels (mean = 745.79 ng/L) than dry-season levels (mean = 186.71 ng/L); most wet-season samples fell within the mild-to-moderate contamination category. Compositional shifts indicated increased levels of PHEN and ANT during the wet season. No significant spatial differences were found ($p = 0.753$), and high correlations between sites ($r = 0.96$) suggest uniform input sources. Diagnostic ratios, inter-species correlations, and principal component analysis (PCA) consistently indicated a predominant pyrogenic origin, with robust PHEN–ANT correlation ($r = 0.824$) confirming shared combustion-related sources. PCA revealed a clear separation between dry- and wet-season samples, with the latter showing greater variability and stronger associations with NAP and ANT. Ecological risk assessment using hazard quotients (HQ_{water}) indicated negligible acute toxicity risk across all sites and seasons (<0.01); the highest HQ_{water} (0.0095), observed upstream during the wet season, remained within this range. However, benchmark exceedances by PHEN and ANT suggest potential chronic risks not captured by the acute ERA framework. These findings support integrated watershed management practices to mitigate PAH pollution and strengthen long-term ecological health in tropical urban wetlands.

Keywords: polycyclic aromatic hydrocarbons; urban wetland; surface water; seasonal variability; ecological risk assessment

1. Introduction

Polycyclic aromatic hydrocarbons (PAHs) are semi-volatile organic compounds with two or more fused aromatic rings, broadly classified into low molecular weight (LMW, 2–3 rings) and high molecular weight (HMW, ≥4 rings), a distinction that governs their environmental behavior [1]. LMW-PAHs, due to their higher solubility and mobility, are generally more abundant in surface waters, whereas HMW-PAHs are particle-associated, prone to sediment accumulation, and often occur near or below detection limits [2,3]. They originate from natural and anthropogenic sources, mainly incomplete combustion of fossil fuels and biomass (pyrogenic), petroleum refining and spills (petrogenic), and, to a lesser extent, biogenic synthesis by microorganisms and plants [4,5]. PAHs are widely dispersed across atmospheric, terrestrial, and aquatic systems [6–8]. Transport processes such as atmospheric deposition, urban runoff, and wastewater discharge drive their accumulation in sediments and subsequent transfer through food webs [9–11]. Due to their persistence, hydrophobicity, and toxic properties, PAHs are classified as priority pollutants by the United States Environmental Protection Agency (USEPA) and are of global concern for their mutagenic, teratogenic, and carcinogenic effects [12,13]. These risks are especially critical in shallow urban wetlands, where stormwater inflows and limited dilution capacity exacerbate ecological exposure [14–16].

Urban wetlands, located within or near densely populated areas, function as transitional ecosystems that provide critical services such as water filtration, flood control, and habitat for aquatic organisms [17]. Because of their proximity to multiple anthropogenic emission sources, they are highly vulnerable to contamination by PAHs, receiving inputs from traffic emissions, industrial discharges, and unregulated stormwater runoff [14,18]. In this context, they act both as reservoirs and conduits for pollutants, while also serving as natural ecological filters that can decrease PAHs through microbial degradation, sedimentation, and plant uptake under favorable conditions [19].

Elevated concentrations of PAHs have been reported in urban rivers, lakes, and drainage canals across Asia, Africa, and Latin America, frequently surpassing international water quality standards [20–22]. Enclosed or semi-enclosed wetlands—such as urban marshes—often retain higher PAH loads due to restricted hydrological exchange, shallow waters, and organic-rich sediments, which enhance pollutant retention, sedimentary accumulation, and long-term ecological exposure [23,24]. However, in tropical urban wetlands—where environmental dynamics are intensified by climatic variability and human pressures—integrated assessments remain scarce, with most existing studies focused on sediments and biota [25], rather than dissolved LMW-PAHs in surface waters.

For example, coastal sediments in Trinidad showed higher concentrations during the wet season [26], while Venezuelan mangrove soils at Tucacas Bay exhibited moderate-to-high PAH levels, likely enhanced by freshwater inflows [27]. Other subtropical mangrove wetlands in Brazil reported moderate sediment contamination linked to navigation and domestic effluents, with hydrodynamic conditions shaping spatial distribution [28], whereas river sediments in Brazilian urban catchments reflected strong vehicular inputs, particularly naphthalene (NAP) dominance [29]. In Cuba, Santana et al. [30] documented LMW-PAH dominance in Almendares River surface waters, with polluted sites exceeding acute ecological risk thresholds, in contrast to the Gulf of Batabanó where Tolosa et al. [31] found low sediment levels dominated by natural organic matter sources. The present study addresses this gap by providing the first spatial and seasonal assessment of dissolved LMW-PAHs in surface waters of Puerto Rico's largest urban wetland.

Spatial and seasonal assessments of LMW-PAHs such as NAP, phenanthrene (PHEN), and anthracene (ANT) are still limited in tropical urban wetlands [32,33]. Due to their phase behavior and environmental mobility, these compounds are readily transported during rainfall events, which facilitates their dispersion and often results in higher detection frequencies in the dissolved phase compared to sediments, where they may occur at lower concentrations due to their greater solubility and lower particle affinity [34,35]. In tropical urban environments, high temperatures and intense land-use pressures may amplify their ecological risks, through seasonal microbial degradation dynamics and transient contamination pulses [36–39]. However, studies that also quantify the ecological risk associated with LMW-PAH concentrations in tropical island wetlands remain notably absent.

Caño La Malaria is a tidal-influenced urban waterway within the highly urbanized Cucharillas Marsh watershed in Cataño, Puerto Rico, which plays a critical role in connecting upland runoff with the San Juan Bay Estuary [40–42]. It receives runoff from roads, informal settlements, and industrial zones before discharging into the San Juan Bay Estuary. Prior studies have reported elevated levels of fecal coliforms, heavy metals, and untreated sewage [42–44], but PAHs have never been characterized in its surface waters. Given its location within a protected natural reserve, Caño La Malaria offers a relevant case study for evaluating the spatial and seasonal variability of these compounds in an ecologically vulnerable wetland system.

Although PAHs are widely recognized as significant aquatic pollutants, field assessments in tropical island environments remain scarce. Particularly in Puerto Rico, existing research has focused primarily on estuarine sediments and amphibian habitats [45–48], with no studies addressing the occurrence or seasonal variability of PAHs in waters of urban channels like Caño La Malaria, despite previous recommendations [49] and recent evidence of human exposure [50,51]. This urban waterway represents a critical yet understudied conduit of pollutant transfer into a protected tropical urban wetland. Within this context, this study addresses a critical knowledge gap by evaluating the occurrence, spatial distribution, seasonal trends, and potential sources of three LMW-PAHs—NAP, PHEN, and ANT—in Caño La Malaria. These compounds were prioritized due to their higher solubility and relevance as indicators of dissolved-phase contamination, while acknowledging that regulatory frameworks such as the USEPA's 16 priority PAHs also include HMW congeners more relevant to sediments and biota [52–54]. In addition, it integrates an ecological risk assessment (ERA) based on measured surface water concentrations and toxicity thresholds to estimate potential impacts on aquatic organisms. These findings support future monitoring and restoration efforts in tropical urban wetlands under similar anthropogenic pressures. This need is further emphasized by the lack of region-specific studies and long-term environmental assessments of PAH contamination in Caribbean urban wetlands, underscoring the urgency of sustained monitoring and management efforts [55].

2. Materials and Methods

2.1. Study Area and Sampling Design

Caño La Malaria is a tidal-influenced estuarine channel located in the municipality of Cataño, on the northern coast of Puerto Rico (18.4450° N, 66.1178° W). It functions as the primary hydrological conduit for surface water flow from the Cucharillas Marsh watershed into the San Juan Bay Estuary [42]. The channel receives year-round surface water inputs from upland tributaries such as the Santa Catalina and San Diego streams [56], which contribute to a continuous outflow into San Juan Bay. Its hydrological regime is artificially regulated by tide gates and a pump system located at the outlet, which limit marine–terrestrial exchange, block tidal inflow, and promote freshwater retention under seasonal rainfall conditions [40,41]. The region follows a bimodal precipitation pattern characteristic of Puerto Rico and the wider Caribbean, with a dry season (December–March), early rainfall (April–July), a brief midsummer drought, and late rainfall season (August–November) [57,58].

Surface water samples were collected at four fixed sampling sites—upstream (U), midstream (M), downstream (D), and outlet (O)—representing spatial variation from upstream headwaters to the channel's outlet. Site selection was based on proximity to stormwater discharge points, urban infrastructure, and accessibility during both dry and wet seasons. A total of four sampling campaigns were conducted in March, September, and November 2022, and February 2023, corresponding to two dry- and wet-season events. At each sampling station and campaign, duplicate surface water samples were collected. Figure 1 illustrates the geographic context of the study area, including watershed boundaries, drainage features, and the spatial distribution of surface water sampling sites along Caño La Malaria.

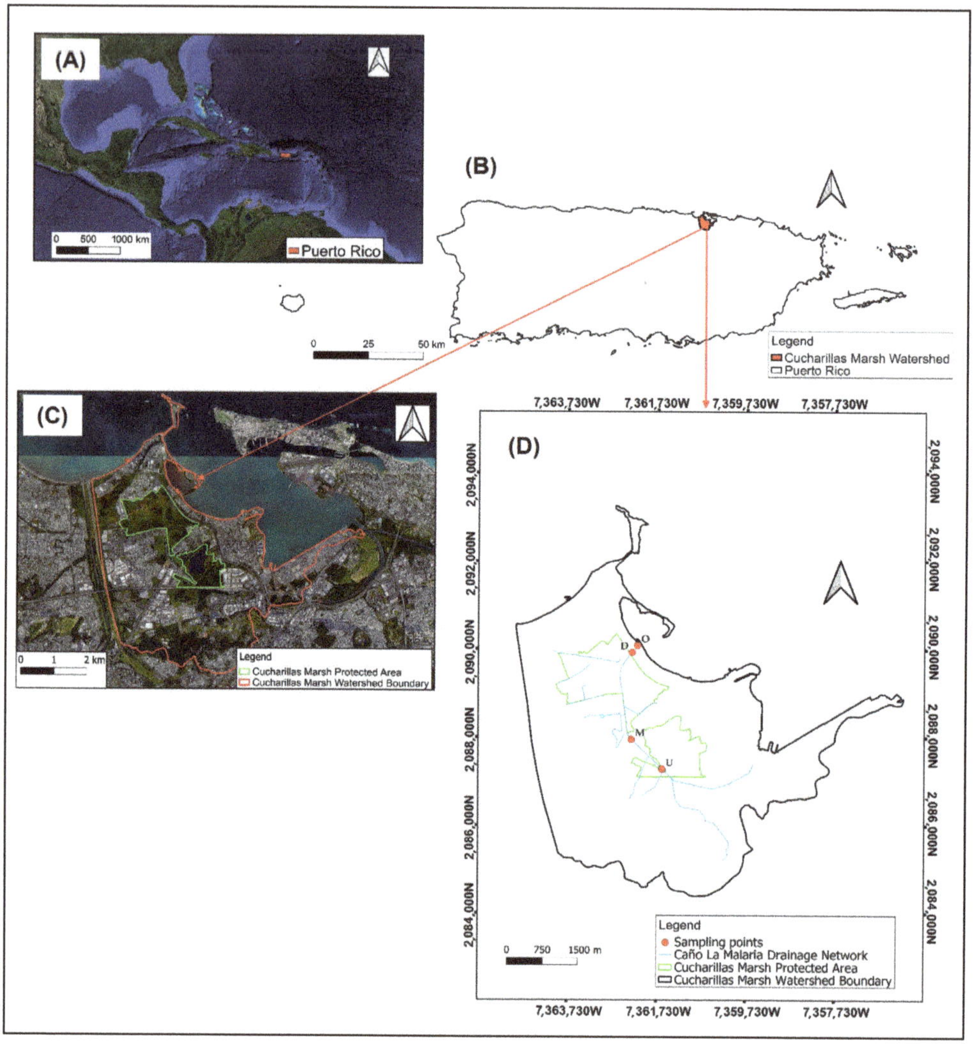

Figure 1. Geographic location and spatial context of the study area within the Cucharillas Marsh watershed, Cataño, Puerto Rico. Panel (**A**) shows the location of Puerto Rico within the Caribbean region. Panel (**B**) highlights the position of the Cucharillas Marsh watershed on the island of Puerto Rico. Panel (**C**) presents the watershed boundary (red) and protected wetland area (green) over satellite imagery. Panel (**D**) displays the drainage network of Caño La Malaria (blue), the Cucharillas Marsh Protected Area boundary (green), and the Cucharillas Marsh watershed boundary (black). Red dots indicate surface water sampling stations corresponding to U, M, D, and O segments along the Caño La Malaria system. Maps prepared using publicly available data from GIS Puerto Rico [59].

2.2. Materials and Reagents

Surface water samples were collected in pre-assembled amber wide-mouth glass bottles (LabForce®, Thomas Scientific, Swedesboro, NJ 08085, USA), equipped with PTFE-lined caps to prevent analyte adsorption and minimize photodegradation of PAHs during transport and storage. All laboratory glassware used in sample collection, extraction, and concentration was thoroughly cleaned following a multi-stage protocol. This included sequential washing with 1% (w/v) Alconox® detergent solution (White Plains, NY 10603, USA), rinsing with distilled water, repeated rinses with deionized water, acid rinsing with 50% nitric acid to remove metal and organic residues, followed by additional deionized water rinses, and a final rinse with acetone. The glassware was subsequently dried in a laboratory oven at 100 °C for a minimum of 2 h prior to use. The liquid–liquid extraction procedure employed certified HPLC-grade dichloromethane (DCM; CAS No. 75-09-2; GFS Chemicals, Powell, OH 43065, USA). Anhydrous sodium sulfate (CAS No. 7757-82-6; Thermo Fisher Scientific, Waltham, MA 02452, USA), was used to remove residual water from the organic extracts. Analytical standards of PAHs, including NAP, PHEN, and ANT, were obtained as a certified solution at 1000 μg/mL in DCM (Agilent Technologies, Santa Clara, CA 95051, USA, Method 610-compliant), and were used for calibration and quality control of gas chromatography-mass spectrometry (GC-MS) quantification procedures.

2.3. Sample Preparation and Extraction

PAHs were extracted from 1-L surface water samples within 48 h of collection to minimize degradation, using a liquid–liquid extraction adapted from the USEPA standard method [52]. Each sample was transferred to a 2000 mL separatory funnel and extracted with three portions of 60 mL of DCM. The combined organic phases were dried with anhydrous sodium sulfate, added incrementally until no residual water remained, as indicated by freely moving granules, followed by gravity filtration through Whatman #1 filter paper. The dried extract was concentrated to dryness using a rotary evaporator equipped with a water bath maintained at 30–35 °C. The residue was then reconstituted in 5 mL of DCM and transferred to amber glass vials. Extracts were stored at 4–6 °C in a laboratory-grade refrigerator and analyzed within 7 days. Each extract was injected in triplicate into the GC–MS.

2.4. GC-MS Instrumental Analysis

LMW-PAHs were quantified by GC–MS using a Shimadzu GC-2010 Plus–QP2020 system (Shimadzu, Kyoto, Japan). Separation of the target compounds was achieved using a Restek Rxi-5Sil MS capillary column (30 m × 0.25 mm i.d., 0.25 μm film thickness; Restek Corporation, Bellefonte, PA, USA) with high-purity helium (99.999%) as the carrier gas at a constant linear velocity of 43.7 cm/s. The injection was conducted in splitless mode with an injector temperature of 300 °C, and an injection volume of 1 μL was delivered using an AOC-20i autosampler (Shimadzu, Kyoto, Japan). The oven temperature program was set as follows: initial temperature of 90 °C (held for 2 min), ramped at 5 °C/min to 320 °C, with a final hold time of 12 min, for a total runtime of 60 min. The MS was operated in electron ionization (EI) mode at 70 eV, with an ion source temperature of 230 °C and an interface temperature of 300 °C. The solvent cut time was set at 3.5 min. Data acquisition was performed in Selected Ion Monitoring (SIM) mode, targeting characteristic m/z values of 128 (NAP), 178 (PHEN and ANT). Isomeric differentiation between PHEN and ANT was achieved by comparing retention times to those of certified standards.

2.5. Data Processing and Statistical Analysis

All GC-MS chromatograms were processed using the Postrun Analysis program of GC-MS solution software, version 2.6 (Shimadzu Corporation, Kyoto, Japan), to quantify the peak areas and retention times of the target PAHs. External calibration curves were constructed using certified PAH standards (NAP, PHEN, and ANT) at known concentrations in DCM. The coefficient of determination (R^2) for each compound exceeded 0.99. Quantified concentrations were compared against national and international water quality guidelines to assess the ecological significance. The Shapiro–Wilk test indicated that ∑3PAHs concentrations were not normally distributed ($W = 0.76$, $p < 0.01$, $n = 16$), supporting the use of non-parametric statistical tests. Accordingly, the Friedman test was applied to assess spatial and seasonal differences in PAH concentrations, using a significance level of $\alpha = 0.05$. All statistical analyses were performed using Microsoft Excel (Microsoft 365; Microsoft Corporation, Redmond, WA, USA).

2.6. Quality Assurance and Quality Control

External calibration curves for each PAH were prepared daily using at least seven concentration levels (0.10–100.0 ng/L), achieving high linearity ($R^2 > 0.99$). The limit of detection (LOD) and limit of quantification (LOQ) were determined for each analyte and batch using the method based on the standard deviation of the intercept divided by the slope of the calibration curve [60]. LOD and LOQ were calculated using the formulas: $LOD = 3.3 \times SE/m$ and $LOQ = 10 \times SE/m$, where SE is the standard error of the y-intercept and m is the slope of the calibration curve. All reported concentrations above LOQ were considered quantifiable. LOD values ranged from 0.02 to 0.59 ng/L, and LOQ values ranged from 0.06 to 1.80 ng/L across the four sampling campaigns (see Supplementary Table S2).

Blanks consisting of HPLC-grade DCM were injected into GC-MS throughout each run to confirm the absence of carryover and instrument background contamination. No target PAHs were detected in the blanks. Analytical precision was assessed through duplicate sample acquisition and triplicate injections in the GC-MS. Relative standard deviation (RSD) and coefficient of variation values were calculated for each sample (see Supplementary Table S3). Nondetected (ND) values in individual injections were treated as zero for mean concentration calculations.

2.7. Ecological Risk Assessment

To estimate the potential ecological risk associated with NAP in surface waters of Caño La Malaria, we applied the Hazard Quotient (HQ) method [61–64]. The HQ was calculated as follows:

$$HQ_{water} = EC_{water}/PNEC_{water} \quad (1)$$

$$PNEC_{water} = L(E)C_{50}/AF \quad (2)$$

EC_{water} represents the maximum concentration of NAP detected in surface water (ng/L). The predicted no-effect concentration in water ($PNEC_{water}$) was derived from acute toxicity data for *Callinectes sapidus* (blue crab), a native estuarine decapod crustacean widely distributed across Puerto Rico's coastal wetlands [65]. Based on the 96-h LC_{50} (lethal concentration for 50% of test organisms), the lowest observed $L(E)C_{50}$ was 0.68 mg/L, applying an assessment factor (AF) of 10, and yielding a $PNEC_{water}$ of 0.068 mg/L (68,000 ng/L) [66]. Toxicity data for NAP, were obtained from the ECOTOX database [67].

HQ_{water} values were then calculated for each site and sampling date using the maximum observed NAP concentrations to reflect worst-case exposure scenarios. Ecological risk categories were assigned following the classification criteria proposed by Li et al. [64]. The categories used are summarized in Table 1.

Table 1. Ecological risk classification based on HQ_{water} values.

HQ_{water} Range	Ecological Risk Category
$HQ_{water} < 0.01$	Insignificant
$0.01 \leq HQ_{water} < 0.1$	Low
$0.1 \leq HQ_{water} < 1.0$	Moderate
$HQ_{water} \geq 1.0$	High

3. Results

3.1. Spatial and Seasonal Variability in PAH Concentrations

Table 2 presents the concentrations of NAP, PHEN, and ANT across four seasonal sampling campaigns and four stations along Caño La Malaria. PAH concentrations exhibited marked variability, with higher levels recorded during the wet season (September and November 2022) compared to the dry season (March 2022 and February 2023). For instance, Σ3PAHs during the dry season ranged from 7.40 ng/L (U, March) to 363.11 ng/L (O, February), while wet season values spiked up to 929.57 ng/L (M, September) and peaked at 2198.83 ng/L (U, November), driven by extreme ANT concentrations (up to 1313.60 ng/L). NAP was the most abundant compound, reaching 485.10 ng/L (U, September), although its dominance shifted to ANT in November. PHEN varied broadly (ND–557.43 ng/L), with notable peaks at M and U sites. The O consistently registered Σ3PAHs between 43.43–712.54 ng/L across all campaigns, underscoring its role as the final receptor before discharge into San Juan Bay.

Table 2. Mean concentrations (±standard deviation) and concentration ranges of NAP, PHEN, and ANT measured in surface water from four sites along Caño La Malaria, Puerto Rico, during four seasonal sampling campaigns conducted between March 2022 and February 2023.

Date	Site	NAP Mean (ng/L) ± SD	Range	PHEN Mean (ng/L) ± SD	Range	ANT Mean (ng/L) ± SD	Range	Σ3PAHs (ng/L)
March 2022	U	6.20 ± 0.00	ND–6.20	1.20 ± 0.00	ND–1.20	ND	ND	7.40
	M	30.13 ± 28.46	10.00–50.25	8.10 ± 1.91	6.75–9.45	ND	ND	38.23
	D	30.50 ± 26.38	11.85–49.15	1.25 ± 0.00	ND–1.25	0.50 ± 0.00	ND–0.50	32.25
	O	34.73 ± 32.99	11.40–58.05	8.70 ± 0.00	ND–8.70	ND	ND	43.43
September 2022	U	485.10 ± 231.93	321.10–649.10	135.87 ± 48.11	85.65–181.55	203.40 ± 0.00	114.50–292.30	824.37
	M	337.60 ± 32.39	308.60–372.55	364.72 ± 47.34	321.55–415.35	227.25 ± 109.39	149.90–304.60	929.57
	D	232.12 ± 29.02	203.80–261.80	240.47 ± 33.11	218.50–278.55	171.95 ± 116.67	100.75–306.60	644.54
	O	281.62 ± 67.63	230.50–358.30	297.17 ± 81.43	224.35–385.10	133.75 ± 36.98	108.90–176.25	712.54
November 2022	U	327.80 ± 183.92	197.75–457.85	557.43 ± 0.00	554.05–560.80	1313.60 ± 0.00	1285.15–1342.05	2198.83
	M	41.48 ± 38.36	14.35–68.60	183.30 ± 167.02	65.20–301.40	301.08 ± 383.29	30.05–572.10	525.86
	D	34.35 ± 0.99	33.65–35.05	5.00 ± 0.49	4.65–5.35	27.68 ± 3.22	25.40–29.95	67.03
	O	6.98 ± 4.56	3.75–10.20	37.13 ± 12.48	28.30–45.95	19.48 ± 2.02	18.05–20.90	63.59
February 2023	U	223.66 ± 185.24	43.00–399.15	183.49 ± 191.92	5.60–359.70	213.04 ± 213.39	2.25–412.20	620.19
	M	43.76 ± 1.29	41.75–45.75	46.04 ± 9.44	31.95–54.90	20.57 ± 17.34	2.45–39.10	110.37
	D	26.73 ± 20.12	<LOQ–52.80	165.11 ± 61.73	115.35–255.00	86.87 ± 15.61	72.20–112.95	278.71
	O	16.12 ± 11.70	3.60–28.80	228.98 ± 24.01	204.50–251.00	118.01 ± 4.33	112.30–123.50	363.11

Sites: upstream (U); midstream (M); downstream (D), and outlet to San Juan Bay (O). Values in the first column of each PAH represent mean ± standard deviation across replicates for each site and date. ND: Not detected; below limit of detection (LOD). <LOQ–xx: Range of replicates where at least one result was <LOQ and the maximum replicate yielded the stated value (xx). Σ3PAHs = sum of NAP, PHEN, and ANT concentrations. Range = minimum–maximum concentrations measured per site.

Figure 2 presents the seasonal and spatial distribution of Σ3PAHs concentrations in surface waters of Caño La Malaria. Panel A illustrates that mean Σ3PAHs concentrations were elevated during the wet season, reaching 777.75 ng/L in September and 713.82 ng/L in November, in contrast to lower values observed in the dry season—31.44 ng/L in March and 343.09 ng/L in February. Panel B further underscores this seasonal pattern, showing that overall mean Σ3PAHs concentrations during the wet season (745.79 ng/L) were more than threefold higher than those recorded during the dry season (186.71 ng/L). Panel C depicts the spatial variation across the four sampling sites, with site U consistently exhibiting the highest concentrations, particularly in November 2022, whereas sites D and O showed persistently lower levels throughout the sampling period.

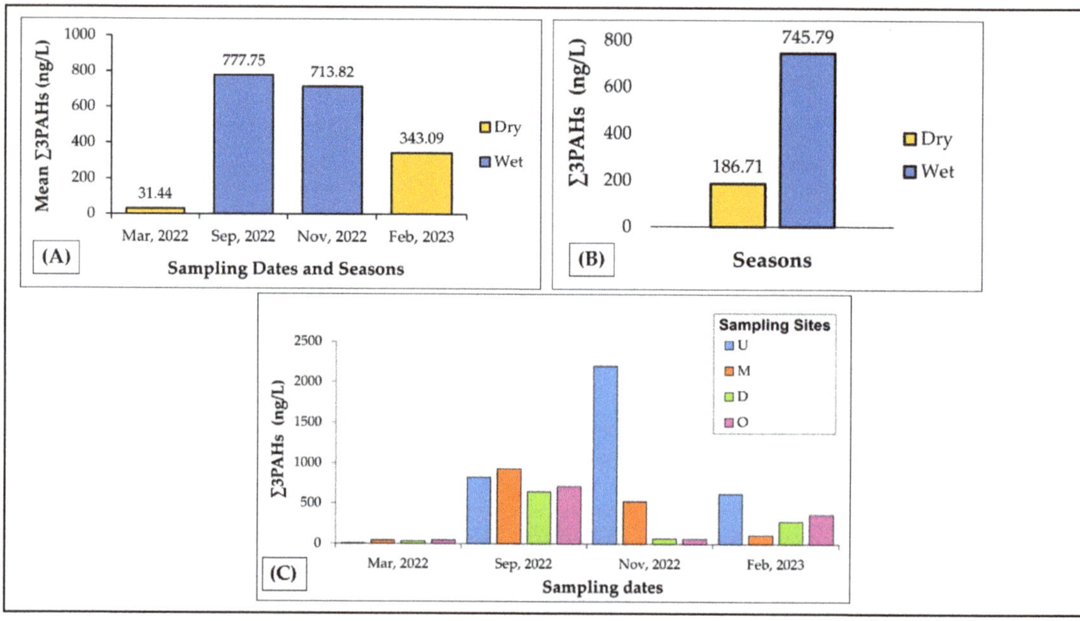

Figure 2. Seasonal and spatial variation of total LMW-PAHs (Σ3PAHs: NAP, PHEN, and ANT) in surface water samples collected from Caño La Malaria. (**A**) Mean Σ3PAHs concentrations (ng/L) per sampling campaign. (**B**) Comparison of overall mean Σ3PAHs concentrations by season. Overall means were calculated as the arithmetic average of all Σ3PAHs values measured across the four sampling sites within each season (n = 8 values per season). (**C**) Spatial and seasonal distribution of Σ3PAHs across four sampling sites (U, M, D, O) and four campaigns.

A Friedman test confirmed seasonal differences in Σ3PAHs concentrations across the four sampling campaigns ($\chi^2 = 21.00$, $p = 4.59 \times 10^{-6}$), supporting the trend of elevated PAH levels during the wet season. In contrast, no significant differences were found among the sampling sites ($\chi^2 = 1.20$, $p = 0.753$), suggesting a homogeneous spatial distribution of PAHs across Caño La Malaria, despite marked seasonal variability.

3.2. PAH Sources and Compositional Patterns

3.2.1. Compositional Patterns

Figure 3 illustrates the relative percentage composition of the three LMW-PAHs across all sampling sites and dates. During the first sampling campaign (March 2022), NAP dominated the ∑3PAHs composition at all sites, representing 80–100% of the total concentration. PHEN contributed at sites M and O (16–20%), while ANT was either absent or accounted for less than 10%. By September 2022, the compositional profile became more balanced, especially at sites U and M, where NAP accounted for approximately 59% and 36%. PHEN and ANT showed increased contributions at these sites during the wet season (e.g., PHEN: 16–39%; ANT: 25–24%).

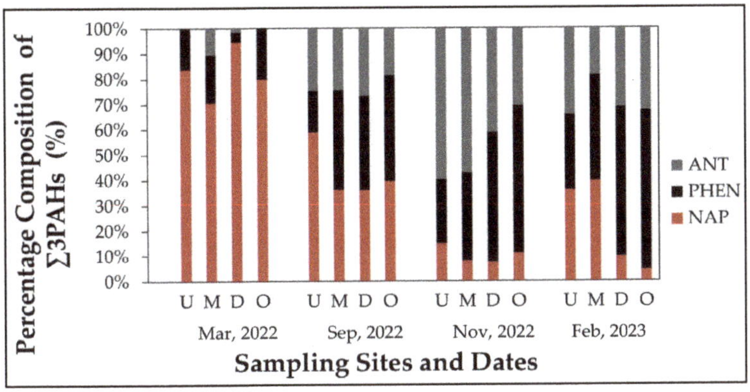

Figure 3. Percentage composition of individual LMW-PAHs—NAP, PHEN, and ANT—in surface water samples collected from Caño La Malaria. Data is shown for four sites: U, M, D, O, across four sampling campaigns. Each bar represents the relative contribution (%) of each PAH to the total Σ3PAHs concentration.

In November 2022, ANT became the dominant compound at all sites, notably at U and M, where it represented approximately 60% and 57% of ∑3PAHs. PHEN also remained elevated (25–58%), while NAP accounted for less than ~20% of the total. In February 2023, PHEN was the most abundant PAH at sites D and O, contributing approximately 59% and 63% of ∑3PAHs, followed by ANT with 31% and 32%. NAP contributions were highly variable: low in November and February (6–15%), but much higher in March (70–98%) and September (28–59%). Across all campaigns, NAP represented between 4% and 98% of Σ3PAHs, dominating during the March 2022 campaign. PHEN and ANT showed greater seasonal and spatial variability, with ANT being notably higher during the peak wet season and PHEN becoming more dominant at downstream sites in the final sampling campaign.

3.2.2. Sources by Diagnostic Ratios

PAH source apportionment was evaluated using two diagnostic ratios: ANT/(ANT + PHEN) and PHEN/ANT. The former distinguishes petrogenic sources (<0.10) from pyrogenic ones (>0.10), while the latter indicates pyrogenic inputs when values are below 10. As illustrated in Figure 4, all ANT/(ANT + PHEN) values exceeded the 0.10 threshold, indicating a consistent pyrogenic signature across Caño La Malaria's surface waters. While PHEN/ANT ratios were also below 10, with most values under 2.50. Elevated values in both ratios aligned with periods of higher Σ3PAHs concentrations. In March 2022, anthracene was not detected at sites U and O (Table 2), which prevented the calculation of diagnostic ratios requiring both PHEN and ANT. For this reason, ratios are only interpreted for sites and campaigns where both compounds were quantifiable.

3.2.3. Sources by Principal Component Analysis

Principal Component Analysis (PCA) is a widely applied multivariate approach in PAH source apportionment, as it reduces complex datasets into a few principal components that explain most of the variance and reflect groups of correlated compounds [68]. The biplot of PC 1 and PC 2 (Figure 5), which together accounted for 96.7% of the total variance (84.8% and 11.9%, respectively), revealed a separation between samples collected during the dry and wet seasons. Dry-season samples clustered tightly in the lower-left quadrant, reflecting uniform PAH compositions and lower mean concentrations. In contrast, samples from the wet season exhibited greater dispersion and were oriented along the NAP and ANT vector directions. Component loadings (Table S9) indicate that ANT had the strongest influence on PC1 (loading = 0.87794), followed by PHEN and NAP. Meanwhile, PC2 was primarily influenced by NAP (loading = 0.89129).

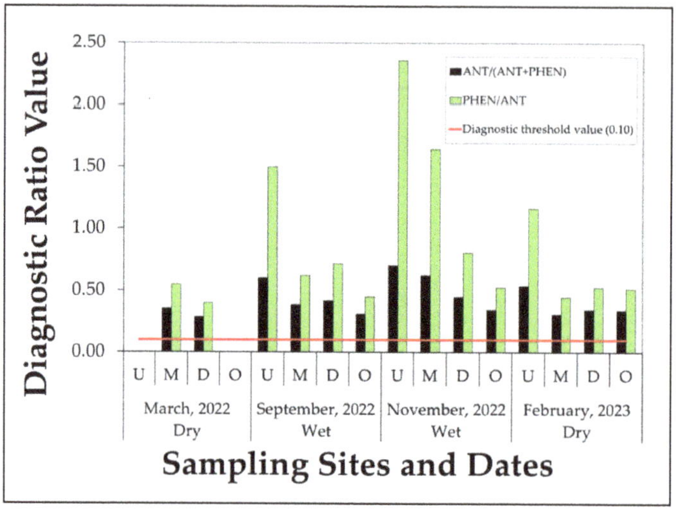

Figure 4. Diagnostic ratios of PAHs used for emission source attribution in surface water samples from Caño La Malaria, collected at four sites—U, M, D, O—during the seasonal campaigns (March, September, November 2022, and February 2023). Bars represent the value generated by the diagnostic ratios used: ANT/(ANT + PHEN) (black) and PHEN/ANT (green). The red line indicates the threshold value of 0.10 for ANT/(ANT + PHEN), above which pyrogenic sources are inferred. PHEN/ANT values below 10 also suggest a pyrogenic origin.

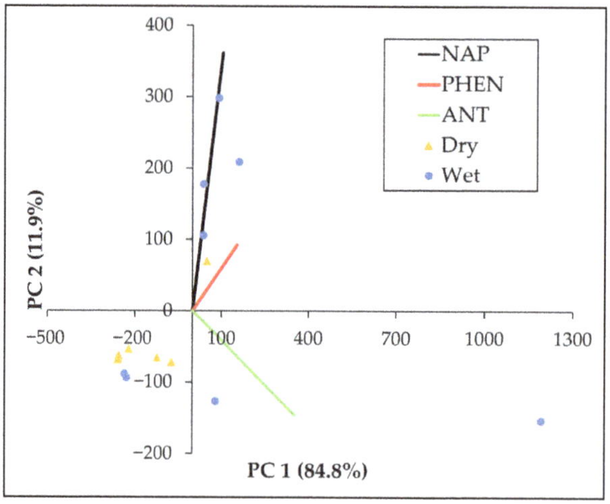

Figure 5. PCA biplot of surface water samples from Caño La Malaria based on concentrations of three LMW-PAHs: NAP, PHEN, and ANT. The plot displays the distribution of 16 samples grouped by season (Dry vs. Wet) along the first two principal components (PC 1 and PC 2), which explain 84.8% and 11.9% of the total variance, respectively. Vectors represent the directional contribution of each PAH to the multivariate pattern. Samples from the wet season exhibit greater variability and are more strongly associated with NAP and ANT concentrations.

3.3. Correlation Patterns

Pearson correlation analysis between sampling sites supported the spatial and seasonal patterns observed. A positive correlation was found between sites U and M ($r = 0.964$), while moderate associations between U–O and M–O suggest similar contamination sources or hydrological connectivity among most sites. In contrast, weak or negative correlations with D indicate local variability, driven by distinct inputs or site-specific environmental processes (Supplementary Table S7). Inter-species correlation analysis also revealed strong relationships between individual PAHs. In particular, PHEN and ANT were strongly correlated ($r = 0.824$), while NAP showed moderate correlation with both compounds (Supplementary Table S8).

3.4. Results of Ecological Risk Assessment

HQ_{water} values were calculated for each site and sampling event using the maximum observed NAP concentrations. Table 3 summarizes the HQ_{water} values calculated for each site–season combination. All HQ_{water} values were below 0.01 (Table 1), ranging from 9.12×10^{-5} to 9.55×10^{-3}, indicating insignificant ecological risk across the study area and seasonal sampling campaigns (Table 3). The highest HQ_{water} value (9.55×10^{-3}), was observed at the upstream site during the September 2022 sampling campaign—corresponding to the wet season and the peak NAP concentrations. Although, this value approached the upper limit of the "insignificant" ecological risk category, it did not cross into the "low ecological risk" range.

Table 3. Hazard Quotient Summary by Site and Sampling Date. PNEC value: 68,000 ng/L, based on acute toxicity data for *Callinectes sapidus* [67].

Date	Site	Max. NAP (ng/L)	HQ_{water} (ng/L)	Ecological Risk Classification
March 2022	U	6.20	9.12×10^{-5}	Insignificant
	M	50.25	7.39×10^{-4}	Insignificant
	D	49.15	7.23×10^{-4}	Insignificant
	O	58.05	8.54×10^{-4}	Insignificant
September 2022	U	649.10	9.55×10^{-3}	Insignificant
	M	372.55	5.48×10^{-3}	Insignificant
	D	261.80	3.85×10^{-3}	Insignificant
	O	358.30	5.27×10^{-3}	Insignificant
November 2022	U	457.85	6.73×10^{-3}	Insignificant
	M	68.60	1.01×10^{-3}	Insignificant
	D	35.05	5.15×10^{-4}	Insignificant
	O	10.20	1.50×10^{-4}	Insignificant
February 2023	U	399.15	5.87×10^{-3}	Insignificant
	M	45.75	6.73×10^{-4}	Insignificant
	D	52.80	7.76×10^{-4}	Insignificant
	O	28.80	4.24×10^{-4}	Insignificant

Spatial patterns showed that HQ_{water} values were higher at upstream locations (U and M), particularly during wet seasons. In contrast, downstream sites (D and O), consistently exhibited lower HQ_{water} values. Seasonally, wet season events (September and November) show elevated HQ_{water} relative to dry season campaigns (March and February). However, none of the HQ_{water} values exceeded the threshold for "low risk".

4. Discussion

4.1. Influence of Seasonal Hydrology on PAH Dynamics

The pronounced seasonal variation in Σ3PAHs concentrations observed in Caño La Malaria (mean = 745.79 ng/L during the wet season vs. 186.71 ng/L in the dry season) underscores the dominant role of rainfall-driven hydrological pulses in mobilizing PAHs within tropical urban wetlands. This enrichment during the wet period is consistent with findings from other tropical and subtropical freshwater systems [16,69,70], where elevated PAH concentrations were linked to runoff-driven inputs during rainy seasons. Across these systems, LMW-PAHs—particularly NAP, PHEN, and ANT—were consistently dominant, reflecting the compositional pattern found in our study.

Contrasting seasonal trends have been reported. For example, Na et al. [71] and Jiang et al. [72] observed higher PAH concentrations during the dry season in the East Liao River and coal-mining-impacted groundwater systems, attributing the lower wet-season levels to dilution and hydrodynamic flushing. Similar dry-season peaks were reported in the Olt River [73] and the Han River [74], suggesting that regional climate regimes, flow variability, and land-use characteristics can lead to divergent seasonal behaviors. These patterns suggest that regional climate, hydrology, and land use drive seasonal PAH dynamics, with the extensive impervious cover and limited riparian vegetation in Caño La Malaria amplifying runoff-driven transport of LMW-PAHs during intense tropical rainfall.

The predominance of LMW-PAHs—especially NAP—during the wet season in Caño La Malaria reflects a trend reported in other tropical aquatic systems, including the Subarnarekha Estuary [75], Gaoqiao wetland sediments [76], and the Sombreiro River Estuary [77]. Elevated temperatures during warmer months may further enhance the dissolution and water-phase partitioning of lighter PAHs [78], contributing to the seasonal increase in aqueous concentrations observed in our study. Dominance of LMW-PAHs has also been reported in highly industrialized watersheds, such as the Kolo Creek [79] and the Kuye River [80], reinforcing their utility as indicators of recent pyrogenic input and their susceptibility to hydrological mobilization. In Caño La Malaria, Σ3PAHs concentrations reached up to 2198.8 ng/L—exceeding values reported in many comparable tropical freshwater systems—highlighting the effects of sustained anthropogenic pressure, deficient stormwater infrastructure, and direct runoff inputs in this densely urbanized wetland catchment.

Compared to other global wetland systems, PAH levels in Caño La Malaria are elevated. In the Anzali Wetland, Iran, total PAH concentrations (Σ16PAHs) in surface waters ranged from 5.14 to 253.37 ng/L, with a mean of 78.31 ng/L, dominated by LMW congeners [81], values that overlap with our measured Σ3PAHs despite the narrower compound scope of this study. In the Hoor Al-Azim Wetland, Σ11PAHs ranged from 15.3 to 160.15 ng/L [82], while in the Shadegan Wetland, Σ16PAHs ranged from 42 to 136 ng/L with a mean of 78 ng/L, indicating low to slightly polluted conditions [83]. All of these reported values remain well below our recorded peak concentration of 2198.8 ng/L for Σ3PAHs in Caño La Malaria. These comparisons reveal significant human impact in Cucharillas Marsh and support the need for pollution control strategies.

In summary, seasonal hydrology plays a pivotal role in shaping PAH dynamics in Caño La Malaria, primarily by facilitating the mobilization and aqueous partitioning of LMW-PAHs during periods of intense rainfall. These findings not only corroborate patterns of rainfall-driven PAH mobilization observed in other tropical systems [32,84], but also contribute novel data from a Caribbean urban wetland. While variability across studies reflects differences in regional hydrology, emission profiles, and land use, the strong influence of wet-season runoff on the transport of pyrogenic and petrogenic PAHs emerges as a consistent feature of tropical and subtropical aquatic systems. It should be noted that these reported values correspond to Σ11–16PAHs, whereas our findings are limited to three LMW compounds (NAP, PHEN, ANT); thus, the comparison is illustrative of relative seasonal dynamics rather than magnitude-equivalent.

4.2. Spatial Distribution and Attenuation of PAHs

Although the Friedman test revealed no statistically significant differences in Σ3PAHs concentrations across the four sampling sites (p = 0.753). Elevated concentrations at U and M sites—particularly during high-rainfall events—suggest local inputs from adjacent roadways and potential illicit discharges, in addition to diffuse runoff [38,85]. In contrast, D and O sites consistently showed lower concentrations, likely due to hydrological dilution, sorption to suspended particulates, and gradual deposition along the canal's flow path [86].

This U-to-D gradient aligns with patterns observed in other urban aquatic systems. For example, elevated PAHs have been reported in the Imiringi River due to vehicular emissions and localized combustion [87], and along the Nile River downstream of wastewater and industrial discharge zones [68]. Similar upstream enrichment has also been observed in the Lipu River [88], the Damodar River Basin [89], and the Guanzhong River in the Danjiangkou Reservoir, where fossil fuel combustion dominates [64]. These spatial trends are further corroborated by findings in Ho Chi Minh City and the East Liao River, where PAH concentrations were higher in densely urbanized, high-traffic areas [70,71]. A consistent pattern emerges in which upstream and highly urbanized segments function as hotspots of PAH contamination, whereas downstream reaches reflect the combined influences of hydrological dilution, sedimentation, and pollutant attenuation, particularly within aquatic coastal environments.

Beyond hydrological controls, land-use patterns exert influence on the spatial distribution of PAHs in Caño La Malaria. As demonstrated in the Sombreiro Estuary and Kuye River, where extreme PAH concentrations were detected near oil activity [77,80]. Although Σ3PAHs in Caño La Malaria were lower than in oil-impacted rivers, the system still exhibited spatial heterogeneity linked to anthropogenic pressure. This pattern parallels observations in Baiyangdian Lake, where coal and biomass combustion contributed to PAH variability modulated by both source proximity and hydrological retention [90]. Comparable trends in Dong and Tangxun Lakes, Wang Lake Wetland, and Brazilian estuarine systems further support that spatial PAH heterogeneity in urbanized tropical wetlands is shaped by a combination of land use and flow dynamics [91–93]. These studies indicate that the spatial heterogeneity observed in Caño La Malaria reflects broader patterns in which land-use intensity and hydrological processes modulate PAH distributions in urbanized wetland systems.

In summary, the spatial distribution of Σ3PAHs in Caño La Malaria reflects localized contamination at U and M sites, driven by road runoff, stormwater discharges, and surrounding land use. Although concentrations gradually decline toward D and O sites—suggesting attenuation through hydrological dilution, sedimentation, or sorption—elevated upstream levels, particularly during the wet season, highlight the persistent influence of urban inputs near the canal headwaters. These findings may inform the prioritization of upstream pollution controls and stormwater management strategies in similarly urbanized wetland systems.

4.3. Source Apportionment of PAHs

Compositional profiles of PAHs in Caño La Malaria revealed distinct seasonal trends. NAP was the most abundant compound during the dry season, while PHEN and ANT increased notably during the wet season. This is consistent with our observations, where NAP contributed up to 98% of Σ3PAHs in March 2022, while PHEN and ANT together accounted for more than 70% at several sites in September and November. These shifts reflect changes in transport mechanisms: volatilization and solubility-driven movement dominate in dry periods [94], whereas particle-bound inputs and surface runoff prevail during rainfall events [16]. Diagnostic ratios [ANT/(ANT + PHEN) > 0.10; PHEN/ANT < 10] consistently pointed to pyrogenic sources, implicating vehicular emissions and biomass combustion [95]. The correlation between PHEN and ANT ($r = 0.824$) reinforces this common origin and behavior under combustion-influenced regimes. In late-season samples (February 2023), PHEN overtook ANT as the predominant PAH at downstream sites, indicating shifting dominance even within the wet-to-dry seasonal transition. Seasonal compositional shifts observed in Caño La Malaria suggest the combined influence of transport mechanisms and combustion-derived sources.

PCA was applied to differentiate seasonal variability in PAH profiles and support source apportionment. PC1 was primarily influenced by ANT, followed by PHEN and NAP, together explaining 96.7% of the total variance. ANT and PHEN are associated with pyrogenic emissions from industrial and residential combustion processes [96], while NAP is often considered a general marker of petrogenic inputs such as oil spills and petroleum leaks [81,97]. Comparable PCA-based approaches in other riverine systems have similarly identified combustion-derived PAHs as major drivers, with additional contributions from petroleum-related inputs depending on local activities and hydrological settings [38,87]. In Caño La Malaria, the loadings of ANT and NAP suggest pyrogenic and petrogenic influences, in which the combined impact of combustion and petroleum sources influence PAH composition.

NAP dominated PC2 in our analysis, a pattern consistent with studies that linked strong NAP loadings to petroleum-derived inputs such as oil leaks and refined fuel residues [76,81]. Grmasha et al. [68] further emphasized that NAP is a major constituent of diesel fuels and gasoline, produced through incomplete combustion [98], and in their reported PCA, NAP appeared as a dominant component of PC1. The presence of NAP in PC2 indicates petroleum-derived sources, while the contributions of ANT and PHEN in PC1 reflect combustion markers, together accounting for the observed seasonal variations in PAHs. Overall, these findings demonstrate that PCA not only separates dry and wet season profiles but also highlights the dual influence of combustion and petroleum-related sources in shaping PAH dynamics in Caño La Malaria.

These diagnostic ratios, compositional profiles, and PCA-derived source attributions are consistent with broader reports from tropical and subtropical aquatic systems. In the Damodar River Basin and the East Liao River, 3- and 4-ring PAHs—including PHEN, fluoranthene, and pyrene—were associated with coal combustion and urban activity [71,89]. Similarly, PHEN and NAP were dominant in the Nile River and Ho Chi Minh City, reflecting urban runoff and motor vehicle emissions [70,99]. NAP was the most abundant PAH in the Danjiangkou Reservoir and the Taige Canal, attributed to petroleum combustion and diesel emissions, closely to the profiles observed in Caño La Malaria [64,100]. The compositional similarities between Caño La Malaria and other urban aquatic systems reveal that combustion-derived inputs are the dominant and persistent drivers of PAH behavior in this wetland.

Urban infrastructure facilitates the mobilization of combustion-derived PAHs, particularly in densely populated tropical catchments. Roads and stormwater systems promote the transfer of particle-bound compounds into aquatic systems, a dynamic also observed near PR-5 and PR-165 sampling sites in our study area [101]. This process is further amplified during the wet season, when increased runoff remobilizes PAHs previously adsorbed to soils and sediments [102]. Similar seasonal enrichment of LMW-PAHs—especially NAP—in urban watersheds suggest the influence of rainfall-driven inputs on PAH dynamics [103,104]. These patterns align with our findings in Caño La Malaria, where combustion-derived LMW-PAHs consistently dominated across both dry and wet seasons. The persistence of pyrogenic markers throughout the study period indicates chronic anthropogenic pressure from vehicular traffic, industrial activity, and stormwater discharge [38]. Altogether, the compositional and multivariate patterns observed in Caño La Malaria underscore the predominance of combustion-derived PAHs, with episodic petroleum inputs amplified by rainfall and stormwater infrastructure. These dynamics highlight the vulnerability of this tropical urban wetland to chronic anthropogenic pressure and establish the importance of source-specific management strategies.

4.4. Risk Implications and Management Recommendations

The concentrations of PAHs measured in Caño La Malaria frequently exceed established national and international environmental guidelines. The maximum concentrations for NAP (0.649 μg/L) remained below its benchmark of 1.1 μg/L, while PHEN (0.557 μg/L) and ANT (1.314 μg/L) notably exceeded their respective freshwater screening benchmarks (0.4 and 0.012 μg/L) [105]. Although widely used classification frameworks define ∑PAH contamination levels as slight (0–100 ng/L), mild (100–1000 ng/L), moderate (1000–5000 ng/L), and severe (>5000 ng/L) [106,107], these thresholds are based on the full suite of 16 priority PAHs and therefore cannot be directly applied to our dataset of three LMW PAHs. Nevertheless, the maximum concentrations observed here (up to 2198.8 ng/L) fall within the numerical range that in other systems corresponds to mild-to-moderate contamination, suggesting that our partial dataset may still reflect ecologically relevant pollution levels. When compared to chronic freshwater quality criteria, PHEN concentrations exceeded the 0.3 μg/L benchmark, reaching up to 0.561 μg/L, while NAP and ANT remained below their respective thresholds of 1 μg/L and 4 μg/L [108]. According to EU guidelines for inland surface waters, NAP concentrations did not exceed the threshold of 2000 ng/L, whereas ANT surpassed the 100 ng/L limit, particularly during the wet season [109,110]. Several wet season samples also exhibited ANT concentrations well above the 120 ng/L threshold set for marine organisms, with maximum values reaching 1342.05 ng/L [111]. These elevated concentrations indicate potential ecological risk under multiple regulatory frameworks, primarily reflecting chronic or sublethal protection thresholds established by international guidelines, and therefore do not necessarily imply acute toxicity under present conditions.

Beyond regulatory benchmarks, the ecological significance of the observed concentrations is further supported by thresholds derived from the target lipid model [112] and guidelines for human health protection [113]. Σ16PAHs concentrations in the Euphrates River ranged from 464 to 992 ng/L, with higher levels downstream of urban areas [68]. In Dong and Tangxun Lakes (China), Yao et al. [93] reported Σ16PAHs ranging from 42.9 to 434.7 ng/L. Σ3PAHs in Caño La Malaria exceeded 2000 ng/L, even though only three compounds were analyzed, compared to the full set of USEPA priority Σ16PAHs in other studies [68]. However, it should be noted that while the cited studies reported Σ16PAHs, our findings are based on three LMW compounds. Thus, the comparison is not magnitude-equivalent but rather illustrative of relative enrichment in dissolved-phase PAHs in tropical urban wetlands. Also, surface water concentrations often exceeded 1000 ng/L, likely due to the canal proximity to urban areas upstream [114,115], shallow depth [116], organic-rich sediments [117], and hydrological connectivity [118] to the San Juan Bay Estuary. Together, these comparisons reinforce that current environmental benchmarks may underestimate site-specific risks in tropical urban wetlands and highlight the need for more comprehensive water quality standards for PAHs.

Effective mitigation of PAH pollution in Caño La Malaria requires a watershed-scale management approach that integrates urban planning, regulatory oversight, and ecosystem-based solutions. In highly urbanized watersheds, persistent PAH contamination has been linked to vehicular emissions and stormwater runoff, which are also prominent sources in the Cucharillas Marsh area [70]. Nevertheless, despite improvements in water quality, legacy pollutants and diffuse PAH sources remain challenging to control without long-term strategies [88]. Among recommended strategies, nature-based solutions offer promising low-cost and sustainable alternatives [119]. Constructed wetlands have proven effective in attenuating PAHs through plant uptake, microbial degradation, and sedimentation [19]. Complementary stormwater management strategies, such as bio-swales [120] and infiltration basins [121], can further reduce runoff-borne PAHs before they enter the aquatic system. When implemented as small ponds or detention features along tributaries or upstream stormwater channels draining into Caño La Malaria, these interventions could intercept runoff during high-intensity rainfall events and significantly mitigate PAH inputs to the canal.

Findings from Caño La Malaria reveal that partial compliance with existing benchmarks does not guarantee ecological safety, because the observed concentrations reveal potential ecological risks that existing benchmarks may underestimate. While our study was limited to three parent LMW-PAHs, we note that traditional PAH assessments focusing solely on parent compounds may underestimate total toxicity in aquatic systems by excluding substituted derivatives, such as nitrated and oxygenated PAHs, which the literature shows can occur at higher concentrations and exhibit greater toxicity than their parent analogues [96]. Thus, expanding the analytical scope to include these compounds in future studies—together with applying updated classification systems for PAH contamination [107,122]—would provide a more comprehensive assessment of ecological risk. In parallel with improved monitoring, management interventions should not be limited to isolated hotspots [123,124]. Instead, efforts should also prioritize upstream pollution controls [125], the restoration of riparian buffers [126], and enhancements to stormwater infrastructure [127]. Ensuring long-term resilience in Cucharillas Marsh will require integrated monitoring and watershed-scale actions—rather than hotspot-only interventions—to mitigate chronic PAH pollution and protect ecological and hydrological functions.

4.5. Ecological Effects

The ecological risk associated with NAP in Caño La Malaria was found to be negligible across all sampling sites and seasons, with HQ_{water} values ranging from 9.1×10^{-5} to 9.5×10^{-3}. These values fall below the 0.01 threshold for "insignificant risk" suggesting negligible acute ecological threat under current environmental conditions. Even the highest HQ_{water}—recorded at the upstream site during the wet season (September 2022)—remained within the insignificant risk range. These findings are consistent with assessments conducted in subtropical systems, where PAH-related risks increased during wet-season runoff events due to pollutant mobilization and elevated concentrations [64]. Similarly, in coastal environments like tidal creeks, hydrological pulses during wet periods facilitated contaminant transport and led to higher ecological risk values [62].

Although HQ_{water} values in Caño La Malaria were consistently low, a clear spatial gradient was observed, with higher values upstream than downstream. This pattern aligns with findings from other river systems, where elevated upstream ecological risks are attributed to limited dilution capacity and closer proximity to pollutant sources [71]. Similarly, PAH-related risks in urbanized tidal creeks influenced by industrial and domestic discharges underscore the role of human activities in amplifying contaminant exposure [62]. Upstream sectors of urban wetlands serve as key vulnerability hotspots, where targeted ecological monitoring and management efforts are needed [128]. The elevated HQ_{water} values observed at the upstream site correspond to the most urbanized section of the watershed, located in the upper reach of Caño La Malaria, where dense residential development and major roadway infrastructure are present.

In the context of each study's ERA framework, findings across wetlands and estuarine systems under strong urban and industrial influence indicate a consistent pattern. In Shadegan Wetland (Iran), the risk of PAHs in water was classified as moderate for several compounds, with benzo[a]anthracene (BaA) reaching high risk levels at most sites, while overall sediment risk was lower [83]. Similarly, in the Hoor Al-Azim wetland (Iran), the authors reported high risk for BaA in water but generally low to moderate risk for most other PAHs, while sediments indicated low biological risk [82]. In Anzali Wetland (Iran), sediments overall showed no harmful biological effects, although site-specific exceedances occurred for fluorene and pyrene, and several compounds, including NAP and ANT, were frequently above negligible-effect thresholds [81]. Consistent with our finding of insignificant acute risk for dissolved NAP, Edku Wetland (Egypt) showed no observed ecological hazard during spring [129]. By contrast, in estuarine systems, several PAHs, including PHEN, pyrene, and fluorene, frequently exceeded ecological risk thresholds, with clear seasonal variability [130]. More recently, studies demonstrated that, even when PAH water risk was negligible [129], sediment-bound PAHs can pose major ecological hazards in areas with dense urban infrastructure [131]. These comparisons indicate that while Caño La Malaria is not pristine, its present surface-water PAH levels imply insignificant ecological risk ($HQ_{water} < 0.01$) relative to those reported for other impacted wetlands; nonetheless, future risk assessments should integrate sediments and biota for a complete evaluation.

The ecological implications are particularly relevant when considering vulnerable benthic species such as the *Callinectes sapidus*, which serves as a key ecological indicator in Puerto Rico's coastal wetlands. The negligible HQ_{water} values obtained in this study, calculated using toxicity thresholds derived from *Callinectes sapidus*, studies confirm a low acute ecological risk under current PAH concentrations [64]. However, evidence indicates that this species bioaccumulates PAHs in its tissues under long-term exposure conditions—particularly during wet seasons—with PHEN and ANT showing persistence that suggests slow metabolic degradation [132]. Other studies found no physiological or molecular effects after short-term exposure to PAH-contaminated sediments, highlighting the importance of exposure duration and compartment in ecological risk assessment [133]. Although *Callinectes sapidus* does not biomagnify PAHs across trophic levels, it is known to bioaccumulate these compounds from sediments, porewater, and diet in estuarine environments, showing its role in shaping PAH bioavailability within aquatic food webs [134]. Additionally, comparative assessments in rivers have shown that crustaceans and mollusks often exhibit higher ecological risk than fish, reinforcing their importance in monitoring frameworks [135]. These findings reaffirm the utility of *Callinectes sapidus* as a sentinel species and highlight the need to consider long-term exposure, seasonal variability, and sediment-phase contamination in future assessments.

Despite the insights gained from this study, several limitations must be acknowledged to contextualize the ERA more accurately. The risk assessment was restricted to dissolved NAP and focused solely on acute toxicity endpoints. It did not account for sediment-associated PAHs, high molecular weight congeners, or sublethal and chronic toxic effects—factors which have been shown to elevate ecological risk in similar wetland systems [91,136]. However, LMW-PAHs can induce oxidative stress and endocrine disruption even at concentrations below acute toxicity benchmarks, particularly under chronic exposure scenarios [137,138]. Similarly, in estuarine studies, PAHs such as fluoranthene, pyrene, and BaA were major contributors to total ecological risk, particularly when sediment-phase concentrations were considered [130]. These limitations underscore the need for integrative ERA approaches in Caño La Malaria. Exceedances of guideline values for PHEN and ANT suggest potential chronic concerns, while acute HQ_{water} results for NAP indicate insignificant risk—complementary rather than contradictory frameworks.

5. Conclusions

This study presents the first spatial–seasonal assessment of three LMW-PAHs (NAP, PHEN, and ANT) in surface waters of Caño La Malaria, the main hydrological conduit of the Cucharillas Marsh urban wetland (Cataño, Puerto Rico). Seasonal variability reflected rainfall-driven runoff and upstream urban inputs, while downstream attenuation was consistent with dilution and sediment–sorption processes. Compositional shifts were evident across campaigns: NAP dominated in March 2022, ANT prevailed during the peak wet season, and PHEN became most abundant at several downstream sites. Source apportionment analyses indicated a predominance of pyrogenic inputs—principally vehicular emissions and biomass combustion—with wet-season samples more strongly associated with NAP and ANT, with episodic contributions from petroleum-related sources. Correlation patterns revealed strong upstream connectivity but local variability at site D, while the strong association between PHEN and ANT suggests a common combustion-derived origin.

Ecological risk characterization showed negligible acute risk from NAP. However, PHEN and ANT frequently exceeded international benchmarks, underscoring the potential for chronic or sublethal effects not captured by the ERA, which was restricted to acute endpoints for NAP. Peak wet-season concentrations reached levels comparable to moderate pollution categories. NAP and PHEN regularly exceeded freshwater guideline values, while ANT occasionally neared guideline values for marine organisms. These suggest that current benchmark values may underestimate site-specific risks in Caño La Malaria. This highlights the need for comprehensive frameworks that integrate sediments, biota, and substituted PAHs to more accurately capture long-term ecological hazards.

From a management perspective, the results emphasize the necessity of integrated, watershed-scale interventions, including upstream pollution control, restoration of riparian buffers, and improvements to stormwater infrastructure, complemented by nature-based solutions. Future monitoring should include sediments and biota, the full suite of 16 USEPA priority PAHs and their derivatives, and chronic toxicity endpoints. Given that this study was limited to surface waters and three parent LMW-PAHs, a logical next step is to incorporate additional PAHs and seasonal sediment sampling. Collectively, these insights provide a transferable framework to strengthen the assessment and management of PAH pollution in tropical urban wetlands.

Supplementary Materials: The following supporting information can be downloaded at: https://www.mdpi.com/article/10.3390/toxics13100860/s1, Table S1. Physical properties and chemical structure of the 16 USEPA priority PAHs [138–141]; Table S2: Calibration curve parameters and calculated limits of detection (LOD) and quantification (LOQ) for naphthalene (NAP), phenanthrene (PHEN), and anthracene (ANT) in surface water samples; Table S3: Relative standard deviation (RSD%) and coefficient of variation (CV) for naphthalene (NAP), phenanthrene (PHEN), and anthracene (ANT) across all sites and sampling dates; Table S4: Shapiro–Wilk test for normality of Σ3PAHs concentrations; Table S5: Friedman test for seasonal differences in Σ3PAHs concentrations; Table S6: Friedman test for spatial differences in Σ3PAHs concentrations among sampling sites across all campaigns; Table S7: Pearson correlation matrix of Σ3PAHs concentrations between sampling sites in Caño La Malaria; Table S8: Pearson correlation matrix of naphthalene (NAP), phenanthrene (PHEN), and anthracene (ANT) concentrations in Caño La Malaria surface water; Table S9: Loadings of PAH compounds on the first three principal components derived from principal component analysis (PCA) of PAH concentrations in surface water samples from Caño La Malaria.

Author Contributions: Conceptualization, P.J.B.-R., M.C.C. and F.M.; methodology, P.J.B.-R., M.C.C. and F.M.; writing—original draft preparation, P.J.B.-R.; writing—review and editing, P.J.B.-R., M.C.C.

and F.M.; supervision, M.C.C. The authors have reviewed and edited the output and take full responsibility for the content of this publication. All authors have read and agreed to the published version of the manuscript.

Funding: Financial support from NSF Center for the Advancement of Wearable Technologies-CAWT (Grant 1849243) and from the Consortium of Hybrid Resilient Energy Systems CHRES (DE-NA0003982), are gratefully acknowledged.

Institutional Review Board Statement: The study was approved by the Biosecurity Committee (Comité de Bioseguridad, IBC) of the Universidad Ana G. Méndez, Gurabo Campus, School of Natural Sciences and Technology (protocol code B03-111-19, approved on 20 February 2020).

Informed Consent Statement: No applicable.

Data Availability Statement: The data presented in this study are contained within the article and are available from the corresponding author upon reasonable request.

Acknowledgments: The authors express special thanks to Mariangely Colón-Cruz for her assistance with field sampling and her valuable input on methodological design and statistical analysis. Gratitude is extended to El Corredor del Yaguazo Inc., Pedro Carrión, and the staff who provided access to key sampling areas near the Cucharillas Marsh Nature Reserve.

Conflicts of Interest: The authors declare no conflicts of interest.

Abbreviations

The following abbreviations are used in this manuscript:

PAHs	polycyclic aromatic hydrocarbons
LMW-PAHs	low molecular weight polycyclic aromatic hydrocarbons
HMW-PAHs	high molecular weight polycyclic aromatic hydrocarbons
NAP	naphthalene
PHEN	phenanthrene
ANT	anthracene
BaA	benzo[a]anthracene
Σ3PAHs	sum of three low molecular weight PAHs (naphthalene, phenanthrene, anthracene)
DCM	dichloromethane
PCA	Principal Component Analysis
USEPA	United States Environmental Protection Agency
GC-MS	Gas Chromatography–Mass Spectrometry
ng/L	Nanograms per Liter
U	upstream
M	midstream
D	downstream
O	outlet
RSD	relative Standard Deviation
EU	European Union
ND	nondetected
ERA	ecological risk assessment
HQ	hazard quotient
EC_{water}	environmental concentration in water
$PNEC_{water}$	predicted no-effect concentration in water
$L(E)C_{50}$	lethal (or effect) concentration for 50% of organisms
AF	assessment factor

References

1. Harvey, R.G. Environmental Chemistry of PAHs. In *PAHs and Related Compounds: Chemistry*; Neilson, A.H., Ed.; The Handbook of Environmental Chemistry; Springer: Berlin/Heidelberg, Germany, 1998; pp. 1–54, ISBN 978-3-540-49697-7.

2. Amodu, O.S.; Ojumu, T.V.; Ntwampe, S.K.O.; Amodu, O.S.; Ojumu, T.V.; Ntwampe, S.K.O. Bioavailability of High Molecular Weight Polycyclic Aromatic Hydrocarbons Using Renewable Resources. In *Environmental Biotechnology—New Approaches and Prospective Applications*; IntechOpen: London, UK, 2013; ISBN 978-953-51-0972-3.
3. Rabodonirina, S.; Net, S.; Ouddane, B.; Merhaby, D.; Dumoulin, D.; Popescu, T.; Ravelonandro, P. Distribution of Persistent Organic Pollutants (PAHs, Me-PAHs, PCBs) in Dissolved, Particulate and Sedimentary Phases in Freshwater Systems. *Environ. Pollut.* **2015**, *206*, 38–48. [CrossRef]
4. Alaekwe, I.O.; Abba, O. Polycyclic Aromatic Hydrocarbons in Water: A Review of the Sources, Properties, Exposure Pathways, Bionetwork and Strategies for Remediation. *J. Geosci. Environ. Prot.* **2022**, *10*, 137–144. [CrossRef]
5. Lourenço, R.A.; Lube, G.V.; Jarcovis, R.D.L.M.; da Silva, J.; de Souza, A.C. Navigating the PAH Maze: Bioaccumulation, Risks, and Review of the Quality Guidelines in Marine Ecosystems with a Spotlight on the Brazilian Coastline. *Mar. Pollut. Bull.* **2023**, *197*, 115764. [CrossRef] [PubMed]
6. Dat, N.-D.; Chang, M.B. Review on Characteristics of PAHs in Atmosphere, Anthropogenic Sources and Control Technologies. *Sci. Total Environ.* **2017**, *609*, 682–693. [CrossRef]
7. Behera, B.K.; Das, A.; Sarkar, D.J.; Weerathunge, P.; Parida, P.K.; Das, B.K.; Thavamani, P.; Ramanathan, R.; Bansal, V. Polycyclic Aromatic Hydrocarbons (PAHs) in Inland Aquatic Ecosystems: Perils and Remedies through Biosensors and Bioremediation. *Environ. Pollut.* **2018**, *241*, 212–233. [CrossRef] [PubMed]
8. Campos, I.; Abrantes, N. Forest Fires as Drivers of Contamination of Polycyclic Aromatic Hydrocarbons to the Terrestrial and Aquatic Ecosystems. *Curr. Opin. Environ. Sci. Health* **2021**, *24*, 100293. [CrossRef]
9. Al Ali, S.; Debade, X.; Chebbo, G.; Béchet, B.; Bonhomme, C. Contribution of Atmospheric Dry Deposition to Stormwater Loads for PAHs and Trace Metals in a Small and Highly Trafficked Urban Road Catchment. *Environ. Sci. Pollut. Res.* **2017**, *24*, 26497–26512. [CrossRef]
10. Awonaike, B.; Lei, Y.D.; Parajulee, A.; Mitchell, C.P.J.; Wania, F. Polycyclic Aromatic Hydrocarbons and Quinones in Urban and Rural Stormwater Runoff: Effects of Land Use and Storm Characteristics. *ACS EST Water* **2021**, *1*, 1209–1219. [CrossRef]
11. Wang, H.; Shu, Y.; Kuang, Z.; Han, Z.; Wu, J.; Huang, X.; Song, X.; Yang, J.; Fan, Z. Bioaccumulation and Potential Human Health Risks of PAHs in Marine Food Webs: A Trophic Transfer Perspective. *J. Hazard. Mater.* **2025**, *485*, 136946. [CrossRef]
12. Keith, L.H. The Source of U.S. EPA's Sixteen PAH Priority Pollutants. *Polycycl. Aromat. Compd.* **2015**, *35*, 147–160. [CrossRef]
13. Abdel-Shafy, H.I.; Mansour, M.S.M. A Review on Polycyclic Aromatic Hydrocarbons: Source, Environmental Impact, Effect on Human Health and Remediation. *Egypt. J. Pet.* **2016**, *25*, 107–123. [CrossRef]
14. Lettoof, D.C.; Bateman, P.W.; Aubret, F.; Gagnon, M.M. The Broad-Scale Analysis of Metals, Trace Elements, Organochlorine Pesticides and Polycyclic Aromatic Hydrocarbons in Wetlands Along an Urban Gradient, and the Use of a High Trophic Snake as a Bioindicator. *Arch. Environ. Contam. Toxicol.* **2020**, *78*, 631–645. [CrossRef]
15. Stephansen, D.A.; Arias, C.A.; Brix, H.; Fejerskov, M.L.; Nielsen, A.H. Relationship between Polycyclic Aromatic Hydrocarbons in Sediments and Invertebrates of Natural and Artificial Stormwater Retention Ponds. *Water* **2020**, *12*, 2020. [CrossRef]
16. Oyo-Ita, I.; Nkom, P.Y.; Ugim, S.U.; Bassey, F.I.; Oyo-Ita, O.E. Seasonal Changes of PAHs in Water and Suspended Particulate Matter from Cross River Estuary, SE Nigeria in Response to Human-Induced Activity and Hydrological Cycle. *Polycycl. Aromat. Compd.* **2022**, *42*, 5456–5473. [CrossRef]
17. Alikhani, S.; Nummi, P.; Ojala, A. Urban Wetlands: A Review on Ecological and Cultural Values. *Water* **2021**, *13*, 3301. [CrossRef]
18. Chen, Z.; Ren, G.; Ma, X.; Zhou, B.; Yuan, D.; Liu, H.; Wei, Z. Presence of Polycyclic Aromatic Hydrocarbons among Multi-Media in a Typical Constructed Wetland Located in the Coastal Industrial Zone, Tianjin, China: Occurrence Characteristics, Source Apportionment and Model Simulation. *Sci. Total Environ.* **2021**, *800*, 149601. [CrossRef] [PubMed]
19. Zhao, C.; Xu, J.; Shang, D.; Zhang, Y.; Zhang, J.; Xie, H.; Kong, Q.; Wang, Q. Application of Constructed Wetlands in the PAH Remediation of Surface Water: A Review. *Sci. Total Environ.* **2021**, *780*, 146605. [CrossRef]
20. Alves, R.N.; Mariz, C.F., Jr.; de Melo Alves, M.K.; Cavalcanti, M.G.N.; de Melo, T.J.B.; de Arruda-Santos, R.H.; Zanardi-Lamardo, E.; Carvalho, P.S.M. Contamination and Toxicity of Surface Waters Along Rural and Urban Regions of the Capibaribe River in Tropical Northeastern Brazil. *Environ. Toxicol. Chem.* **2021**, *40*, 3063–3077. [CrossRef] [PubMed]
21. Lee, C.-C.; Chen, C.S.; Wang, Z.-X.; Tien, C.-J. Polycyclic Aromatic Hydrocarbons in 30 River Ecosystems, Taiwan: Sources, and Ecological and Human Health Risks. *Sci. Total Environ.* **2021**, *795*, 148867. [CrossRef] [PubMed]
22. Makobe, S.; Seopela, M.P.; Ambushe, A.A. Seasonal Variations, Source Apportionment, and Risk Assessment of Polycyclic Aromatic Hydrocarbons (PAHs) in Sediments from Klip River, Johannesburg, South Africa. *Environ. Monit. Assess.* **2025**, *197*, 257. [CrossRef]
23. Caballero-Gallardo, K.; Olivero-Verbel, J.; Corada-Fernández, C.; Lara-Martín, P.A.; Juan-García, A. Emerging Contaminants and Priority Substances in Marine Sediments from Cartagena Bay and the Grand Marsh of Santa Marta (Ramsar Site), Colombia. *Environ. Monit. Assess.* **2021**, *193*, 596. [CrossRef] [PubMed]

24. Huang, R.; Zhang, C.; Xu, X.; Jin, R.; Li, D.; Christakos, G.; Xiao, X.; He, J.; Agusti, S.; Duarte, C.M.; et al. Underestimated PAH Accumulation Potential of Blue Carbon Vegetation: Evidence from Sedimentary Records of Saltmarsh and Mangrove in Yueqing Bay, China. *Sci. Total Environ.* **2022**, *817*, 152887. [CrossRef]
25. Nunes, B.Z.; Zanardi-Lamardo, E.; Choueri, R.B.; Castro, Í.B. Marine Protected Areas in Latin America and Caribbean Threatened by Polycyclic Aromatic Hydrocarbons. *Environ. Pollut.* **2021**, *269*, 116194. [CrossRef]
26. Balgobin, A.; Ramroop Singh, N. Source Apportionment and Seasonal Cancer Risk of Polycyclic Aromatic Hydrocarbons of Sediments in a Multi-Use Coastal Environment Containing a Ramsar Wetland, for a Caribbean Island. *Sci. Total Environ.* **2019**, *664*, 474–486. [CrossRef] [PubMed]
27. López, L.; Barreto, M.B.; Peralba, M.d.C.R.; Barreto-Pittol, E.; Mónaco, S.L.; Díaz, R.; López, L.; Barreto, M.B.; Peralba, M.d.C.R.; Barreto-Pittol, E.; et al. Analysis of Polycyclic Aromatic Hydrocarbons (Pah) in Soil of Mangrove Forest of the Venezuelan Caribbean Coast. *Rev. Int. Contam. Ambient.* **2020**, *36*, 677–687. [CrossRef]
28. Garcia, M.R.; Martins, C.C. A Systematic Evaluation of Polycyclic Aromatic Hydrocarbons in South Atlantic Subtropical Mangrove Wetlands under a Coastal Zone Development Scenario. *J. Environ. Manag.* **2021**, *277*, 111421. [CrossRef]
29. dos Santos, P.R.S.; Moreira, L.F.F.; Moraes, E.P.; de Farias, M.F.; Domingos, Y.S. Traffic-Related Polycyclic Aromatic Hydrocarbons (PAHs) Occurrence in a Tropical Environment. *Environ. Geochem. Health* **2021**, *43*, 4577–4587. [CrossRef]
30. Santana, J.L.; Massone, C.G.; Valdés, M.; Vazquez, R.; Lima, L.A.; Olivares-Rieumont, S. Occurrence and Source Appraisal of Polycyclic Aromatic Hydrocarbons (PAHs) in Surface Waters of the Almendares River, Cuba. *Arch. Environ. Contam. Toxicol.* **2015**, *69*, 143–152. [CrossRef]
31. Tolosa, I.; Mesa-Albernas, M.; Alonso-Hernández, C.M. Assessing the Sources of Organic Matter in Sediments from the Gulf of Batabanó (Cuba) Using Stable Isotopes, Aliphatic and Polycyclic Aromatic Hydrocarbons. *Cont. Shelf Res.* **2025**, *285*, 105405. [CrossRef]
32. Billah, M.M.; Bhuiyan, M.K.A.; Amran, M.I.U.A.; Cabral, A.C.; Garcia, M.R.D. Polycyclic Aromatic Hydrocarbons (PAHs) Pollution in Mangrove Ecosystems: Global Synthesis and Future Research Directions. *Rev. Environ. Sci. Bio/Technol.* **2022**, *21*, 747–770. [CrossRef]
33. Wang, Z.; Meng, Q.; Sun, K.; Wen, Z. Spatiotemporal Distribution, Bioaccumulation, and Ecological and Human Health Risks of Polycyclic Aromatic Hydrocarbons in Surface Water: A Comprehensive Review. *Sustainability* **2024**, *16*, 10346. [CrossRef]
34. Masih, J.; Singhvi, R.; Kumar, K.; Jain, V.K.; Taneja, A. Seasonal Variation and Sources of Polycyclic Aromatic Hydrocarbons (PAHs) in Indoor and Outdoor Air in a Semi Arid Tract of Northern India. *Aerosol Air Qual. Res.* **2012**, *12*, 515–525. [CrossRef]
35. Liu, K.; Wang, X.; Zhang, H.; Wei, Y.; Zhao, G.; Liu, X.; Cui, H.; Han, J. Sources, Transport and Fate of Polycyclic Aromatic Hydrocarbons (PAHs) in a Typical River-Estuary System in the North China: From a New Perspective of PAHs Loading. *Mar. Pollut. Bull.* **2025**, *214*, 117692. [CrossRef]
36. Jesus, F.; Pereira, J.L.; Campos, I.; Santos, M.; Ré, A.; Keizer, J.; Nogueira, A.; Gonçalves, F.J.M.; Abrantes, N.; Serpa, D. A Review on Polycyclic Aromatic Hydrocarbons Distribution in Freshwater Ecosystems and Their Toxicity to Benthic Fauna. *Sci. Total Environ.* **2022**, *820*, 153282. [CrossRef]
37. Li, H.; Wang, X.; Peng, S.; Lai, Z.; Mai, Y. Seasonal Variation of Temperature Affects HMW-PAH Accumulation in Fishery Species by Bacterially Mediated LMW-PAH Degradation. *Sci. Total Environ.* **2022**, *853*, 158617. [CrossRef]
38. Mu, G.; Bian, D.; Zou, M.; Wang, X.; Chen, F. Pollution and Risk Assessment of Polycyclic Aromatic Hydrocarbons in Urban Rivers in a Northeastern Chinese City: Implications for Continuous Rainfall Events. *Sustainability* **2023**, *15*, 5777. [CrossRef]
39. Nahar, A.; Akbor, M.A.; Sarker, S.; Siddique, M.A.B.; Shaikh, M.A.A.; Chowdhury, N.J.; Ahmed, S.; Hasan, M.; Sultana, S. Dissemination and Risk Assessment of Polycyclic Aromatic Hydrocarbons (PAHs) in Water and Sediment of Buriganga and Dhaleswari Rivers of Dhaka, Bangladesh. *Heliyon* **2023**, *9*, e18465. [CrossRef]
40. Webb, R.M.; Gómez-Gómez, F. *Synoptic Survey of Water Quality and Bottom Sediments, San Juan Bay Estuary System, Puerto Rico, December 1994–July 1995*; U.S. Geological Survey; Branch of Information Services: Guaynabo, PR, USA, 1998.
41. Branoff, B.; Cuevas, E.; Hernández, E. *Assessment of Urban Coastal Wetlands Vulnerability to Hurricanes in Puerto Rico*; Department of Natural Resources of Puerto Rico: San Juan, PR, USA, 2018.
42. Martínez-Cintrón, A. Assessment of Land Use Land Cover Patterns Influence on Toxic Metals Distribution in Freshwater Sediments a Case Study—Ciénaga Las Cucharillas, Puerto Rico. Ph.D Thesis, Universidad de Puerto Rico, San Juan PR, USA, 2025.
43. United States Environmental Protection Agency 02/03/2003: EPA Proposes to Fine the Municipality of Cataño for Raw Sewage Discharges. Available online: https://www.epa.gov/archive/epapages/newsroom_archive/newsreleases/d568d21c3a4bd717852571630060f20d.html (accessed on 30 June 2025).
44. Mejías, C.L.; Carlos Musa, J.; Otero, J. Exploratory Evaluation of Retranslocation and Bioconcentration of Heavy Metals in Three Species of Mangrove at Las Cucharillas Marsh, Puerto Rico. *J. Trop. Life Sci.* **2013**, *3*, 14–22. [CrossRef]
45. Aldarondo-Torres, J.X.; Samara, F.; Mansilla-Rivera, I.; Aga, D.S.; Rodríguez-Sierra, C.J. Trace Metals, PAHs, and PCBs in Sediments from the Jobos Bay Area in Puerto Rico. *Mar. Pollut. Bull.* **2010**, *60*, 1350–1358. [CrossRef]

46. Gjeltema, J.; Stoskopf, M.; Shea, D.; De Voe, R. Assessment of Polycyclic Aromatic Hydrocarbon Contamination of Breeding Pools Utilized by the Puerto Rican Crested Toad, Peltophryne Lemur. *Int. Sch. Res. Not.* **2012**, *2012*, 309853. [CrossRef]
47. Pait, A.S.; Whitall, D.R.; Dieppa, A.; Newton, S.E.; Brune, L.; Caldow, C.; Mason, A.L.; Apeti, D.A.; Christensen, J.D. Characterization of Organic Chemical Contaminants in Sediments from Jobos Bay, Puerto Rico. *Environ. Monit. Assess.* **2012**, *184*, 5065–5075. [CrossRef]
48. Whitall, D.; Mason, A.; Pait, A.; Brune, L.; Fulton, M.; Wirth, E.; Vandiver, L. Organic and Metal Contamination in Marine Surface Sediments of Guánica Bay, Puerto Rico. *Mar. Pollut. Bull.* **2014**, *80*, 293–301. [CrossRef]
49. Marengo-Santiago, J.I. *Evaluación de Plantas Con Potencial Fitoremediador de Hidrocarburos Aromaticos Policíclicos de La Ciénaga Las Cucharrillas*; Universidad Metropolitana: San Juan, PR, USA, 2008.
50. Sturla Irizarry, S.M.; Cathey, A.L.; Zimmerman, E.; Rosario Pabón, Z.Y.; Huerta Montañez, G.; Vélez Vega, C.M.; Alshawabkeh, A.N.; Cordero, J.F.; Meeker, J.D.; Watkins, D.J. Prenatal Polycyclic Aromatic Hydrocarbon Exposure and Neurodevelopment among Children in Puerto Rico. *Chemosphere* **2024**, *366*, 143468. [CrossRef]
51. Watkins, D.J.; Zayas, H.R.T.; Welton, M.; Vega, C.M.V.; Pabón, Z.R.; Arroyo, L.D.A.; Cathey, A.L.; Cordero, N.R.C.; Alshawabkeh, A.; Cordero, J.F.; et al. Changes in Exposure to Environmental Contaminants in the Aftermath of Hurricane Maria among Pregnant Women in Northern Puerto Rico. *Heliyon* **2024**, *10*, e39767. [CrossRef] [PubMed]
52. United States Environmental Protection Agency. *Method 610: Polynuclear Aromatic Hydrocarbons*; United States Environmental Protection Agency: Washington, DC, USA, 1984.
53. Vistnes, H.; Sossalla, N.A.; Røsvik, A.; Gonzalez, S.V.; Zhang, J.; Meyn, T.; Asimakopoulos, A.G. The Determination of Polycyclic Aromatic Hydrocarbons (PAHs) with HPLC-DAD-FLD and GC-MS Techniques in the Dissolved and Particulate Phase of Road-Tunnel Wash Water: A Case Study for Cross-Array Comparisons and Applications. *Toxics* **2022**, *10*, 399. [CrossRef]
54. Semenov, M.Y.; Marinaite, I.I.; Silaev, A.V.; Begunova, L.A. Composition, Concentration and Origin of Polycyclic Aromatic Hydrocarbons in Waters and Bottom Sediments of Lake Baikal and Its Tributaries. *Water* **2023**, *15*, 2324. [CrossRef]
55. Gramlich, K.C.; Monteiro, F.C.; Carreira, R.d.S. Polycyclic Aromatic Hydrocarbons on the Atlantic Coast of South America and the Caribbean: A Systematic Literature Review on Biomonitoring Coastal Regions Employing Marine Invertebrates. *Reg. Stud. Mar. Sci.* **2024**, *78*, 103792. [CrossRef]
56. Mendoza Pantojas, G.M.; Vélez, V.; Zayas, B.; Malavé Zayas, K. Evaluación de la Calidad Microbiológica del Agua del Caño la Malaria y el Riesgo a Las Comunidades. *Perspect. Asun. Ambient.* **2014**, *3*, 83–98.
57. Taylor, M.A.; Enfield, D.B.; Chen, A.A. Influence of the Tropical Atlantic versus the Tropical Pacific on Caribbean Rainfall. *J. Geophys. Res. Ocean.* **2002**, *107*, 10-1–10-14. [CrossRef]
58. Torres-Valcárcel, Á.; Harbor, J.; González-Avilés, C.; Torres-Valcárcel, A. Impacts of Urban Development on Precipitation in the Tropical Maritime Climate of Puerto Rico. *Climate* **2014**, *2*, 47–77. [CrossRef]
59. GIS Puerto Rico Geodatos—Sistemas de Información Geográfica (GIS). Available online: https://gis.pr.gov/ (accessed on 1 September 2025).
60. European Union Reference Laboratories. *Guidance Document on the Estimation of LOD and LOQ for Measurements in the Field of Contaminants in Feed and Food*; European Union Reference Laboratory: Freiburg, Germany, 2016.
61. Gao, X.; Li, J.; Wang, X.; Zhou, J.; Fan, B.; Li, W.; Liu, Z. Exposure and Ecological Risk of Phthalate Esters in the Taihu Lake Basin, China. *Ecotoxicol. Environ. Saf.* **2019**, *171*, 564–570. [CrossRef]
62. Zhao, Y.; Li, J.; Qi, Y.; Guan, X.; Zhao, C.; Wang, H.; Zhu, S.; Fu, G.; Zhu, J.; He, J. Distribution, Sources, and Ecological Risk Assessment of Polycyclic Aromatic Hydrocarbons (PAHs) in the Tidal Creek Water of Coastal Tidal Flats in the Yellow River Delta, China. *Mar. Pollut. Bull.* **2021**, *173*, 113110. [CrossRef]
63. Li, W.; Wang, L.; Wang, X.; Liu, R. Derivation of Predicted No Effect Concentration and Ecological Risk Assessment of Polycyclic Musks Tonalide and Galaxolide in Sediment. *Ecotoxicol. Environ. Saf.* **2022**, *229*, 113093. [CrossRef]
64. Li, R.; Pang, H.; Guo, Y.; Zhou, X.; Fu, K.; Zhang, T.; Han, J.; Yang, L.; Zhou, B.; Zhou, S. Distribution of Polycyclic Aromatic Hydrocarbons and Pesticides in Danjiangkou Reservoir and Evaluation of Ecological Risk. *Toxics* **2024**, *12*, 859. [CrossRef] [PubMed]
65. Windsor, A.M.; Moore, M.K.; Warner, K.A.; Stadig, S.R.; Deeds, J.R. Evaluation of Variation within the Barcode Region of Cytochrome c Oxidase I (COI) for the Detection of Commercial Callinectes Sapidus Rathbun, 1896 (Blue Crab) Products of Non-US Origin. *PeerJ* **2019**, *7*, e7827. [CrossRef] [PubMed]
66. Sabourin, T.D. Respiratory and Circulatory Responses of the Blue Crab to Naphthalene and the Effect of Acclimation Salinity. *Aquat. Toxicol.* **1982**, *2*, 301–318. [CrossRef]
67. United States Environmental Protection Agency ECOTOX Knowledgebase. Available online: https://cfpub.epa.gov/ecotox/ (accessed on 1 September 2025).
68. Grmasha, R.A.; Abdulameer, M.H.; Stenger-Kovács, C.; Al-sareji, O.J.; Al-Gazali, Z.; Al-Juboori, R.A.; Meiczinger, M.; Hashim, K.S. Polycyclic Aromatic Hydrocarbons in the Surface Water and Sediment along Euphrates River System: Occurrence, Sources, Ecological and Health Risk Assessment. *Mar. Pollut. Bull.* **2023**, *187*, 114568. [CrossRef]

69. de Souza Bery, C.C.; dos Santos Gois, A.R.; Silva, B.S.; da Silva Soares, L.; Santos, L.G.G.V.; Fonseca, L.C.; da Silva, G.F.; Freitas, L.S.; Santos, E.; Alexandre, M.R.; et al. Polycyclic Aromatic Hydrocarbons in Surface Water of Rivers in Sergipe State, Brazil: A Comprehensive Analysis of Sources, Spatial and Temporal Variation, and Ecotoxicological Risk. *Mar. Pollut. Bull.* **2024**, *202*, 116370. [CrossRef]
70. Nguyen, N.T.; Nhon, N.T.T.; Hai, H.T.N.; Chi, N.D.T.; Hien, T.T. Characteristics of Microplastics and Their Affiliated PAHs in Surface Water in Ho Chi Minh City, Vietnam. *Polymers* **2022**, *14*, 2450. [CrossRef]
71. Na, M.; Zhao, Y.; Rina, S.; Wang, R.; Liu, X.; Tong, Z.; Zhang, J. Residues, Potential Source and Ecological Risk Assessment of Polycyclic Aromatic Hydrocarbons (PAHs) in Surface Water of the East Liao River, Jilin Province, China. *Sci. Total Environ.* **2023**, *886*, 163977. [CrossRef] [PubMed]
72. Jiang, C.; Zhao, D.; Chen, X.; Zheng, L.; Li, C.; Ren, M. Distribution, Source and Ecological Risk Assessment of Polycyclic Aromatic Hydrocarbons in Groundwater in a Coal Mining Area, China. *Ecol. Indic.* **2022**, *136*, 108683. [CrossRef]
73. Teodora Ciucure, C.; Geana, E.I.; Lidia Chitescu, C.; Laurentiu Badea, S.; Elena Ionete, R. Distribution, Sources and Ecological Risk Assessment of Polycyclic Aromatic Hydrocarbons in Waters and Sediments from Olt River Dam Reservoirs in Romania. *Chemosphere* **2023**, *311*, 137024. [CrossRef]
74. Dong, L.; Lin, L.; He, J.; Pan, X.; Wu, X.; Yang, Y.; Jing, Z.; Zhang, S.; Yin, G. PAHs in the Surface Water and Sediments of the Middle and Lower Reaches of the Han River, China: Occurrence, Source, and Probabilistic Risk Assessment. *Process Saf. Environ. Prot.* **2022**, *164*, 208–218. [CrossRef]
75. Ambade, B.; Sethi, S.S.; Giri, B.; Biswas, J.K.; Bauddh, K. Characterization, Behavior, and Risk Assessment of Polycyclic Aromatic Hydrocarbons (PAHs) in the Estuary Sediments. *Bull. Environ. Contam. Toxicol.* **2022**, *108*, 243–252. [CrossRef] [PubMed]
76. Yan, Y.; Bao, K.; Zhao, K.; Neupane, B.; Gao, C. A Baseline Study of Polycyclic Aromatic Hydrocarbons Distribution, Source and Ecological Risk in Zhanjiang Mangrove Wetlands, South China. *Ecotoxicol. Environ. Saf.* **2023**, *249*, 114437. [CrossRef]
77. Iloma, R.U.; Okpara, K.E.; Tesi, G.O.; Techato, K. Spatio-Temporal Distribution, Source Apportionment, Ecological and Human Health Risks Assessment of Polycyclic Aromatic Hydrocarbons in Sombreiro River Estuary, Niger Delta, Nigeria. *Int. J. Environ. Health Res.* **2025**, *35*, 2244–2265. [CrossRef]
78. Li, A.; der Beek, T.A.; Zhang, J.; Schmid, C.; Schüth, C. Characterizing Spatiotemporal Variations of Polycyclic Aromatic Hydrocarbons in Taihu Lake, China. *Environ. Monit. Assess.* **2022**, *194*, 713. [CrossRef] [PubMed]
79. Ezekwe, C.; Onwudiegwu, C.A.; Uzoekwe, S.A. Risk Assessment and Distribution of Polycyclic Aromatic Hydrocarbons in Water and Sediments Along Ibelebiri Axis of Kolo Creek, Ogbia Local Government Area, Niger Delta Region of Nigeria. *Chem. Afr.* **2025**, *8*, 1199–1213. [CrossRef]
80. Wu, X.; Dong, Y.; Zhao, J.; Liu, J.; Zhang, Y. Distribution, Sources, and Ecological Risk Assessment of Polycyclic Aromatic Hydrocarbons in Surface Water in the Coal Mining Area of Northern Shaanxi, China. *Environ. Sci. Pollut. Res.* **2023**, *30*, 50496–50508. [CrossRef]
81. Cheshmvahm, H.; Keshavarzi, B.; Moore, F.; Zarei, M.; Esmaeili, H.R.; Hooda, P.S. Investigation of the Concentration, Origin and Health Effects of PAHs in the Anzali Wetland: The Most Important Coastal Freshwater Wetland of Iran. *Mar. Pollut. Bull.* **2023**, *193*, 115191. [CrossRef]
82. Fakhradini, S.S.; Moore, F.; Keshavarzi, B.; Lahijanzadeh, A. Polycyclic Aromatic Hydrocarbons (PAHs) in Water and Sediment of Hoor Al-Azim Wetland, Iran: A Focus on Source Apportionment, Environmental Risk Assessment, and Sediment-Water Partitioning. *Environ. Monit. Assess.* **2019**, *191*, 233. [CrossRef]
83. Ashayeri, N.Y.; Keshavarzi, B.; Moore, F.; Kersten, M.; Yazdi, M.; Lahijanzadeh, A.R. Presence of Polycyclic Aromatic Hydrocarbons in Sediments and Surface Water from Shadegan Wetland—Iran: A Focus on Source Apportionment, Human and Ecological Risk Assessment and Sediment-Water Exchange. *Ecotoxicol. Environ. Saf.* **2018**, *148*, 1054–1066. [CrossRef]
84. Mojiri, A.; Zhou, J.L.; Ohashi, A.; Ozaki, N.; Kindaichi, T. Comprehensive Review of Polycyclic Aromatic Hydrocarbons in Water Sources, Their Effects and Treatments. *Sci. Total Environ.* **2019**, *696*, 133971. [CrossRef]
85. Zhu, Z.; Li, L.; Yu, Y.; Tan, L.; Wang, Z.; Suo, S.; Liu, C.; Qin, Y.; Peng, X.; Lu, H.; et al. Distribution, Source, Risk and Phytoremediation of Polycyclic Aromatic Hydrocarbons (PAHs) in Typical Urban Landscape Waters Recharged by Reclaimed Water. *J. Environ. Manag.* **2023**, *330*, 117214. [CrossRef]
86. Koh, C.-H.; Khim, J.S.; Kannan, K.; Villeneuve, D.L.; Senthilkumar, K.; Giesy, J.P. Polychlorinated Dibenzo-*p*-Dioxins (PCDDs), Dibenzofurans (PCDFs), Biphenyls (PCBs), and Polycyclic Aromatic Hydrocarbons (PAHs) and 2,3,7,8-TCDD Equivalents (TEQs) in Sediment from the Hyeongsan River, Korea. *Environ. Pollut.* **2004**, *132*, 489–501. [CrossRef]
87. Tarawou, T.; Erepamowei, Y.; Aigberua, A. Determination of Sources, Spatial Variability, and Concentration of Polycyclic Aromatic Hydrocarbons in Surface Water and Sediment of Imiringi River. *World J. Adv. Res. Rev.* **2021**, *9*, 309–318. [CrossRef]
88. Luo, J.; Huang, G.-B.; Wang, M.; Zhang, Y.-N.; Liu, Z.-X.; Zhang, Q.; Bai, S.-Y.; Xu, D.-D.; Liu, H.-L.; Mo, S.-P.; et al. Composition Characteristics, Source Analysis and Risk Assessment of PAHs in Surface Waters of Lipu. *J. Hazard. Mater.* **2025**, *490*, 137733. [CrossRef] [PubMed]

89. Ambade, B.; Sethi, S.S.; Kurwadkar, S.; Kumar, A.; Sankar, T.K. Toxicity and Health Risk Assessment of Polycyclic Aromatic Hydrocarbons in Surface Water, Sediments and Groundwater Vulnerability in Damodar River Basin. *Groundw. Sustain. Dev.* **2021**, *13*, 100553. [CrossRef]
90. Kong, J.; Han, M.; Cao, X.; Cheng, X.; Yang, S.; Li, S.; Sun, C.; He, H. Sedimentary Spatial Variation, Source Identification and Ecological Risk Assessment of Parent, Nitrated and Oxygenated Polycyclic Aromatic Hydrocarbons in a Large Shallow Lake in China. *Sci. Total Environ.* **2023**, *863*, 160926. [CrossRef]
91. Santos, E.; Souza, M.R.R.; Vilela Junior, A.R.; Soares, L.S.; Frena, M.; Alexandre, M.R. Polycyclic Aromatic Hydrocarbons (PAH) in Superficial Water from a Tropical Estuarine System: Distribution, Seasonal Variations, Sources and Ecological Risk Assessment. *Mar. Pollut. Bull.* **2018**, *127*, 352–358. [CrossRef] [PubMed]
92. Shi, C.; Qu, C.; Sun, W.; Zhou, J.; Zhang, J.; Cao, Y.; Zhang, Y.; Guo, J.; Zhang, J.; Qi, S. Multimedia Distribution of Polycyclic Aromatic Hydrocarbons in the Wang Lake Wetland, China. *Environ. Pollut.* **2022**, *306*, 119358. [CrossRef]
93. Yao, K.; Xie, Z.; Zhi, L.; Wang, Z.; Qu, C. Polycyclic Aromatic Hydrocarbons in the Water Bodies of Dong Lake and Tangxun Lake, China: Spatial Distribution, Potential Sources and Risk Assessment. *Water* **2023**, *15*, 2416. [CrossRef]
94. Areguamen, O.I.; Calvin, N.N.; Gimba, C.E.; Okunola, O.J.; Elebo, A. Seasonal Assessment of the Distribution, Source Apportionment, and Risk of Water-Contaminated Polycyclic Aromatic Hydrocarbons (PAHs). *Environ. Geochem. Health* **2023**, *45*, 5415–5439. [CrossRef] [PubMed]
95. Yunker, M.B.; Macdonald, R.W.; Vingarzan, R.; Mitchell, R.H.; Goyette, D.; Sylvestre, S. PAHs in the Fraser River Basin: A Critical Appraisal of PAH Ratios as Indicators of PAH Source and Composition. *Org. Geochem.* **2002**, *33*, 489–515. [CrossRef]
96. Fan, J.; Jing, Z.; Guo, F.; Jia, J.; Jiang, Y.; Cai, X.; Wang, S.; Zhao, H.; Song, X. Spatial Variation Characteristics of Polycyclic Aromatic Hydrocarbons and Their Derivatives in Surface Water of Suzhou City: Occurrence, Sources, and Risk Assessment. *Toxics* **2025**, *13*, 403. [CrossRef]
97. Li, Y.; Liu, M.; Hou, L.; Li, X.; Yin, G.; Sun, P.; Yang, J.; Wei, X.; He, Y.; Zheng, D. Geographical Distribution of Polycyclic Aromatic Hydrocarbons in Estuarine Sediments over China: Human Impacts and Source Apportionment. *Sci. Total Environ.* **2021**, *768*, 145279. [CrossRef]
98. Soltani, N.; Keshavarzi, B.; Moore, F.; Tavakol, T.; Lahijanzadeh, A.R.; Jaafarzadeh, N.; Kermani, M. Ecological and Human Health Hazards of Heavy Metals and Polycyclic Aromatic Hydrocarbons (PAHs) in Road Dust of Isfahan Metropolis, Iran. *Sci. Total Environ.* **2015**, *505*, 712–723. [CrossRef]
99. Grmasha, R.A.; Stenger-Kovács, C.; Bedewy, B.A.H.; Al-sareji, O.J.; Al-Juboori, R.A.; Meiczinger, M.; Hashim, K.S. Ecological and Human Health Risk Assessment of Polycyclic Aromatic Hydrocarbons (PAH) in Tigris River near the Oil Refineries in Iraq. *Environ. Res.* **2023**, *227*, 115791. [CrossRef]
100. Kong, J.; Ma, T.; Cao, X.; Li, W.; Zhu, F.; He, H.; Sun, C.; Yang, S.; Li, S.; Xian, Q. Occurrence, Partition Behavior, Source and Ecological Risk Assessment of Nitro-PAHs in the Sediment and Water of Taige Canal, China. *J. Environ. Sci.* **2023**, *124*, 782–793. [CrossRef]
101. Azah, E.; Kim, H.; Townsend, T. Source of Polycyclic Aromatic Hydrocarbon in Roadway and Stormwater System Maintenance Residues. *Environ. Earth Sci.* **2015**, *74*, 3029–3039. [CrossRef]
102. Deka, J.P.; Dash, S.; Sandil, S.; Chaminda, T.; Mahlknecht, J.; Kumar, M. Long-Range Transport of Polycyclic Aromatic Hydrocarbons and Metals in High Altitude Lacustrine Environments of the Eastern Himalayas: Speciation, and Source Apportionment Perspectives. *ACS EST Water* **2024**, *4*, 3400–3411. [CrossRef]
103. Astrahan, P.; Lupu, A.; Leibovici, E.; Ninio, S. BTEX and PAH Contributions to Lake Kinneret Water: A Seasonal-Based Study of Volatile and Semi-Volatile Anthropogenic Pollutants in Freshwater Sources. *Environ. Sci. Pollut. Res.* **2023**, *30*, 61145–61159. [CrossRef] [PubMed]
104. Chen, C.; Lin, T.; Sun, X.; Wu, Z.; Tang, J. Spatiotemporal Distribution and Particle–Water Partitioning of Polycyclic Aromatic Hydrocarbons in Bohai Sea, China. *Water Res.* **2023**, *244*, 120440. [CrossRef] [PubMed]
105. United States Environmental Protection Agency. *Freshwater Screening Benchmarks*; United States Environmental Protection Agency: Washington, DC, USA, 2006.
106. Baumard, P.; Budzinski, H.; Michon, Q.; Garrigues, P.; Burgeot, T.; Bellocq, J. Origin and Bioavailability of PAHs in the Mediterranean Sea from Mussel and Sediment Records. *Estuar. Coast. Shelf Sci.* **1998**, *47*, 77–90. [CrossRef]
107. Liu, S.; Yao, X.; Zang, S.; Wan, L.; Sun, L. A National-Scale Study of Polycyclic Aromatic Hydrocarbons in Surface Water: Levels, Sources, and Carcinogenic Risk. *Water* **2024**, *16*, 3027. [CrossRef]
108. European Parliament and Council of the European Union. Directive 2013/39/EU of 12 August 2013: Amending Directives 2000/60/EC and 2008/105/EC as Regards Priority Substances in the Field of Water Policy. *Off. J. Eur. Union* **2013**; *L* 226, 1–17. Available online: https://eur-lex.europa.eu/eli/dir/2013/39/oj/eng (accessed on 8 October 2025).
109. Sajid, M.; Nazal, M.K.; Ihsanullah, I. Novel Materials for Dispersive (Micro) Solid-Phase Extraction of Polycyclic Aromatic Hydrocarbons in Environmental Water Samples: A Review. *Anal. Chim. Acta* **2021**, *1141*, 246–262. [CrossRef]

110. Nagpal, N.K. *Ambient Water Quality Criteria for Polycyclic Aromatic Hydrocarbons (PAHs)*; Ministry of Environment, Lands and Parks Province of British Columbia: Victoria, BC, Canada, 1993.
111. Tillmanns, A.R.; McGrath, J.A.; Di Toro, D.M. International Water Quality Guidelines for Polycyclic Aromatic Hydrocarbons: Advances to Improve Jurisdictional Uptake of Guidelines Derived Using the Target Lipid Model. *Environ. Toxicol. Chem.* **2024**, *43*, 686–700. [CrossRef]
112. Chen, J.; Fan, B.; Li, J.; Wang, X.; Li, W.; Cui, L.; Liu, Z. Development of Human Health Ambient Water Quality Criteria of 12 Polycyclic Aromatic Hydrocarbons (PAH) and Risk Assessment in China. *Chemosphere* **2020**, *252*, 126590. [CrossRef]
113. Han, M.; Liu, F.; Kang, Y.; Zhang, R.; Yu, K.; Wang, Y.; Wang, R. Occurrence, Distribution, Sources, and Bioaccumulation of Polycyclic Aromatic Hydrocarbons (PAHs) in Multi Environmental Media in Estuaries and the Coast of the Beibu Gulf, China: A Health Risk Assessment through Seafood Consumption. *Environ. Sci. Pollut. Res.* **2022**, *29*, 52493–52506. [CrossRef]
114. Hoang, A.Q.; Tran, K.H.; Vu, Y.H.T.; Nguyen, T.P.T.; Nguyen, H.D.; Nguyen, H.T.; Hoang, N.; Van Vu, T.; Tran, T.M. Comprehensive Investigation of Polycyclic Aromatic Hydrocarbons in Multiple Water Types from Hanoi, Vietnam: Contamination Characteristics, Influencing Factors, and Ecological Risks. *Environ. Toxicol. Chem.* **2025**, *44*, 2889–2899. [CrossRef]
115. Liu, S.; Liu, X.; Liu, M.; Yang, B.; Cheng, L.; Li, Y.; Qadeer, A. Levels, Sources and Risk Assessment of PAHs in Multi-Phases from Urbanized River Network System in Shanghai. *Environ. Pollut.* **2016**, *219*, 555–567. [CrossRef]
116. Mundo, R.; Matsunaka, T.; Iwai, H.; Ochiai, S.; Nagao, S. Environmental Processes and Fate of PAHs at a Shallow and Enclosed Bay: West Nanao Bay, Noto Peninsula, Japan. *Mar. Pollut. Bull.* **2022**, *184*, 114105. [CrossRef]
117. Zhang, S.; Xing, X.; Yu, H.; Du, M.; Zhang, Y.; Li, P.; Li, X.; Zou, Y.; Shi, M.; Liu, W.; et al. Fate of Polycyclic Aromatic Hydrocarbon (PAHs) in Urban Lakes under Hydrological Connectivity: A Multi-Media Mass Balance Approach. *Environ. Pollut.* **2025**, *366*, 125556. [CrossRef]
118. Franchi, E.; Cardaci, A.; Pietrini, I.; Fusini, D.; Conte, A.; De Folly D'Auris, A.; Grifoni, M.; Pedron, F.; Barbafieri, M.; Petruzzelli, G.; et al. Nature-Based Solutions for Restoring an Agricultural Area Contaminated by an Oil Spill. *Plants* **2022**, *11*, 2250. [CrossRef]
119. Wang, Y.; Yin, H.; Liu, Z.; Wang, X. A Systematic Review of the Scientific Literature on Pollutant Removal from Stormwater Runoff from Vacant Urban Lands. *Sustainability* **2022**, *14*, 12906. [CrossRef]
120. Jafarzadeh, A.; Matta, A.; Moghadam, S.V.; Dessouky, S.; Hutchinson, J.; Kapoor, V. Field Performance of Two Stormwater Bioretention Systems for Treating Heavy Metals and Polycyclic Aromatic Hydrocarbons from Urban Runoff. *J. Environ. Manag.* **2024**, *370*, 123080. [CrossRef] [PubMed]
121. Zoveidadianpour, Z.; Doustshenas, B.; Alava, J.J.; Savari, A.; Karimi Organi, F. Environmental and Human Health Risk Assessment of Polycyclic Aromatic Hydrocarbons in the Musa Estuary (Northwest of Persian Gulf), Iran. *J. Sea Res.* **2023**, *191*, 102335. [CrossRef]
122. Chapman, D.V.; Sullivan, T. The Role of Water Quality Monitoring in the Sustainable Use of Ambient Waters. *One Earth* **2022**, *5*, 132–137. [CrossRef]
123. Zerga, B. Integrated Watershed Management: A Review. *Discov. Sustain.* **2025**, *6*, 657. [CrossRef]
124. Yoon, T.; Rhodes, C.; Shah, F.A. Upstream Water Resource Management to Address Downstream Pollution Concerns: A Policy Framework with Application to the Nakdong River Basin in South Korea. *Water Resour. Res.* **2015**, *51*, 787–805. [CrossRef]
125. Pillay, V.; Moodley, B. Assessment of the Impact of Reforestation on Soil, Riparian Sediment and River Water Quality Based on Polyaromatic Hydrocarbon Pollutants. *J. Environ. Manag.* **2022**, *324*, 116331. [CrossRef]
126. Gómez-Ávila, C.; Rao, B.; Hussain, T.; Zhou, H.; Pitt, R.; Colvin, M.; Hayman, N.; DeMyers, M.; Reible, D. Particle Size-Based Evaluation of Stormwater Control Measures in Reducing Solids, Polycyclic Aromatic Hydrocarbons (PAHs) and Polychlorinated Biphenyls (PCBs). *Water Res.* **2025**, *277*, 123299. [CrossRef]
127. Sarkar, S.K.; Binelli, A.; Chatterjee, M.; Bhattacharya, B.D.; Parolini, M.; Riva, C.; Jonathan, M.P. Distribution and Ecosystem Risk Assessment of Polycyclic Aromatic Hydrocarbons (PAHs) in Core Sediments of Sundarban Mangrove Wetland, India. *Polycycl. Aromat. Compd.* **2012**, *32*, 1–26. [CrossRef]
128. Okbah, M.A.; Nassar, M.; Ibrahim, M.S.; El-Gammal, M.I. Environmental Study of Some Polycyclic Aromatic Hydrocarbons in Edku Wetland Waters, Egypt. *Egypt. J. Aquat. Biol. Fish.* **2024**, *28*, 211–233. [CrossRef]
129. Yan, J.; Liu, J.; Shi, X.; You, X.; Cao, Z. Polycyclic Aromatic Hydrocarbons (PAHs) in Water from Three Estuaries of China: Distribution, Seasonal Variations and Ecological Risk Assessment. *Mar. Pollut. Bull.* **2016**, *109*, 471–479. [CrossRef]
130. Nozarpour, R.; Bakhtiari, A.R.; Gorabi, F.G.; Azimi, A. Assessing Ecological and Health Risks of PAH Compounds in Anzali Wetland: A Weight of Evidence Perspective. *Mar. Pollut. Bull.* **2025**, *220*, 118428. [CrossRef] [PubMed]
131. Ololade, I.A.; Lajide, L.; Amoo, I.A. Occurrence and Toxicity of Hydrocarbon Residues in Crab (Callinectes Sapidus) from Contaminated Site. *J. Appl. Sci. Environ. Manag.* **2008**, *12*, 19–23. [CrossRef]
132. Pie, H.V.; Schott, E.J.; Mitchelmore, C.L. Investigating Physiological, Cellular and Molecular Effects in Juvenile Blue Crab, *Callinectus Sapidus*, Exposed to Field-Collected Sediments Contaminated by Oil from the Deepwater Horizon Incident. *Sci. Total Environ.* **2015**, *532*, 528–539. [CrossRef] [PubMed]

133. Khairy, M.A.; Weinstein, M.P.; Lohmann, R. Trophodynamic Behavior of Hydrophobic Organic Contaminants in the Aquatic Food Web of a Tidal River. *Environ. Sci. Technol.* **2014**, *48*, 12533–12542. [CrossRef]
134. Feng, J.; Hu, P.; Li, X.; Liu, S.; Sun, J. Ecological and Health Risk Assessment of Polycyclic Aromatic Hydrocarbons (PAHs) in Surface Water from Middle and Lower Reaches of the Yellow River. *Polycycl. Aromat. Compd.* **2016**, *36*, 656–670. [CrossRef]
135. Bhatti, S.S.; Bhatia, A.; Bhagat, G.; Singh, S.; Dhaliwal, S.S.; Sharma, V.; Verma, V.; Yin, R.; Singh, J. PAHs in Terrestrial Environment and Their Phytoremediation. In *Bioremediation for Sustainable Environmental Cleanup*; CRC Press: Boca Raton, FL, USA, 2024; ISBN 978-1-003-27794-1.
136. Umeh, C.T.; Nduka, J.K.; Omokpariola, D.O.; Morah, J.E.; Mmaduakor, E.C.; Okoye, N.H.; Lilian, E.-E.I.; Kalu, I.F. Ecological Pollution and Health Risk Monitoring Assessment of Polycyclic Aromatic Hydrocarbons and Heavy Metals in Surface Water, Southeastern Nigeria. *Environ. Anal. Health Toxicol.* **2023**, *38*, e2023007-0. [CrossRef] [PubMed]
137. Ndwabu, S.; Malungana, M.; Mahlambi, P. Efficiency Comparison of Extraction Methods for the Determination of 11 of the 16 USEPA Priority Polycyclic Aromatic Hydrocarbons in Water Matrices: Sources of Origin and Ecological Risk Assessment. *Integr. Environ. Assess. Manag.* **2024**, *20*, 1598–1610. [CrossRef] [PubMed]
138. National Center for Biotechnology Information PubChem Dataset 2023. Available online: https://pubchem.ncbi.nlm.nih.gov/ (accessed on 2 September 2025).
139. Shen, H. *Polycyclic Aromatic Hydrocarbons*; Springer Theses; Springer: Berlin/Heidelberg, Germany, 2016; ISBN 978-3-662-49678-7.
140. Agency for Toxic Substances and Disease Registry Toxicological Profile for Polycyclic Aromatic Hydrocarbons 1996. Available online: https://www.ncbi.nlm.nih.gov/books/NBK598190/ (accessed on 2 September 2025).
141. Mackay, D.; Shiu, W.-Y.; Shiu, W.-Y.; Lee, S.C. *Handbook of Physical-Chemical Properties and Environmental Fate for Organic Chemicals*; CRC Press: Boca Raton, FL, USA, 2006; ISBN 978-0-429-15007-4.

toxics

MDPI

Article

Cytotoxicity of Typical Diiodoalkanes from Shale Gas Wastewater in HepG2 Cells

Maoyuan Xu [1], Yusheng Wu [1], Yunmei Cai [2,*], Ruijie Wang [1] and Guofa Ren [1,*]

[1] Institute of Environmental Pollution and Health, School of Environmental and Chemical Engineering, Shanghai University, Shanghai 200444, China

[2] School of Environmental Monitoring, Guangdong Polytechnic of Environmental Protection Engineering, Foshan 528216, China

* Correspondence: 18702030877@163.com (Y.C.); rgf2008@shu.edu.cn (G.R.); Tel./Fax: +86-757-81773127 (Y.C.); +86-21-66137753 (G.R.)

Abstract: Shale gas extraction releases significant quantities of organic iodides of "unknown origin", which generally pose high ecological and health risks, yet their toxic mechanisms remain unclear. In this study, the human hepatocellular carcinoma (HepG2) cell line was employed as an in vitro cell model to assess the cytotoxic effects of three typical organic iodides (1,2-diiodoethane, 1,3-diiodopropane, and 1,4-diiodobutane) identified in shale gas extraction wastewater from Chongqing, China. The results demonstrated that all three diiodoalkanes exhibited significant toxic effects on HepG2 cells at a concentration of 25 μM, and this effect demonstrated a dose-dependent pattern. As the concentration of diiodoalkanes increased, the viability of HepG2 cells decreased significantly, while cell mortality increased markedly. The transcriptomic analysis indicated that exposure to these three diiodoalkanes induced abnormal expression of genes associated with the extracellular space, extracellular matrix (ECM), and endoplasmic reticulum (ER) in HepG2 cells, which was presumed to be linked to the disruption of the intracellular redox-antioxidant system homeostasis by the diiodoalkanes. Furthermore, assays of intracellular reactive oxygen species (ROS) and antioxidant enzyme/molecule levels suggested that diiodoalkane exposure triggered excessive intracellular ROS production, induced oxidative stress, and ultimately resulted in cell death.

Keywords: shale gas; diiodoalkanes; HepG2 cells; oxidative stress; transcriptomic analysis

1. Introduction

Iodine, as a ubiquitous trace element, has consistently been a research hotspot due to its significance in both biological and geochemical fields [1,2]. The forms of iodine in the environment are diverse. Currently, most research primarily focuses on inorganic iodine, mainly iodide ions (I^-) and iodate ions (IO_3^-), with relatively limited attention paid to iodine-containing disinfection byproducts (I-DBPs) generated during drinking water disinfection [3–5]. However, research on other organic iodides in water remains relatively scarce.

The distribution of iodine in the natural environment exhibits significant heterogeneity: apart from marine systems (where seawater contains relatively high iodine concentrations), iodine is also closely associated with natural gas extraction, with shale gas extraction as a typical example [6]. This association leads to sustained elevated iodine levels in the surroundings of shale gas extraction sites. In recent years, halogenated organic compounds

https://doi.org/10.3390/toxics13110943

(HOCs) have been frequently detected in shale gas wastewater [7–10]. Notably, several studies have identified organic iodides as the dominant class of detected HOCs. For instance, Luek et al. reported that in flowback wastewater from the Marcellus Shale, organic compounds containing iodine, bromine, chlorine, and mixed halogens (two distinct halogens) accounted for 52%, 20%, 9%, and 19% of total HOCs, respectively [10]. Furthermore, the concentration of organic iodides showed a gradual increase during wastewater discharge and remained relatively high even 10 months after gas well production began [8]. In our previous study, 21 HOCs were identified from the discharged wastewater of the shale gas wastewater treatment plant in Chongqing, China [11]. The composition included 5 chlorinated, 10 iodinated, 3 brominated, and 3 mixed-halogen (bromo/chloro or iodo/chloro) organic compounds. Consistent with prior findings, iodinated organic compounds constituted the most abundant HOC class detected. Quantitative analysis further revealed that diiodoalkanes (including 1,2-diiodoethane, 1,3-diiodopropane, and 1,4-diiodobutane) were present at relatively high concentrations. Among these, 1,2-diiodoethane reached a maximum concentration of 72.6 μg/L (0.26 μM) in surface water [12]. The formation mechanism of organic iodides in shale gas extraction may resemble that of DBPs in drinking water: during hydraulic fracturing or wastewater treatment, dissolved inorganic iodine is oxidized by strong oxidants to generate reactive halogen species (e.g., molecular halogens and halogen-free radicals). These reactive species then react with natural organic matter (NOM) in geological formations to form "unknown-origin" organic iodides [9,11].

HOCs are typically persistent organic pollutants, characterized by chemical stability and low biodegradability, which pose severe threats to the ecological environment and human health. Consequently, HOCs remain a core focus in environmental science. As transformation products of inorganic iodine, organic iodides exhibit far more complex environmental behaviors and toxic effects than their inorganic counterparts. For example, organic iodides generated via atmospheric photochemical processes (e.g., iodomethane and iodoethane) have low intrinsic toxicity but trigger cascading environmental reactions: these volatile compounds undergo long-range atmospheric transport, decompose under ultraviolet (UV) radiation, and release reactive iodine atoms that deplete stratospheric ozone [13]. Additionally, limited cellular and genetic studies on DBPs have demonstrated that brominated and iodinated DBPs exhibit significantly higher cytotoxicity and genotoxicity than their chlorinated congeners [4,14]. Based on these findings, it can be inferred that organic iodides formed during shale gas extraction may also possess substantial toxicity. However, the mechanisms underlying their toxic effects remain poorly understood, and relevant research is extremely limited. Given that shale gas has become a key focus of global energy development, and considering the "high complexity, strong toxicity, and persistence" of organic iodine pollutants derived from shale gas extraction, comprehensive ecological and human health risk assessments prior to shale gas development are of great significance.

Transcriptome analysis unravels the molecular mechanisms by which toxins disrupt cellular gene expression by systematically deciphering dynamic changes in cellular transcriptional states and gene regulatory networks. Within this framework, RNA sequencing (RNA-seq) stands as a core, foundational transcriptomic technology [15]. By leveraging the advantages of high-throughput sequencing, this technique comprehensively captures cellular transcriptomic profiles and precisely identifies differentially expressed genes (DEGs) upon toxic exposure. Upon integration with bioinformatics analyses—such as gene functional annotation and pathway enrichment analysis—RNA-seq can further delineate the mechanisms by which toxic substances perturb core cellular pathways, including those governing metabolism, stress responses, and apoptotic regulation, thereby providing critical molecular insights into the toxic mechanisms of the target substances. Given the high detection frequency and potential high toxicity of iodoalkanes in shale gas wastewater, clarifying the key mechanisms underlying their biological damage is critical. In this study, we evaluated the toxic effects of three typical diiodoalkanes, namely 1,2-diiodoethane, 1,3-diiodopropane, and 1,4-diiodobutane, on HepG2 cells. Transcriptomic analysis was further performed on HepG2 cells exposed to these pollutants, with the aim of identifying key signaling pathways involved in the toxic effects of typical diiodoalkanes.

2. Materials and Methods

2.1. Chemicals and Reagents

1,2-diiodoethane ($C_2H_4I_2$, CAS:624-73-7, 98% purity), 1,3-diiodopropan ($C_3H_6I_2$, CAS:627-31-6, 98% purity), and 1,4-diiodobutane ($C_4H_8I_2$, CAS:628-21-7, 98% purity) were purchased from Tokyo Chemical Industry Co., Ltd (Tokyo, Japan). The Dulbecco's modified Eagle medium (High Glucose DMEM) was bought from Datahill Biotechnology Co., Ltd. (Shanghai, China). The Fetal Bovine Serum (FBS) was obtained from Biological Industries Israel Beit Haemek Ltd (Beit Haemek, Israel). Serum-free animal protein-free cell freezing medium was derived from Nell Cell & Molecular Biotech Co., Ltd. (Jiangsu, China). 2′,7′-dichlorofluorescein diacetate (DCFH-DA) and 3-(4,5-dimethylthiazol-2)-2,5-diphenyltetrazolium bromide (MTT) were bought from Sigma-Aldrich (St. Louis, MO, USA). Other unspecified reagents were purchased from Beyotime Biotechnology (Shanghai, China).

2.2. Cell Culture

The human hepatocellular carcinoma (HepG2) cell line was purchased from the American Type Culture Collection (ATCC). Cells are routinely cultured in Dulbecco's Modified Eagle Medium (DMEM) supplemented with 10% fetal bovine serum (FBS) and 1% penicillin-streptomycin (P/S), in a humidified incubator maintained at 37 °C with 5% carbon dioxide (CO_2). Cells should be passaged at least twice. When cells exhibited stable morphology and reached the logarithmic growth phase (prior to confluence), they were seeded into 96-well or 6-well plates. Following 24 h of adherent culture to allow cell attachment, treatment with diiodoalkanes was initiated. Diiodoalkanes were dissolved in dimethyl sulfoxide (DMSO) for preparation. A solvent control group was included in the experiment, and the final concentration of DMSO in the medium across all groups was maintained at 0.1% to rule out potential solvent-induced interference with cell viability or function.

2.3. Cell Viability Assay

MTT assay was employed to evaluate the toxicity and cell viability of diiodoalkanes on HepG2 cells: HepG2 cells were seeded at a density of 4×10^3 cells/well in a 96-well plate. After 24 h of pre-incubation at 37 °C in a humidified incubator with 5% CO_2 to allow cells to adhere and stabilize, cells were treated with 25 μM, 50 μM, 100 μM, and 200 μM concentrations of diiodoalkanes, and 0.1% DMSO (solvent control). After 24 h exposure, 100 μL of MTT reagent (final concentration ~5 mg/mL) was added to each well and incubated in the dark for 4 h to form MTT crystals. After removing the supernatant, 150 μL of DMSO was added to each well, and the mixture was incubated on a shaking platform in the dark for 3 min to completely dissolve the crystals. Finally, absorbance at 570 nm was measured using a Spark® 20M multimode microplate reader (Tecan, Switzerland). Absorbance values indirectly reflected live cell numbers and cell viability.

2.4. Cellular Morphology Analysis

Cells were gently rinsed 1–2 times with phosphate-buffered saline (PBS) to remove residual medium, and the sample was then mounted on the optical microscope stage.

During observation, the cell distribution area was first located using the 10× low-power objective, after which the objective was switched to the 20× and 40× medium-to-high power ones. The overall cell morphology was examined and photographed using the microscope's imaging system, with a focus on recording typical damage features such as cell shrinkage, membrane rupture, abnormal aggregation, or detachment. At least three representative fields per sample were selected to ensure the reliability of the analysis.

2.5. RNA-Sequencing and Bioinformatics Analysis

HepG2 cells were exposed to 50 μM 1,2-diiodoethane, 1,3-diiodopropane, and 1,4-diiodobutane for 12 h. Total cellular RNA was then extracted using TRIZOL reagent (Life Technologies, Carlsbad, CA, USA). RNA quality was assessed with an Agilent Bioanalyzer 2100 (Agilent Technologies, Santa Clara, CA, USA) to ensure suitability for subsequent experimental procedures. To guarantee the reliability of results, three biological replicates were included throughout the experiment. The quality assurance and quality control (QA/QC) information for the raw sequencing data and omics analysis of RNA-seq are provided in the Supplementary Materials. cDNA library sequencing was performed by OE Biotech (Shanghai, China) using an Illumina HiSeqTM 2000 sequencer (Illumina, Inc., San Diego, CA, USA). Subsequently, DESeq software (Bioconductor 3.12) was utilized to standardize gene count data across all samples and calculate fold changes in gene expression. The significance of differential expression was evaluated using negative binomial distribution tests. Differentially expressed genes (DEGs) were identified based on the criteria of "absolute fold change (FC) ≥ 1.5 and $p < 0.05$". Finally, functional annotation and enrichment analysis of the identified DEGs were conducted using the Gene Ontology (GO; database link: http://geneontology.org/ (accessed on 28 October 2025)) and the Kyoto Encyclopedia of Genes and Genomes (KEGG; database link: https://www.kegg.jp/ (accessed on 28 October 2025), aiming to explore their biological functions and potential involvement in signaling pathways.

2.6. Detection of Reactive Oxygen Species

DCFH-DA fluorescent probe was used to detect intracellular ROS levels. HepG2 cell suspensions were seeded at a density of 1.5×10^5 cells per well in a 6-well plate. After the cells reached 70~80% confluence, they were exposed to different concentrations of diiodoalkanes (25, 50, 100, and 200 μM) and 0.1% dimethyl sulfoxide (DMSO, solvent control) for 4 h and 24 h, respectively. After exposure, remove the culture medium from the wells. Add 10 μM DCFH-DA probe solution to each well. Incubate the cells in a 37 °C dark incubator for 30 min to ensure complete probe loading. After incubation, discard unloaded probe solution and gently wash cells three times with phosphate-buffered saline (PBS) to remove residual probe and minimize background fluorescence interference. Subsequently, observe and capture cellular fluorescence images using the ZOE™ Fluorescence Cell Imager (BIO-RAD, Hercules, CA, USA). Analyze the fluorescence images with Image-Pro Plus software (Medical Cybernetics, Inc., Albemarle, NC, USA, Version 6.0) to quantitatively calculate intracellular fluorescence intensity—the level of fluorescence indirectly reflects changes in intracellular ROS levels.

2.7. Determination of Catalase, Glutathione, and Cysteine

Commercially available assay kits were used to measure the levels of catalase (CAT), glutathione (GSH), and cysteine (Cys) in HepG2 cells following exposure to the toxicant. As described above, HepG2 cells were first treated with different concentrations of diiodoalkanes (including a 50 μM iodinated alkane group and a 0.1% DMSO solvent control group) for 24 h. Following treatment, cells were collected via trypsin digestion and gently washed 2~3 times with phosphate-buffered saline (PBS) to remove residual medium and digestion solution. Subsequently, intracellular levels of the three antioxidant molecules were measured using the CAT assay kit, GSH assay kit, and Cys assay kit, respectively. All procedures strictly followed the manufacturers' instructions for each kit. After detection, the absorbance values corresponding to each indicator were recorded using a Spark® 20M Multifunction Microplate Reader (Tecan, Switzerland). All assays were independently replicated three times to ensure the reliability of the experimental data.

2.8. Statistical Analysis

GraphPad Prism 9.0 software (San Diego, CA, USA) was used for graphing and statistical analysis. All experiments were independently replicated three times with at least three parallel samples. The data were presented as mean ± standard deviation (SD). Multiple comparisons between groups were analyzed by using one-way ANOVA and post hoc Tukey's Test. The significance level of the data differences was determined by the p-values: p-value < 0.05 (*), p-value < 0.01 (**).

3. Results

3.1. Diiodoalkanes Affect the Cell Viability and Morphology of HepG2

The MTT cell proliferation assay was employed to evaluate the effects of 24 h exposure to 1,2-diiodoethane, 1,3-diiodopropene, and 1,4-diiodoobutane on the proliferation ability of HepG2 cells. As shown in Figure 1A, compared with the control group, exposure to any of the three diiodoalkanes at concentrations of 25 μM or 50 μM led to a significant decrease in HepG2 cell viability ($p < 0.01$). These concentrations differ by less than 100-fold from those detected in actual surface water, suggesting potential ecological risks from diiodoalkanes in shale gas wastewater discharge. When the exposure concentration increased to 100 μM, all three diiodoalkanes significantly inhibited HepG2 cell viability ($p < 0.05$), reducing it to 79.9 ± 6.2%, 65.7 ± 7.5%, and 72.7 ± 8.4% of the control group, respectively. With a further concentration increase to 200 μM, cell viability continued to decrease significantly, reaching 51.6 ± 6.0%, 35.3 ± 10.3%, and 47.2 ± 9.5% of the control group, respectively. Quantitative analysis revealed the toxic potency of the three diiodoalkanes against HepG2 cells in the following order: 1,3-diiodopropane > 1,4-diiodobutane ≥ 1,2-diiodoethane.

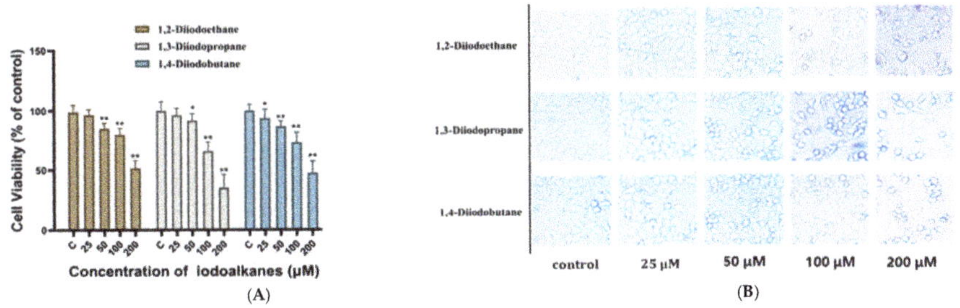

Figure 1. Effects of 1,2-diiodoethane, 1,3-diiodopropane and 1,4-diiodobutane exposure on HepG2. (**A**) The viability measured by the MTT assay. (**B**). The cell morphology observed by microscopic. (The control was treated with medium containing 0.1% DMSO (v/v), * $p < 0.05$, ** $p < 0.01$).

The results of microscopic observations (Figure 1B) revealed significant morphological changes in HepG2 cells following 24 h exposure to the three diiodoalkanes. Compared with the control group, all treated groups exhibited typical toxicological characteristics, such as cell shrinkage, reduced cell numbers, and increased cellular debris. The extent of cell loss showed a positive correlation with diiodoalkanes concentration. At the high exposure concentration of 200 μM, specifically, a marked reduction in cell area was observed. The most pronounced decrease in cell number was observed in the 1,3-diiodopropane-treated group, followed by the 1,4-diiodobutane-treated group. This phenomenon aligns with the aforementioned reported toxicity ranking, further suggesting that the toxic effects of these three diiodoalkanes may be associated with their chemical structure.

3.2. Effects of Diiodoalkanes on Gene Expression in HepG2 Cells

To elucidate the molecular mechanisms underlying the cytotoxicity induced by diiodoalkanes on HepG2 cells, transcriptomic analysis was performed in this study to investigate differential gene expression profiles following exposure to 50 μM diiodoalkanes. The goal was to identify potential regulatory pathways mediating their toxic effects. Statistical results for differentially expressed genes (DEGs) are shown in Figure 2A. Compared with the control group, a total of 158 DEGs were identified in the 1,2-diiodoethane-treated group, including 63 upregulated genes and 95 downregulated genes; the number of DEGs in the 1,3-diiodopropane-treated group increased significantly to 491, comprising 348 upregulated and 143 downregulated genes; and 221 DEGs were detected in the 1,4-diiodobutane-treated group (105 upregulated and 116 downregulated). Figure 2B further presents the distribution characteristics of DEGs across the three treatment groups, clearly demonstrating the quantitative relationship between group-specific and shared DEGs. Specifically, 85 unique DEGs were identified in the 1,2-diiodoethane-exposed group, 319 in the 1,3-diiodopropane-exposed group, and 86 in the 1,4-diiodobutane-exposed group. Regarding shared DEGs, the 1,3-diiodopropane and 1,4-diiodobutane groups shared 132 DEGs, while the 1,2-diiodoethane group shared 70 and 33 DEGs with the 1,3-diiodopropane and 1,4-diiodobutane groups, respectively. These statistical data further indicate that the three diiodoalkanes have significantly different intensities of perturbation on the gene expression of HepG2 cells: 1,2-diiodoethane has a relatively weak regulatory effect on gene expression, while 1,3-diiodopropane has the most prominent perturbation effect on the cell transcriptome.

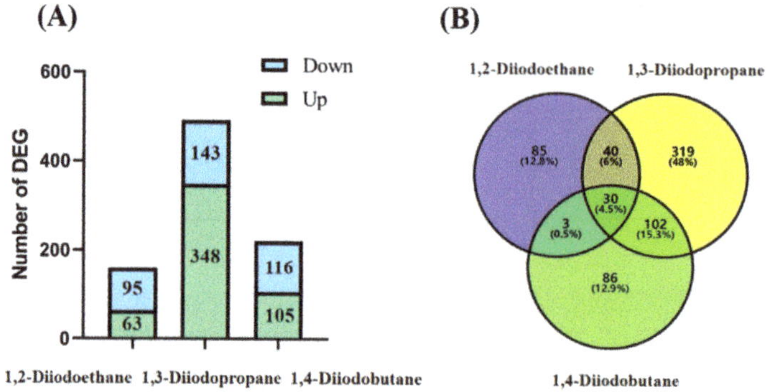

Figure 2. Analysis of differentially expressed genes (DEGs) after treatment of HepG2 cells with 1,2-diiodoethane, 1,3-diiodopropene, and 1,4-diiodobutane. (**A**) Number and distribution of DEGs in each group; (**B**) Venn diagram of DEGs.

3.3. Gene Ontology Enrichment Analysis

To clarify the biological functions associated with DEGs, Gene Ontology (GO) enrichment analysis was performed on the DEGs identified in this study. Figure 3 display the top 30 most significantly enriched biological function terms associated with DEGs in the 1,2-diiodoethane, 1,3-diiodopropane, and 1,4-diiodobutane treatment groups, respectively. These terms span three categories: biological process, cellular component, and molecular function, with the top 10 terms selected from each category. Analysis showed that nearly all significantly enriched GO terms were associated with the extracellular space and extracellular matrix. The core GO terms that are common to all three treatment groups and related to these extracellular components include: extracellular matrix organization (GO:0030198), extracellular space (GO:0005615), extracellular matrix (GO:0031012), extracellular exosome (GO:0070062), extracellular matrix structural constituent (GO:0005201), and collagen-containing extracellular matrix (GO:0062023). In the top GO enrichment results for the 1,3-diiodopropane and 1,4-diiodobutane exposure groups, both showed enrichment in GO terms related to endoplasmic reticulum function, specifically: endoplasmic reticulum lumen (GO:0005788), chaperone cofactor-dependent protein refolding (GO:0051085), and unfolded protein binding (GO:0051082). Additionally, GO terms enriched for 1,2-diiodoethane exposure, enriched GO terms were related to the negative regulation of the apoptotic process (GO:0043066), while for 1,3-diiodopropane exposure, terms related to the positive regulation of the apoptotic process (GO:0043065) were identified.

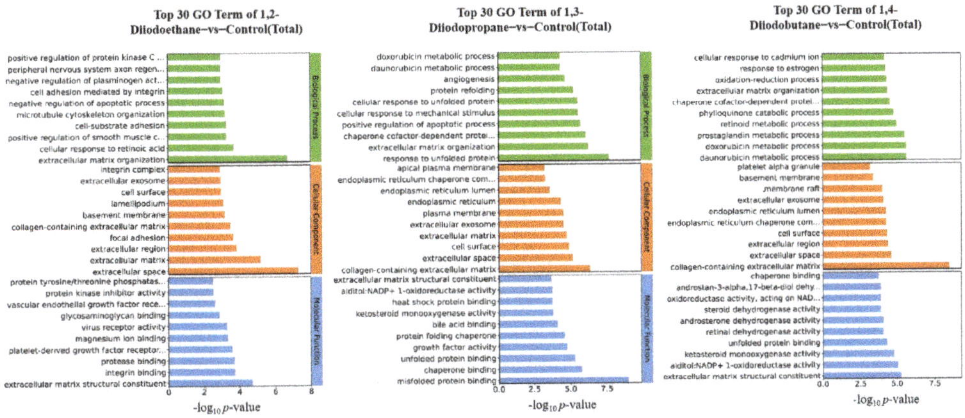

Figure 3. GO functional enrichment analysis of differentially expressed genes.

3.4. Kyoto Encyclopedia of Genes and Genomes Pathway Enrichment Analysis

To elucidate the potential molecular mechanisms underlying the effects of 1,2-diiodoethane, 1,3-diiodopropene, and 1,4-diiodobutane on HepG2 cells, this study further performed Kyoto Encyclopedia of Genes and Genomes (KEGG) pathway annotation and enrichment analysis on differentially expressed genes (DEGs). The top 20 signaling pathways significantly altered by the three diiodoalkanes were presented in Figure 4, respectively. Pathway analysis revealed that 1,2-diiodoethane primarily disrupted two key pathways: the ECM-receptor interaction pathway and the PI3K-Akt signaling pathway. In contrast, 1,3-diiodopropane and 1,4-diiodobutane shared partially overlapping disrupted pathways, specifically the ferroptosis pathway, TNF signaling pathway, IL-17 signaling pathway, and protein processing in the endoplasmic reticulum. Additionally, 1,3-diiodopropane uniquely regulated the MAPK signaling pathway. Previous studies have suggested that all the aforementioned disrupted pathways are directly or indirectly associated with oxidative stress [16–20]. These findings further suggest that 1,2-diiodoethane, 1,3-diiodopropane, and 1,4-diiodobutane significantly disrupt the homeostasis between oxidative stress-induced damage and antioxidant defense in HepG2 cells.

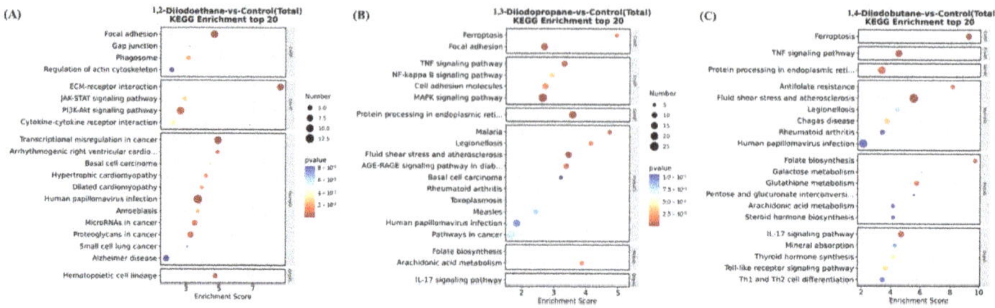

Figure 4. KEGG pathway analysis of differentially expressed genes.

3.5. *Effects of Diiodoalkanes on Reactive Oxygen Species in HepG2 Cells*

When cells undergo oxidative stress, the balance between pro-oxidative and antioxidant systems shifts toward oxidation, thereby inducing neutrophilic inflammatory infiltration, increased protease secretion, and the production of numerous oxidative intermediates [21,22]. Reactive oxygen species (ROS), as core markers of oxidative stress, directly reflect its severity. To investigate the effects of diiiodoalkanes on ROS levels in HepG2 cells under different exposure conditions, this study employed the DCFH-DA fluorescent probe assay. Five concentration gradients of 0, 25, 50, 100, and 200 μM were tested, with exposure durations of 4 h and 24 h, respectively. Results are shown in Figure 5. In the 4 h exposure group, the highest concentration (200 μM) of 1,2-diiodoethane, 1,3-diiodopropane, and 1,4-diiodobutane increased ROS levels in HepG2 cells to 253.2%, 250.2%, and 179.1% of the control group, respectively. In the 24 h exposure group, by contrast, ROS levels induced by these three compounds at the same concentration were 239.8%, 203.5%, and 168.3% of the control group, respectively. These results indicate that all three diiodoalkanes significantly induced an increase in ROS levels in HepG2 cells. The trend of ROS elevation was consistent across both 4 h and 24 h exposure groups, suggesting temporal stability in their oxidative stress-inducing effects.

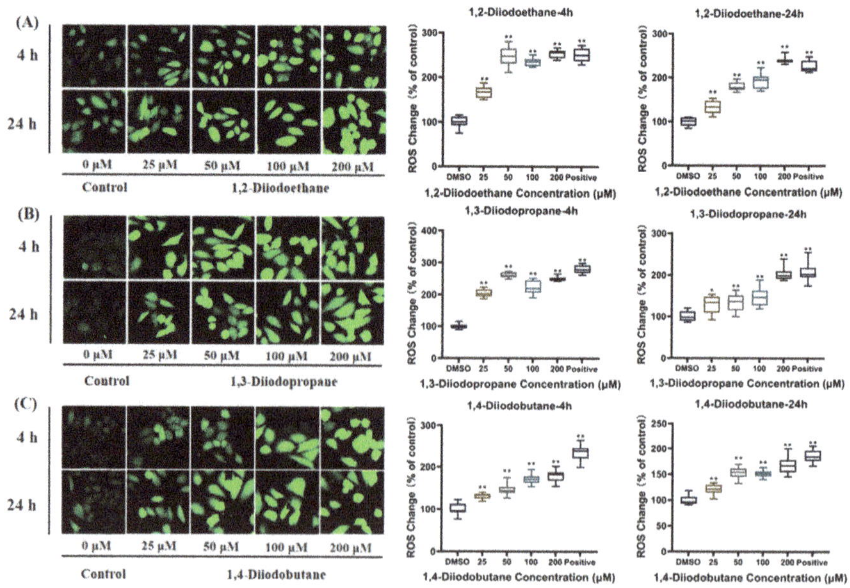

Figure 5. Effect of exposure to different iodinated alkanes on reactive oxygen species (ROS) generation in HepG2 cells. (**A–C**) show the fluorescence images and their corresponding quantification plots of

ROS overproduction induced in cells by the 1,2-diiodoethane, 1,3-diiodopropene, and 1,4-diiodobutane groups, respectively. Values are the mean ± SD of three independent experiments. The control group was treated with medium containing 0.1% (*v*/*v*) DMSO. * $p < 0.05$, ** $p < 0.01$.

3.6. Effects of Diiodoalkanes on Antioxidant Enzymes/Molecules in HepG2 Cells

Cells primarily rely on endogenous antioxidant systems to maintain ROS homeostasis, with cysteine (Cys), glutathione (GSH), and catalase (CAT) serving as core defense factors [23]. This study further examined antioxidant markers in HepG2 cells following 24 h exposure to three diiodoalkanes at a concentration of 50 μM. Compared with the control group, the three diiodoalkanes exhibited significant differences in their effects on CAT, GSH, and Cys in HepG2 cells. Regarding CAT activities (Figure 6A), exposure to 1,2-diiodoethane significantly reduced CAT activity to 57.1% of the control, while exposure to 1,4-diiodobutane significantly increased it to 147.3%. In contrast, exposure to 1,3-diiodopropane did not induce significant changes in CATactivities. For GSH levels (Figure 6B), 1,2-diiodoethane exposure caused an increase, whereas 1,4-diiodobutane exposure led tp a significant decrease. Similarly, 1,3-diiodopropane had no significant effect on GSH levels. With respect to Cys levels (Figure 6C), exposure to 1,2-diiodoethane did not result in significant changes, while both 1,3-diiodopropane and 1,4-diiodobutane exposure significantly elevated HepG2 intracellular Cys levels.

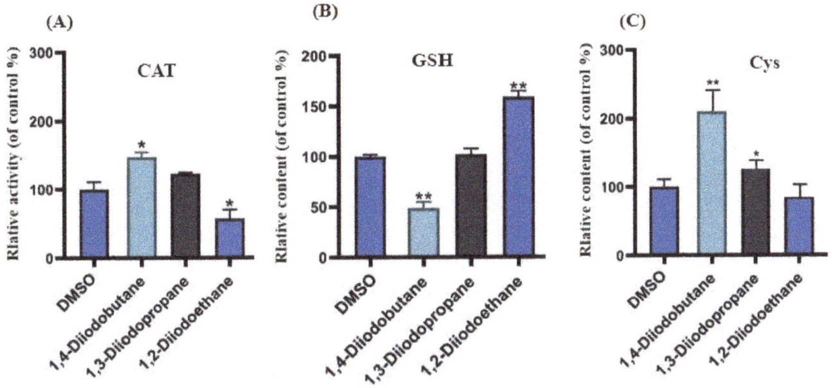

Figure 6. Effects of Diiodoalkanes on Antioxidant Enzymes/Molecules in HepG2 Cells. (**A**–**C**) show the relative activities or contents of catalase (CAT), glutathione (GSH), and cysteine (Cys) detected in HepG2 cells, respectively. Values are the mean ± SD of three independent experiments. The control group was treated with medium containing 0.1% (*v*/*v*) DMSO. * $p < 0.05$, ** $p < 0.01$.

4. Discussions

Shale gas extraction not only releases pollutants into the surrounding environment—including fracturing additives and geogenic compounds—but also generates "unknown source" environmental transformation products, derived from both anticipated and unanticipated sources, in the process. Among these pollutants, organic iodides typically pose more significant health risks than chlorides or bromides; however, their presence and toxicological potential remain unclear. Previous studies have demonstrated that various di iodoalkanes exhibit cytotoxicity [11]. Their toxic effects not only exhibit significant time- and dose-dependence, but also vary in the intensity of toxicity based on the number of iodine substitutions in the molecule. Building on this, the present study further showed that three diiodoalkanes exerted clear effects on HepG2 cells at concentrations ranging from 25 to 200 μM: They significantly inhibited cell proliferation, induced cell morphological abnormalities, and even caused extensive cell death. These results further suggested the dose-dependent characteristics of the diiodoalkanes' toxicity. This difference indicates that the toxicity of diiodoalkanes is closely related to their molecular structure. Mechanistically, the strong electron-withdrawing property of iodine atoms increases the molecular polarity, thereby promoting their binding to the intracellular enzyme systems. Ultimately, differences in molecular structure and iodine atom substitution positions determine the toxicity intensity of these pollutants.

Transcriptomic sequencing results revealed that the number of differentially expressed genes (DEGs) in HepG2 cells exposed to 1,2-diiodopropane, 1,3-diiodobutane, and 1,4-diiodoethane was 158, 491, and 221, respectively (Figure 2). Notably, the 1,3-diiodopropane-exposed group exhibited significantly more DEGs than the 1,2-diiodoethane and 1,4-diiodobutane groups, suggesting that 1,3-diiodopropane may exert a stronger molecular perturbation effect on HepG2 cells than the other two diiodoalkanes. This finding aligns with the established toxicity hierarchy of the three diiodoalkanes, further validating the correlation between molecular structure and toxic intensity. To elucidate the biological functions of identified DEGs, Gene Ontology (GO) and Kyoto Encyclopedia of Genes and Genomes (KEGG) enrichment analyses were conducted. The enriched GO terms were primarily associated with the extracellular space, extracellular matrix (ECM), and endoplasmic reticulum (ER). The extracellular space acts as an intermediary microenvironment for material exchange and signal transduction between cells and their surrounding environment. It is mainly composed of the ECM and extracellular fluids and also serves as a critical compartment for the diffusion and action of oxidative stress-related molecules, such as reactive oxygen species (ROS) and antioxidants. The ECM not only provides mechanical support and intercellular connectivity but also participates in intercellular signaling. It acts as a crucial binding site for cytokines and growth factors, and the maintenance of its structural and functional integrity is vital for sustaining cellular homeostasis [24]. When cells are subjected to adverse stimuli, the impaired function of the antioxidant defense system results in the overproduction of ROS, which disrupts the oxidant-antioxidant balance and induces oxidative stress. Oxidative stress not only directly damages ECM components (e.g., collagen and elastin) and matrix metalloproteinases (MMPs) but also downregulates the expression levels of ECM-related regulatory factors [25]. Notably, extracellular matrix remodeling is a key process jointly regulated by oxidative stress and DNA damage, exhibiting bidirectional interactions: oxidative stress/DNA damage induces ECM remodeling, while abnormal ECM remodeling conversely affects cellular oxidative stress status.

The ER serves as the primary site for protein folding within cells. The protein folding process mediated by it consumes ATP and involves redox reactions, consequently generating small amounts of ROS continuously. When cells are exposed to external toxins, the ER's protein folding capacity becomes disrupted, leading to the accumulation of unfolded or misfolded proteins within its lumen. This induces endoplasmic reticulum stress (ERS), which further promotes the overproduction of ROS. Excessive ROS not only damages ER structure (e.g., by disrupting membrane integrity and oxidizing membrane phospholipids) but also oxidatively inactivates key functional molecules in the ER (e.g., molecular chaperones and foldases) [19]. This ultimately results in the total collapse of ER homeostasis, potentially inducing apoptosis.

For the 1,2-diiodoethane, 1,3-diiodopropane, and 1,4-diiodobutane exposure groups, the GO enrichment results all included entries related to the ECM. It is well established that oxidative stress can alter the metabolic processes of ECM components and the expression of related regulatory factors, leading to excessive accumulation or degradation of specific ECM components and the disruption of ECM homeostasis. Consistent with this mechanism, the aforementioned GO enrichment results imply that 1,3-diiodopropane and 1,4-diiodobutane may impair cell viability in HepG2 cells by exacerbating oxidative stress-induced damage and disrupting antioxidant function. Furthermore, KEGG enrichment analysis identified multiple signaling pathways related to the ECM and ER, nearly all of which have direct or indirect associations with oxidative stress. This finding further supports the notion that exposure to the aforementioned diiodoalkanes may induce cytotoxicity in HepG2 cells by perturbing the balance between oxidative stress-induced damage and antioxidant defense.

ROS are natural byproducts of normal cellular metabolic activity. Under physiological conditions, the production and clearance (quenching) of intracellular ROS maintain a dynamic equilibrium to ensure stable physiological functions. However, when cells are exposed to toxic stimuli, intracellular redox homeostasis is disrupted. The balance between ROS production and clearance shifts toward elevated ROS levels, leading to cellular dysfunction and the onset of oxidative stress [26]. The results of this study indicated that diiodoalkanes significantly induce increased ROS levels in HepG2 cells across the experimental exposure concentrations (25~200 μM) and exposure time points (4 h and 24 h). This elevation exhibits clear dose-dependent effects (Figure 4). Excessive intracellular ROS can trigger multiple outcomes, including oxidative damage to proteins and lipids, DNA damage, apoptosis, and inflammatory responses, potentially exerting significant adverse effects on HepG2 cells. Cells typically rely on their antioxidant systems to maintain ROS homeostasis. Therefore, to further investigate the regulatory effects of diiodoalkanes on the cellular antioxidant system, this study examined the levels of key enzymes and molecules closely associated with antioxidant function and redox homeostasis in HepG2 cells following 24 h exposure to the three diiodoalkanes. Results showed that: Compared with the control group, the 1,2-diiodoethane-exposed group exhibited significantly elevated GSH levels and markedly reduced CAT activity, with no significant change in Cys levels. The 1,3-diiodopropane-exposed group caused only a significant increase in cellular Cys levels. The 1,4-diiodobutane-exposed group led to a significant decrease in both CAT and Cys levels, along with a significant reduction in GSH levels. These findings indicated that diiodoalkanes can modulate cellular antioxidant systems by regulating key enzymes and molecules to counteract ROS accumulation, yet their regulatory patterns exhibit compound-specific structural dependencies.

5. Conclusions

This study focuses on "unknown-source" organic iodides generated during shale gas extraction—pollutants for which their occurrence, toxic effects, and mechanisms of toxicity remain unclear—and systematically investigates the toxic effects and underlying mechanisms of three typical diiodoalkanes on human hepatocellular carcinoma (HepG2) cells. Results show that at concentrations ranging from 25 to 200 μM, all three diiodoalkanes significantly inhibit HepG2 cell proliferation, induce cellular morphological abnormalities, and trigger extensive cell death; their toxicity exhibits pronounced time- and dose-dependent patterns, and 1,3-diiodopropane demonstrates the strongest toxicity. Transcriptomic analysis revealed that the number of differentially expressed genes (DEGs) in HepG2 cells treated with 1,2-diiodoethane, 1,3-diiodopropane, and 1,4-diiodobutane was 158, 491, and 221, respectively—with the 1,3-diiodopropane group exhibiting the highest DEG count—and this DEG profile perfectly aligns with the aforementioned toxicity hierarchy. Further Gene Ontology (GO) and Kyoto Encyclopedia of Genes and Genomes (KEGG) enrichment analyses showed that DEGs were primarily enriched in functional terms associated with the extracellular matrix (ECM) and endoplasmic reticulum (ER), suggesting that the toxic effects of diiodoalkanes may be linked to the disruption of cellular oxidant-antioxidant system homeostasis, and subsequent validation experiments suggested that diiodoalkanes exposure significantly increased reactive oxygen species (ROS) levels in HepG2 cells while markedly perturbing intracellular antioxidant molecules (glutathione [GSH], cysteine [Cys]) and the antioxidant enzyme catalase (CAT). However, this study has limitations, specifically as follows: First, the experimental model utilized HepG2 cells (a hepatocellular carcinoma cell line) rather than primary liver cells. Their physiological characteristics differ from those of normal liver cells, limiting the applicability of the findings for assessing health risks in real physiological scenarios. Second, the selected pollutants are volatile. Volatilization during experiments may have reduced the actual exposure concentration in the system, potentially leading to overestimated toxicity effect values that do not fully reflect the true toxicity profile under real environmental exposure conditions. Finally, due to technical limitations, we did not measure the actual concentration of pollutants taken up/utilized by the cells. Analysis was based solely on the initial exposure concentration. This hinders in-depth analysis of the metabolic pathways and mechanisms of action of pollutants within cells and also affects the precise interpretation of the dose–response relationship for toxic effects.

Supplementary Materials: The following supporting information can be downloaded at: https://www.mdpi.com/article/10.3390/toxics13110943/s1, Table S1: The physicochemical properties of the studied compounds. Table S2: Genes and their specific primers. Table S3: Preprocessing results of RNA-Sequencing Raw Data. Figure S1: Results of cluster analysis. Figure S2: Volcano plot showing the distribution of DEGs in HepG2 cells treated with 1,2-diiodoethane, 1,3-diiodopropane, and 1,4-diiodobutane. Red, green, and gray represent upregulation, downregulation, and no significant change in DEGs, respectively. A: 1,2-diiodoethane vs. control; B: 1,3-diiodopropane; C: 1,4-diiodobutane vs. control. Table S4: Comparison of gene expression between RT-qPCR and RNA-sequencing. Table S5: Dataset of all DEG.

Author Contributions: Methodology and writing—original draft, M.X.; software and visualization, Y.W.; conceptualization, resources, data curation, and supervision, Y.C.; writing—review and edit, funding acquisition, and supervision, G.R.; data curation and visualization, R.W. All authors have read and agreed to the published version of the manuscript.

Funding: We are grateful for the support from the following funding projects: the National Natural Science Foundation of China (4247071761), the National Key R&D Plan (2019YFC1805501), and the Innovative Research Team in University (No. IRT13078).

Institutional Review Board Statement: Not applicable.

Informed Consent Statement: Not applicable.

Data Availability Statement: The data that support the findings of this study are available from the corresponding author upon reasonable request.

Conflicts of Interest: The authors declare no conflicts of interest.

References

1. Jiang, Z.; Huang, M.; Jiang, Y.; Dong, Y.; Shi, L.; Li, J.; Wang, Y. Microbial Contributions to Iodide Enrichment in Deep Groundwater in the North China Plain. *Environ. Sci. Technol.* **2023**, *57*, 2625–2635. [CrossRef] [PubMed]
2. Ma, R.; Yan, M.; Han, P.; Wang, T.; Li, B.; Zhou, S.; Zheng, T.; Hu, Y.; Borthwick, A.G.L.; Zheng, C.; et al. Deficiency and excess of groundwater iodine and their health associations. *Nat. Commun.* **2022**, *13*, 7354. [CrossRef] [PubMed]
3. Harkness, J.S.; Dwyer, G.S.; Warner, N.R.; Parker, K.M.; Mitch, W.A.; Vengosh, A. Iodide, Bromide, and Ammonium in Hydraulic Fracturing and Oil and Gas Wastewaters: Environmental Implications. *Environ. Sci. Technol.* **2015**, *49*, 1955–1963. [CrossRef]
4. Hong, H.; Wu, H.; Chen, J.; Wu, B.; Yu, H.; Yan, B.; Liang, Y. Cytotoxicity induced by iodinated haloacetamides via ROS accumulation and apoptosis in HepG-2 cells. *Environ. Pollut.* **2018**, *242*, 191–197. [CrossRef]
5. Jeong, C.H.; Postigo, C.; Richardson, S.D.; Simmons, J.E.; Kimura, S.Y.; Marinas, B.J.; Barcelo, D.; Liang, P.; Wagner, E.D.; Plewa, M.J. Occurrence and Comparative Toxicity of Haloacetaldehyde Disinfection Byproducts in Drinking Water. *Environ. Sci. Technol.* **2015**, *49*, 13749–13759. [CrossRef]
6. Lu, Z.; Hummel, S.T.; Lautz, L.K.; Hoke, G.D.; Zhou, X.; Leone, J.; Siegel, D.I. Iodine as a sensitive tracer for detecting influence of organic-rich shale in shallow groundwater. *Appl. Geochem.* **2015**, *60*, 29–36. [CrossRef]

7. Chen, M.; Rholl, C.A.; He, T.; Sharma, A.; Parker, K.M. Halogen Radicals Contribute to the Halogenation and Degradation of Chemical Additives Used in Hydraulic Fracturing. *Environ. Sci. Technol.* **2021**, *55*, 1545–1554. [CrossRef]
8. Luek, J.L.; Harir, M.; Schmitt-Kopplin, P.; Mouser, P.J.; Gonsior, M. Temporal dynamics of halogenated organic compounds in Marcellus Shale flowback. *Water Res.* **2018**, *136*, 200–206. [CrossRef]
9. Hoelzer, K.; Sumner, A.J.; Karatum, O.; Nelson, R.K.; Drollette, B.D.; O'Connor, M.P.; D'Arnbro, E.L.; Getzinger, G.J.; Ferguson, P.L.; Reddy, C.M.; et al. Indications of Transformation Products from Hydraulic Fracturing Additives in Shale-Gas Wastewater. *Environ. Sci. Technol.* **2016**, *50*, 8036–8048. [CrossRef]
10. Luek, J.L.; Schmitt-Kopplin, P.; Mouser, P.J.; Petty, W.T.; Richardson, S.D.; Gonsior, M. Halogenated Organic Compounds Identified in Hydraulic Fracturing Wastewaters Using Ultrahigh Resolution Mass Spectrometry. *Environ. Sci. Technol.* **2017**, *51*, 5377–5385. [CrossRef]
11. Wu, K.; Cui, W.; Ren, G.; An, J.; Zheng, K.; Zeng, X.; Ouyang, M.; Yu, Z. Organoiodines in effluents of a shale-fracturing wastewater treatment plant. *Environ. Chem. Lett.* **2023**, *21*, 1943–1949. [CrossRef]
12. Wu, K.; Ouyang, M.; Luo, Y.; Xu, M.; Ren, G.; An, J.; Zheng, K.; Shang, Y.; Zeng, X.; Yu, Z. Characteristics and potential cytotoxicity of halogenated organic compounds in shale gas wastewater-impacted surface waters in Chongqing area, China. *Sci. Total Environ.* **2024**, *912*, 169226. [CrossRef] [PubMed]
13. Bhujel, M.; Marshall, D.L.; Maccarone, A.T.; McKinnon, B.I.; Trevitt, A.J.; da Silva, G.; Blanksby, S.J.; Poad, B.L.J. Gas phase reactions of iodide and bromide anions with ozone: Evidence for stepwise and reversible reactions. *Phys. Chem. Chem. Phys.* **2020**, *22*, 9982–9989. [CrossRef] [PubMed]
14. Yang, Y.; Komaki, Y.; Kimura, S.Y.; Hu, H.-Y.; Wagner, E.D.; Marinas, B.J.; Plewa, M.J. Toxic Impact of Bromide and Iodide on Drinking Water Disinfected with Chlorine or Chloramines. *Environ. Sci. Technol.* **2014**, *48*, 12362–12369. [CrossRef] [PubMed]
15. Hrdlickova, R.; Toloue, M.; Tian, B. RNA-Seq methods for transcriptome analysis. *Wiley Interdiscip. Rev.-RNA* **2017**, *8*, e1364. [CrossRef]
16. He, S.; Li, X.; Li, R.; Fang, L.; Sun, L.; Wang, Y.; Wu, M. Annexin A2 Modulates ROS and Impacts Inflammatory Response via IL-17 Signaling in Polymicrobial Sepsis Mice. *PLoS Pathog.* **2016**, *12*, e1005743. [CrossRef]
17. Blaser, H.; Dostert, C.; Mak, T.W.; Brenner, D. TNF and ROS Crosstal in Inflammation. *Trends Cell Biol.* **2016**, *26*, 249–261. [CrossRef]
18. Niu, B.; Liao, K.; Zhou, Y.; Wen, T.; Quan, G.; Pan, X.; Wu, C. Application of glutathione depletion in cancer therapy: Enhanced ROS-based therapy, ferroptosis, and chemotherapy. *Biomaterials* **2021**, *277*, 121110. [CrossRef]
19. Cao, S.S.; Kaufman, R.J. Endoplasmic Reticulum Stress and Oxidative Stress in Cell Fate Decision and Human Disease. *Antioxid. Redox Signal.* **2014**, *21*, 396–413. [CrossRef]
20. Qi, S.; Xin, Y.; Guo, Y.; Diao, Y.; Kou, X.; Luo, L.; Yin, Z. Ampelopsin reduces endotoxic inflammation via repressing ROS-mediated activation of PI3K/Akt/NF-κB signaling pathways. *Int. Immunopharmacol.* **2012**, *12*, 278–287. [CrossRef]
21. Tang, J.-Y.; Fu, O.-Y.; Hou, M.-F.; Huang, H.-W.; Wang, H.-R.; Li, K.-T.; Fayyaz, S.; Shu, C.-W.; Chang, H.-W. Oxidative stress-modulating drugs have preferential anticancer effects—Involving the regulation of apoptosis, DNA damage, endoplasmic reticulum stress, autophagy, metabolism, and migration. *Semin. Cancer Biol.* **2019**, *58*, 109–117. [CrossRef]
22. Slimen, I.B.; Najar, T.; Ghram, A.; Dabbebi, H.; Ben Mrad, M.; Abdrabbah, M. Reactive oxygen species, heat stress and oxidative-induced mitochondrial damage. A review. *Int. J. Hyperth.* **2014**, *30*, 513–523. [CrossRef]
23. Wang, L.; Ding, L.; Yu, Z.; Zhang, T.; Ma, S.; Liu, J. Intracellular ROS scavenging and antioxidant enzyme regulating capacities of corn gluten meal-derived antioxidant peptides in HepG2 cells. *Food Res. Int.* **2016**, *90*, 33–41. [CrossRef]
24. Rees, M.D.; Kennett, E.C.; Whitelock, J.M.; Davies, M.J. Oxidative damage to extracellular matrix and its role in human pathologies. *Free Radic. Biol. Med.* **2008**, *44*, 1973–2001. [CrossRef]
25. Li, H.; Wang, D.; Chen, Y.; Yang, M. β-Caryophyllene inhibits high glucose-induced oxidative stress, inflammation and extracellular matrix accumulation in mesangial cells. *Int. Immunopharmacol.* **2020**, *84*, 106556. [CrossRef]
26. Prasad, S.; Gupta, S.C.; Tyagi, A.K. Reactive oxygen species (ROS) and cancer: Role of antioxidative nutraceuticals. *Cancer Lett.* **2017**, *387*, 95–105. [CrossRef]

MDPI AG
Grosspeteranlage 5
4052 Basel
Switzerland
Tel.: +41 61 683 77 34

Toxics Editorial Office
E-mail: toxics@mdpi.com
www.mdpi.com/journal/toxics

www.ingramcontent.com/pod-product-compliance
Lightning Source LLC
Chambersburg PA
CBHW040828250926
51850CB00007BA/1290
9783725861330